The Medicine Wheel

The Medicine Wheel

ENVIRONMENTAL DECISION-MAKING PROCESS OF INDIGENOUS PEOPLES

Michael E. Marchand, Kristiina A. Vogt, Rodney Cawston,
John D. Tovey, John McCoy, Nancy Maryboy,
Calvin T. Mukumoto, Daniel J. Vogt, Melody Starya Mobley

Higher Education Press/Michigan State University Press
China/East Lansing

Copyright © 2020 by Higher Education Press. All rights reserved.

♾ The paper used in this publication meets the minimum requirements of ANSI/NISO Z39.48-1992 (R 1997) (Permanence of Paper).

Michigan State University Press
East Lansing, Michigan 48823-5245

Published in China by the Higher Education Press Limited Company
Published in the United States of America by Michigan State University Press

Printed and bound in the United States of America.

29 28 27 26 25 24 23 22 21 20   1 2 3 4 5 6 7 8 9 10

Library of Congress Cataloging-in-Publication Data is available
ISBN 978-1-61186-358-1 (paperback)
ISBN 978-1-60917-633-4 (PDF)
ISBN 9781628953954 (ePub)
ISBN 9781628963960 (Kindle)

Cover art is an artistic rendition of a MedicineWheel found in the Bighorn Mountains of the U.S. state of Wyoming (see https://en.wikipedia.org/wiki/Medicine Wheel/Medicine Mountain National Historic Landmark#/media/File:MedicineWheel.jpg). The Medicine Wheel has been in use for 7,000 years.

Michigan State University Press is a member of the Green Press Initiative and is committed to developing and encouraging ecologically responsible publishing practices. For more information about the Green Press Initiative and the use of recycled paper in book publishing, please visit www.greenpressinitiative.org.

Visit Michigan State University Press at www.msupress.org

Contents

Preface —— xi

Acronyms —— xix

Chapter I Indigenous Knowledge Framework and the Medicine Wheel —— 1
1.1 Bighorn Medicine Wheel Story —— 7
1.2 The Medicine Wheel: Non-linear Knowledge-forming Process —— 9

Chapter II What Is Needed to be a "Leader without Borders"? —— 19
2.1 My People's 9,400 Year Ancestral History —— 20
2.2 Becoming a "Leader without Borders": Interview of Dr. Mike Marchand —— 21

Chapter III How Do You Become "Cultured"? —— 51
3.1 Western European Culture: You Live it, You Wear it and You Eat it —— 54
3.2 Culture According to Indigenous People —— 59
3.2.1 Cultural Resources Defined by Tribes —— 61
3.2.2 What Is a Cultural Resource? —— 64
3.3 Keeping Deep Culture in Two Worlds: Interview of Dr. Mike Tulee —— 66
3.4 Culture Defined by Nation-Level Melting Pots —— 87
3.5 Tribal Peoples' Cultural Context: Interview of JD Tovey —— 89
3.6 Cultural Foods and Food Security —— 104
3.6.1 Loss of Food Security: Chemical Contamination —— 111
3.6.2 Loss of Huckleberries and Tribal Culture: Interview of JD Tovey —— 113
3.6.3 Skokomish Litigation for Rights to Hunt by Indian Tribes —— 115
3.7 Holistic Nature Knowledge not Decoupled from Nature and Religion —— 117
3.8 Languages and Indigenous People —— 119
3.9 What Is Your Real Name? Dr. Mike's Wolverine Encounter —— 126
3.10 Sports and Games Invented by American Indians —— 128

Chapter IV Western Science ≠ Indigenous Forms of Knowledge —— 135
4.1 Knowledge-forming Processes: Western Science ≠ Indigenous Ways of Knowing —— 139

4.2	How Knowledge Frameworks Address Scarcity of Land or Lack of Knowledge —— 142	
4.3	The Challenge of Culture for Western Scientists —— 145	
4.4	Traditional Knowledge: Native Ways of Knowing —— 149	
4.4.1	How Indigenous People Form Knowledge —— 151	
4.4.2	Indigenous Ecological and Spiritual Consciousness —— 153	
4.4.3	Ecological Calendars in Nature Literacy —— 155	
4.5	Juxtaposition of Western and Traditional Knowledge —— 157	
4.6	Who Are Trusted for Their Science Knowledge? —— 160	
4.6.1	How Citizens of the Western World Get Their FACTS —— 161	
4.6.2	How Native Americans Get Western Science FACTS —— 163	
4.7	Women's Role in Passing Indigenous Knowledge Inter-Generationally: Interview of JD Tovey —— 165	
4.8	Role of Environmental Economics in Environmental Justice —— 170	
4.8.1	Natural Capital versus Cultural Values —— 171	
4.8.2	Makah Tribe's Cultural Revitalization: Whaling —— 173	
4.8.3	Lower Elwha Klallam Tribe's Dam Removal —— 175	
4.8.4	Restructuring Environmental Economics to be More Inclusive of Environmental Justice —— 176	
4.8.5	Special Acknowledgements —— 177	

Chapter V   Forestry Lens: Culture-based Planning and Dealing with Climate Change ——179

5.1	PNW U.S. Tribes and Leadership in Climate Change Planning —— 179
5.2	Tribes, Tribal Resources and Forest Losses —— 184
5.2.1	Historical Loss: Manifest Destiny and Loss of Forests —— 185
5.2.2	Fire Cause Loss of Forests, Cultural Resources and Timber from a Shrinking Land Base —— 191
5.3	Today Better Forest Management on Tribal Lands Compared to Their Neighbors —— 198
5.3.1	Good Tribal Forestry under Federally Mandated Assessments (IFMAT): Interview of Dr. John Gordon —— 198
5.3.2	Tribal Forestland Management: A Growing Force in the PNW U.S. —— 215
5.4	Realities in Developing Resources on Reservations —— 223
5.4.1	Making Business Decisions: Interview of Cal Mukumoto and Dr. Mike Marchand —— 224
5.4.2	Challenges: A Boom? or a Dis-economy of Scale for Tribes? —— 242

Chapter VI   Tribes, State and Federal Agencies: Leadership and Knowledge Sharing Dynamics ——249

6.1	Tribal/Federal/State Cultural Resource Policy —— 253
6.2	Tribes and Washington State —— 257
6.2.1	Washington State Policy Process and Tribes: Interview of John McCoy —— 257

6.2.2	Washington Department of Natural Resources and Tribes: Interview of Rodney Cawston —— 267	
6.3	Alaska Natives, Conservation and Policy Process —— 292	
6.3.1	Alaska Native Perspectives on the Governance of Wildlife Subsistence and Conservation Resources in the Arctic —— 293	
6.3.2	Partnership — A Role for Nonprofits and Agencies in Conservation of Native Lands in Alaska —— 309	
6.4	Federal Agency and Tribes: Continuing Challenges to Tribal Rights —— 318	
6.4.1	Indigenous People's Role in National Forest Planning —— 323	
6.4.2	USDA Forest Service Use of Culture in Land and Resource Management Planning Decisions —— 334	
6.4.3	Working as an Individual within a Federal Corporate Culture —— 336	
6.5	Inter-Tribal Collaborations: Increase Tribal Role in Natural Resource Planning —— 337	
6.5.1	The Water Protectors: Protest at Standing Rock —— 341	
6.6	Intra- and Inter-Governmental Affairs and Public Policy Process —— 343	
6.6.1	Preface —— 344	
6.6.2	Intra- and Inter-Governmental Affairs and Public Policy Process —— 345	

Chapter VII   **Native People's Knowledge-Forming Approaches Needed for Nature Literacy to Emerge among Citizens ——355**

7.1	Why We Need New Education Tool for Nature Literacy for the Masses —— 355
7.2	Massive Amounts of Fragmented Data in STEM Sciences —— 358
7.3	Critical Analysis Lacking in Environmental Education —— 359
7.4	Native People's Storytelling Practices to Communicate Holistic Science —— 362

Chapter VIII   **Learning Indigenous People's Way to Tell Circular Stories ——365**

8.1	Technology to Digitize Stories Part of Popular Culture —— 366
8.2	Digital Technologies Part of Popular Culture —— 368
8.3	Challenges in Communicating and Telling Circular Stories —— 371
8.3.1	Science Literacy Needs to be Circular and Not Linear —— 372
8.3.2	Science Literacy Is an Information Problem —— 373
8.3.3	Indigenous Stories Are Not Linear but Cultural and Transdisciplinary Science Knowledge —— 374
8.4	Digitizing Native Stories without Pickling Culture: Interview of JD Tovey —— 375
8.5	Stories in Navajo Lands —— 384

Chapter IX	Medicine Wheel: Moving beyond Nature, People and Business Stereotypes —— 389	
9.1	When I Was a Young Boy —— 391	
9.2	Communicating Indigenous Knowledge to the Masses —— 396	
9.3	Medicine Wheel and Not Case Studies —— 398	
9.4	"Fictional Tribe" as an Educational Tool to Teach How to Form Holistic Knowledge —— 400	

References —— 403

Authors —— 419

Contributing Authors —— 421

Index —— 423

List of Boxes

Box 1: The Story of the Medicine Wheel —— 5
Box 2: Description of a Traditional Navajo Home (Dr. Nancy Maryboy, Dr. David Begay) —— 11
Box 3: Shrinking Land Base of the Colville Tribes (Dr. Mike Marchand) —— 31
Box 4: History of Surplus Food Given to Tribes —— 109
Box 5: Some Chemical Contaminants of Concern in the Puget Sound —— 112
Box 6: Code Talkers from the Choctaw Nation and the Congressional Gold Medal —— 124
Box 7: Members of the 3rd and 4th Division Navajo Code Talker Platoons of World War II —— 124
Box 8: Einstein Quotes on Religion and Science —— 146
Box 9: Why We Need to Think about Natural Capital —— 148
Box 10: IFMAT I Indian Peoples' Visions from Forests —— 201
Box 11: John Gordon's Matrix to Evaluate People's Potential to Lead (Gordon and Berry 2006) —— 207
Box 12: Matrix to Evaluate People's Selection to be on a Complex, Transdisciplinary Committee —— 208
Box 13: Makah Tribe History —— 225
Box 14: Commissioner's Order on Tribal Relations —— 269
Box 15: In 2009, Price/Unit for Huckleberry Fruit —— 275
Box 16: Washington Public Land Harvest Regulations for Wild Huckleberries —— 277
Box 17: Diné (Navajo) Paradigm of *Są'áh Naagháí Bik'eh Hózhóó* (SNBH) —— 385

Preface

Photo Source: Cal Mukumoto

We live in an age where persistent environmental problems are decreasing the health of our lands, water, species and air compared to the impacts of natural processes [1]. The gigantean environmental problems we face today only emerged within the last century. They are a result of technology allowed industrializing societies to utilize fossil carbon compounds for energy production that transformed our lands by altering our carbon and nutrient cycles [2]. These changes also build our synthetic lifestyles of today. **The enormous capacity of humans to impact our environment has even led many geologists to declare we have been ushered into the "Anthropocene Era" where the activities of humans swamp natural cycles and processes.**

These world-wide problems will not be solved by only searching for solutions in academic institutions of higher education. Educational institutions forming knowledge to solve problems in isolation from the rest of society and other disciplinary fields have not worked since the same environmental problems continue to persist for centuries. Today there is a call for science to become part of popular media so the decision-makers are sufficiently knowledgeable in the sciences to understand the problems they need to solve. We think this is important but also that there is a need to engage youth from K-12 grades to learn how to form knowledge and think critically on complex environmental and societal problems. They are our future decision-makers!

Further the most common framework and scientific methods used to form knowledge needs to be completely restructured and envisioned. Current tools are not solving the problems emerging in today's environment, and certainly won't solve tomorrow's problems. Generally, economics is the tool used to resolve environmental and societal problems because decision-makers decontextualize and compartmentalize environmental problems. Economics is a common tool to explore trade-offs since it simplifies the problem-solving process into discrete numbers than can be compared and scored for "value". But this approach is not working for the environment because it misses the holistic, intrinsic, and qualitative aspects of ecological systems that are just not capable of being quantified or assigned a dollar value. There has to be a better way to make decisions that do not increase the risks of impacts on social and environmental health.

Two reasons for writing this book can be summarized as: (1) Scientists, citizens and amateur scientists are unfamiliar with the Indigenous ways of forming holistic knowledge despite its importance for addressing tricky and complex environmental problems. Indigenous knowledge forming processes and communication approaches are an essential tool for non-Tribal communities to learn. It will allow decision-makers to resolve environmental problems in a shorter period of time, so problems do not fester and be left for future generations to figure out; and (2) Tribes are natural-resource dependent communities and are at the forefront of facing the impacts of climate change. They and their ancestors lived resiliently in the natural environment and sustainably managed their resources for thousands of years, often with comparable human populations as present day. They passed this knowledge down through the generations and the entire community was involved in the decision process. Tribes do not compartmentalize their society into those people who form knowledge of nature and the environment, and those who make decisions. In contrast, Western societies mostly educate scientists to form knowledge on nature and non-scientists, especially economists, make the policy decisions. Another important issue is that knowledge in the Western world is not holistic. It is fragmented and decontextualized from the emerging problem and compartmentalized by disciplinary fields. Such an approach is not designed to identify emerging resource problems in contrast to Tribal ways of forming knowledge.

REASON ONE: The Tribal ways of forming holistic and balanced knowledge needs to become the tool used to resolve environmental problems. The Western world process for forming science knowledge are not designed to address complex and interdisciplinary-based environmental problems. Western-trained scientists do research complex environmental problems but typically use a disciplinary lens incapable of mechanistically linking people and nature. The Western world use of the scientific method narrows the scope of the problem and assumes causality based on the current context. Thus, a researcher decontextualizes the problem from its holistic environmental context or the ecosystem within which it is embedded. These approaches result in each environmental problem persisting for decades since an environmental manager identifies the wrong causal factor(s) to monitor for detecting emerging problems. Also, it is impossible to link a specific management intervention as the reason for the successful resolution of an environmental problem.

The Western science approach is not holistic to the temporal and spatial scale of the problem. It develops general principles or paradigms to focus the research problem. These are general principles that are applied everywhere, and not localized to where the problem is occurring. Today, it is common to use technology and a diversity of models to develop knowledge. These tools are not sensitive to the diversity of processes that occur across the landscape through space and time. **Amateur naturalists**, who focus on long-term observations of nature at the local scale, are frequently the first to warn of an emerging environmental problem. Stager wrote a New York Times article entitled "The Silence of the Bugs" how amateur naturalists made the alarming discovery that *"76 percent decline in the total seasonal biomass of flying insects netted at 63 locations in Germany over the last three decades"* [4]. University scientists did not make this discovery because their tools are not sensitive to identifying local changes occurring in the environment and they are not observing nature over long time scales.

Further, when "early warnings" indicators of a negative environment change are not local place-based knowledge, a researcher may conclude there is no emerging environmental catastrophe even though an undetected problem has already emerged. Once the existence of the problem is recognized at a later time, a tipping point may already have been passed where it may be difficult or impossible to mitigate the impacts of the problem. Conniff comments on the use of "total area protected" as an indication of future conservation success and noted this indicator is delusional [5]. As he writes:

> *"Designating protected areas is relatively easy (and with publicity bonus points for politicians), but hardly anyone seems to be bothering with the hard work of actually protecting them. Roughly a third of national*

parks, reserves, refuges and the like now face intensive and increasing human pressure. So many protected areas now face development that there's an acronym for it — Paddd, for protected area downgrading, downsizing and degazettement — and a website for keeping up on the bad news."

"Politicians, like the rest of us, are suckers for numeric targets.... These targets seem simple, objective, easily comparable from one place to the next, and inexpensive to measure. ***Pretending to protect species based purely on the number of acres protected is like managing human health care based on the number of hospital beds* "irrespective of the presence of trained medical staff"** *or* **"whether patients live or die."**

There has to be a better way to protect our environments than creating monitoring tools unable to provide an early warning that a problem is emerging. This is where the Native Peoples' practices should be used to identify and detect environmental problems before they explode.

REASON TWO: Tribes are natural-resource dependent communities already impacted by climate change. They have a history and knowledge developed over several thousand years of living on resources collected from the land and waters. Tribal People continue to practice resilient decision-making and form nature knowledge that is holistic despite their need to adapt and survive from a smaller land area after European colonization. Indigenous forms of knowledge is local-based but also embedded in a regional context since tribes historically managed large areas of land. Lands managed by Tribes were resilient and natural, which is why the early colonialists thought the lands they conquered were "wilderness" areas. They practice a nature-based ethical decision-making process that does not compartmentalize knowledge by professions or practice top-down decision-making. It is not human-egocentric but nature-focused and holistic. Therefore, Tribal approaches to forming nature knowledge are fundamentally different from those used by Western societies. If the goal is to manage environments to be resilient and retain the characteristics of "wilderness" areas, we need to understand the practices of Native People.

In the Western world, scientists and decision-makers rarely make decisions as a community or tribe. Scientists are the reservoirs of disciplinary-based knowledge while decision-makers mostly have little science knowledge. This situation appears to politicize environmental decisions and evidence-based knowledge is not part of the decision-making toolkit. Scientists should not be faulted for all the problems emerging in our environment and which may persist for decades. It's the politicians or policy-makers, with little science education, who make the decisions and determine whether to fund activities that result in environmental problems. No one suggests that scientists alone or decision-makers should make environmental decisions. Each contributes knowledge that is only part of the information needed to plan and make decisions. **Also, the community needs to be involved in the decision process on complex environmental problems. People with holistic knowledge of an issue, and those with no vested interest in the final outcome, need to be "at the table" when decisions are made. Environmental and social justice is only possible when environmental problems are de-politicized and not human centric.** If this does not happen, it will be difficult to build consensus on complex environmental issues.

There is urgency for Western societies to accelerate the rate at which environmental planning becomes holistic and decision-makers use evidence-based knowledge in their decision process. Amateur naturalists do play a very important role in observing and forming knowledge of nature but the community of people capable of forming science knowledge needs to expand beyond to the general public. More lay people need to become scientifically competent, but not by following the traditional "scientist" education track. Science knowledge needs to be part of decision-making, but this is not going to happen if more citizens and future decision-makers need to go to college. We need to focus on youth learning and practicing holistic approaches to environmental management. This needs to start when youth first attend kindergarten and continue into their high school years. Unfortunately, today our youth formally learn about nature and the environment when they matriculate into institutions of higher education. However we suggest that this could be too late. In contrast to that, Tribal youth, for example, learn holistic environmental knowledge throughout their life from their grandparents and the stories they hear throughout their life. We think that non-Tribal youth need to learn nature stories when they are young. They need to learn to tell or digitize stories just like Native people. Youth — our future leaders — can use technology to build applications that transcribes knowledge given as stories into a digital or multi-media format. Thus, youth can create innovative communication tools for complex resource and environmental problems that are challenging for less technologically-skilled adults.

To fully grasp why the general public needs to learn Indigenous knowledge-forming processes to solve complex environmental problems, you need to understand the differences between Western science and Indigenous knowledge. The goal of this book is to provide a context for the attributes of the Western science and Indigenous knowledge frameworks so the reader can build their own holistic and ethical decision framework to provide environmental leadership. Our take-home message is that Indigenous knowledge should be an equal partner with Western Science in environmental assessments under today's climate-changing umbrella. We contend that the Western science knowledge framework is incomplete without this localized intergenerational knowledge introduced by Indigenous people. What is different about our book is that we do not just describe the problems inherent to each knowledge framework but offer new insights for how to connect culture to a science knowledge-forming framework. We also want the reader to think about who they listen to when getting their facts about their knowledge of science and culture, e.g., stories of knowledge passed down through multiple generations via Tribal elders or scientific experts. Today, our science communicators and cultural facts move along parallel tracks that seldom seem to cross.

This story also has to explore how culture gives "tenacity of purpose and guts" to Indigenous people. This character is what sustained the continuity of Tribal members, practices and preservation of their Traditional knowledge and their cultures, despite the numerous road-block they have experienced over the past several hundred years. They continue to be challenged by Western-trained local, state and federal-level agency scientists who do not know how to include culture in their decision-making. These battles continue despite the many published success stories reporting how Indigenous Peoples' practices are increasing conservation efficacy and land health. Some might question why we should listen to Indigenous people since they are "legally recognized as holding only 10 percent" [6] of the global terrestrial land area today. We say that despite Indigenous people having lost their lands to European colonialists, that they care about these lands and have fought and continue to fight many battles to restore nature's health.

This book presents our roadmap of how two knowledge streams bound by different cultural/art/spiritual landscapes may provide a pathway by which decision-makers can use scientists as filters or a lens to interpret large datasets. We offer a second pathway for environmental managers to become "Essential Leaders" capable of making culturally-based decisions for complex environmental problems. However both approaches should be simultaneously considered to effectively address environmental issues, especially with today's climate change impacts are occurring in highly altered environments. A strength of the Indigenous knowledge framework is culture and localized intergenerational knowledge that drives the knowledge-forming process. This process allows Indigenous knowledge practitioners to address "scarcity of knowledge". The obvious strengths of the Western science approach are its tools and methods for assessing scarcity of land and resources under a climate-change scenario, albeit at the "airplane science" scale. In contrast, a weakness in the Western science framework is the lack of functional drivers that link peoples' decisions and land-use activities to ecological systems. Therefore, both knowledge-forming processes can contribute to building a holistic approach to more effectively define and manage emerging environmental problems. A holistic approach would shift problem identification from a reactive assessment of what has already played out to a proactive approach that has a greater potential to diminish the intensity of a problem before it erupts. Today we react to an emerging environmental problems with just a superficial bandage that only addresses the symptom of the problem, not its underlying cause.

In the Western scientific world, a lack of practical methods or tools, to easily integrate fundamentally different forms of knowledge, has been a hindrance to holistic planning for nature. At first glance it appears to be a wide chasm that cannot be easily bridged because of the different space and time scales and the role people play in impacting the process. The Western science process will need to fundamentally change, since it currently follows strict rules on how to practice and communicate science as defined by each disciplinary research group. A Western-trained scientist may find it challenging to incorporate into their science framework the holistic construct and practices conceptualized in this "Medicine Wheel". Also Indigenous knowledge is not easily bounded by disciplinary-based knowledge nor can it be converted into mathematical equations. As part of their training, Western-trained scientists learn strict research protocols that ultimately become encoded into the very fabric of being a "scientist". There are many incentives to follow these practices, since they strongly influence whether you will be hired at a university, and if you are able publish papers or receive grant money. This should be changed so Western scientists have the chance to contribute their holistic knowledge to the planning efforts related to nature!!

Western scientists or decision-makers cannot form holistic knowledge just by memorizing Tribal knowledge. It is the "process" of gaining the holistic knowledge that needs to be experienced by the Western scientists. Further, you also cannot talk to only one Tribe and learn everything you need to know about forming holistic knowledge. Today, there are 562 federally-recognized Indian Tribes in the United States. Each Tribe uses different tools and frameworks to form knowledge and follow different journeys during a planning process. Indigenous people respond to environmental issues using a "nonlinear process" so the endpoint may never be reached and the journey is what is important. The Western scientific method uses a linear process that tests pre-determined solutions supported by scientifically accepted paradigms for how the world works. Reading this book you are not going to learn Tribal knowledge paradigms. Non-Tribal People need to learn to follow a process which will result in different solutions depending on the problem context. This is the focus of this book.

Today, decision-makers need to utilize a framework that makes them think like a "system" and acknowledge that management of "drivers of change" in land/water must include culture/art. The "Medicine Wheel" provides the context for holistic management which is why it is the title of our book. Our mental construct should be a spoked wheel instead of a seesaw (i.e., integrating rather than choosing one side or the other). Our vision is to help today's global citizens to more competently determine how to become Essential Environmental leaders by forming their own mental construct of the natural world — one that is viewed holistically and formed by our relationships with nature. Read this book and learn how you can move beyond stereotypes for connecting with nature. We can't pickle nature and culture so we don't recognize them anymore.

Authors
June 24, 2018

Acronyms

AAAS	American Association for the Advancement of Science
AAC	Annual Allowable Cuts
ABA	Arctic Biodiversity Assessment
ABC	American Broadcasting Company
ACT	American College Testing
AHDR	All Hazards Disaster Response
AIS	American Indian Studies at University of Washington
AML	Aboriginal Media Lab
ANC	Alaska Native Corporations
ANCSA	Alaska Native Claims Settlement Act
ANILCA	Alaska National Interest Lands Conservation Act
AP	Associated Press
apps	short for *application* especially for small wireless computing devices such as a smartphone
ATNI	Affiliated Tribes of the Northwest Indians
BIA	Bureau of Indian Affairs
BLM	Bureau of Land Management
BMW	Bayerische Motoren Werke (in German), or Bavarian Motor Works (in English)
BS	Bull Stuff (in polite form), slang for meaning a mistake, bad idea, wrong, inappropriate
CAT	Caterpillar
CBS	Columbia Broadcast System
CCH	Commerce Clearing House
CCVI	Climate Change Vulnerability Index
CDFI	Community Development Financial Institution
CEA	Council of Economic Advisers
CEO	Chief Executive Officer
CFR	Code of Federal Regulations
D.C.	District of Columbia, typically Washington, D.C. in the U.S.
D.Q.	Deganawidah-Quetzalcoatl University in California, U.S.
DAPL	Dakota Access Pipeline
DNA	deoxyribonucleic acid

DNR	Department of Natural Resources
DOI	U.S. Department of the Interior
DOJ	U.S. Department of Justice
DV	Daniel Vogt
E4	Enlisted rank 4 (sergeant)
EEO	equal employment opportunity
EPA	Environmental Protection Agency
ESP	extra sensory perception
EVCC	Everett Community College
FAQ	frequently asked questions
FDPIR	Food Distribution Program on Indian Reservations
FES	Yale's School of "Forestry and Environmental Studies"
FIT	fire, investment, and transformation
FMP	Forest Management Plan
FORTRAN	Formula Translation, programming language for computing
GAO	Government Accountability Office
GED	General Educational Development
GHGs	Greenhouse Gases, gas in the atmosphere that absorbs and emits radiant energy within the thermal infrared range
GMP	Game Management Plan
GRE	Graduate Record Examination
GSE	Graduate School of Education (Harvard University)
IBM	International Business Machines, American multinational technology company
ICDP	Integrated Conservation and Development Project
IFMAT	Indian Forest Management Assessment Team
IGERT	Integrative Graduate Education and Research Traineeship program in the U.S. National Science Foundation for PhDs
IK	Indigenous Knowledge
IPCC	Intergovernmental Panel on Climate Change
IQ	Intelligence Quotient
IRA	Indian Reorganization Act
IRENA	International Renewable Energy Agency
IRS	Internal Revenue Service
ITARA	Indian Trust Asset Reform Act
ITC	Intertribal Timber Council
ITEDSA	Indian Tribal Energy Development and Self-Determination Act
ITO	Indian Tribal Organizations
JD	John David Tovey III
JLARC	Joint Legislative Audit and Review Committee
LLC	Limited Liability Company
MBA	Master of Business Administration

MEMA	Meaningful Engagement of Indigenous Peoples and Communities in Marine Activities
MIT	Massachusetts Institute of Technology
MM	Melody Mobley
MPA	Master of Public Administration
NAGPRA	Native American Graves Protection and Repatriation Act
NASA	National Aeronautics and Space Administration
NBC	National Broadcasting Company
NCAI	National Congress of American Indians
NEPA	National Environmental Policy Act
NGOs	Non-Governmental Organizations
NIFRMA	National Indian Forest Resource Management Act
NMTC	New Markets Tax Credits
NORC	National Opinion Research Center
NPS	National Park Service
NPT PBS	Nashville Public Television Public Broadcasting Service
NRC	National Research Council
NSF IGERT	National Science Foundation's Integrative Graduate Education and Research Traineeship program
NWIFC	Northwest Indian Fisheries Commission
NYT	New York Times
ODOT	Oregon Department of Transportation
OSB	oriented strand board
PAHs	polycyclic aromatic hydrocarbons
PBDEs	polybrominated diphenyl ethers
PBS	Public Broadcasting Service
PCBs	polychlorinated biphenyls
PEIS	programmatic environmental impact statement
PhC	Candidate of Philosophy
PhD	Doctor of Philosophy
PNW	Pacific Northwest (in the U.S.)
POW WOWS	a social gathering held by many different native American communities to meet and dance, sing, socialize, and honor their cultures
PTSD	Post-Traumatic Stress Disorder
QFC	Quality Food Centers, supermarket chain
RCW	Revised Code of Washington
REI	Recreational Equipment, Inc
RFP	Request for Proposal
S&L	Savings and Loan
SAMBR	State of the Arctic Marine Biodiversity Report
SCADA	Supervisory Control and Data Acquisition system

SISU	Finnish term meaning tenacity of purpose, grit, bravery, resilience and hardiness
SNAP	Supplemental Nutrition Assistance Program
SNBH	Sa'áh Naagháí Bik'eh Hózhóó, Navajo belief system that guides harmonious living
STEAM	a science, technology, engineering, arts and mathematics in science education
STEM	science, technology, engineering, and mathematics in science education
STREAM	a science, technology, writing, engineering, arts, and mathematics in science education
SWOT	strengths, weaknesses, opportunities and threats
TED	Tribal Economic Development
TFPA	Tribal Forest Protection Act
TIMOs	timber investment management organizations
TMI	Thornton Media, Inc.
TV	television
U.S.	United States
U.S.C. (USC)	U.S. Code
U.W.	University of Washington, Seattle, Washington, USA
UNDRIP	United Nations Declaration on the Right of Indigenous Peoples
UNISYS	an American global information technology company based in Pennsylvania, USA.
UNIVAC	Universal Automatic Computer, first commercial computer produced in the U.S.
USDA	United States Department of Agriculture
USFS	United States Forest Service
VA	Veteran Affairs
VIP	very important person
WA	Washington, a state in the U.S.
WDFW	Washington (State) Department of Fish and Wildlife
WSDOT	Washington State Department of Transportation
WWF	World Wildlife Fund
WWII	World War II

Chapter I
Indigenous Knowledge Framework and the Medicine Wheel

> *"The Indian sense of natural law is that nature informs us and it is our obligation to read nature as you would a book, to feel nature as you would a poem, to be part of that and step into its cycles as much as you can."*
>
> ~ John Mohawk, Haudenosaunee Scholar and Author (1945—2006) ~

Photo Source: Cal Mukumoto

Rodney Cawston (Confederated Tribes of the Colville; U.W.-PhD Student) speaks for us when he describes the resiliency of Native American people and how they have adapted over thousands of years to their environment and changing climates. They are not new to these lands or tourists in nature and the rest of the world needs to listen to the knowledge they share. Rodney wrote:

Rodney Cawston.
Source: Confederated Tribes of the Colville.

"Many Tribal communities reside on or near the Pacific Ocean and rivers and streams that feed into the Pacific Ocean. These Tribes are natural-resource dependent communities for commercial, ceremonial and subsistence living which makes them especially vulnerable to climate change. All of these Tribe's usual and accustomed areas are in high-risk areas on islands or low-lying coastal areas. Each of the Tribes in Western Washington has a unique language, culture and legal/political history. Many of these Tribes have lost almost all of their fluent speakers of their traditional language. This is why it is important to seek out the Tribal elders and fluent traditional speakers, traditional practitioners and others who have traditional knowledge about hunting, fishing, gathering traditional use trees and plants and how they once took care of these sacred landscapes. The history of this area has been exposed to many different kinds of environmental changes such as earthquakes, volcanic eruptions, landslides, and tsunami's and tribal villages needed to develop coping strategies to face these phenomena. The resiliency of Native American people's history may offer valuable knowledge to learn from for future adaptation and climate change mitigation measures.

Traditionally, many Tribal villages were located where the fresh water meets the ocean and today's Tribal governments are still situated near their traditional village sites, where sea level rise is expected to have especially serious impacts. Sea level rise due to melting glaciers and sea-ice and the expansion of water because of a rise in temperatures will cause some low-lying coastal areas to become completely submerged, while others will increasingly face short-lived high-water levels. Tribes who reside on islands will be especially vulnerable to the effects of sea level rise. These anticipated changes could have a major impact on the lives of Tribes and their members. They may have to relocate outside of their traditional homes or established reservations. All of the Tribes in the Puget Sound vicinity rely on marine ecosystems and depend on marine resources. Coastal erosion can also impact sea-level rise. Hurricanes and typhoons can lead to a loss of land and property and also dislocate Native people. Salt water intrusion on ground water can cause the salinization of freshwater resources. With climate change the gathering of plants and other resources along these riparian areas will become threatened and the continuation of their traditional practices can become threatened. Mountain glaciers and snow packs feeding lakes and creeks have already declined significantly and continued climate change can lead to changes in land surface characteristics and drainage systems that will impact near shore environments of the ocean. Changes in precipitation are hard to predict but they affect water quality and the loss of biodiversity.

Indigenous people whether they live on or off the reservation participate in traditional ceremonies, which utilize many resources which are gathered from natural areas within their usual and accustomed areas, on both uplands and aquatic lands. Indigenous people hold first food ceremonies to pay respect for traditional foods that are harvested each year. No member of the Tribe is allowed to gather until these ceremonies have been completed. Ceremonial dinners often consist of elk and deer meat, salmon, shell fish, edible roots and berries, teas, and water. These foods are considered sacred. Many Tribes will pray and cleanse themselves both physically and spiritually before harvesting these foods for ceremonial use. Traditional gatherers are often selected to gather for ceremonial dinners and follow traditional protocols in order to conduct this work. Indigenous people also gather many of these foods for their own subsistence throughout the year. Families still harvest large quantities of fish, shellfish, roots and berries and use traditional methods of preparing these foods. Fish, shell fish, deer and elk meat are often cured by smoking with alder wood and then either dried, canned or frozen for future consumption. Families will travel to higher elevations to harvest berries of many different varieties and either dry, can or freeze them. Medicinal and other traditional use plants are also gathered throughout the year. Indigenous people still trade or gift with other Tribes on occasion with traditional foods or cultural articles.

Most Tribes in Western Washington also participate in the annual Canoe Journey. Each year one of the Pacific Northwest Tribes will sponsor this event and send invitations to many other Tribes to paddle to their homelands for a weeklong event of traditional song and dance and potlatch. Many of the canoes are hand carved from old growth cedar and sometimes they need to be replaced. These Tribes will have many small potlatches throughout the week and the host Tribe will have a *large potlatch at the end of the week. During the event, the host Tribe will serve traditional foods to all of their guests, both native and non-native. Salmon, game meat, shell fish and other foods are served all week long. The host Tribe takes much effort in gathering all of these traditional foods throughout the year. They will also gather other traditional materials such as cedar bark, cedar root, and carving materials to make many articles that they will give away to invited guests. Canoes paddle from very far distances such as Alaska, Canada, California, Oregon, Maori, Hawaii, and other locations. This is one of the largest traditional events in Western Washington held each year.*

While there is a growing knowledge about the impacts of climate change on species and ecosystems, the understanding about the potential impacts of climate change on livelihoods and cultures of Indigenous and traditional communities is fragmented. Furthermore, there is a lack of recognition of the importance which traditional people will play in their own future adaptation to climate change.

Today, modern forms of Tribal government rely on Tribal traditions for guidance when dealing with current climate change policy and political challenges. Tribal governments have always been independent nations. Tribes have always had the right to hunt, fish, gather, run casinos, and run their own governments. These rights were reserved through treaties with the federal government. In order for Tribes to perpetuate cultural sovereignty they must have access to their usual and accustomed hunting and gathering sites both on and off their reservations. They must be able to participate in federal and state environmental and habitat restoration programs. Cultural resources, both tangible and intangible are very critical and sacred; and, Indigenous people feel they need to protect what little remains."

The title of this book is portrayed by a stone circle built by North American Indians that symbolized a mental construct of the natural world that includes religious, astronomical, territorial, and calendrical significance (*see* Box 1). It describes a holistic approach to making environmental and human health decisions through our relationships with nature — this is the tribal way to form knowledge on nature. The Medicine Wheel acknowledges that ecosystems experience unpredictable recurring cycles and people/environment/ecology are all interconnected. A Medicine Wheel depicts how the spiritual and physical aspects remain connected to the land, and also the relationship of people to nature through ecological calendars. In the Indigenous knowledge framework, the foundation of people's ethical behavior towards the environment is rooted in the knowledge gained through the inter-generational transfer of holistic information that interconnects local and landscape scales and cultures. This is what you will learn when you read the first part of this book. This is what Western science and societies mostly do not practice and they don't understand what it means to form holistic knowledge of nature.

Box 1: The Story of the Medicine Wheel

"The Medicine Wheel can be called a mental construct. It orients us on a time-space continuum. The Wheel divides our world into different directions and applies specific meaning and significance to each direction. This directional orientation is achieved by simple observation of the natural world. The sun rises in the east and sets in the west. Regardless of where we sit on the globe there are four phases of the moon and typically four recognized seasons. These phases and seasons follow each other in a circular and sequential rotation, because of this, our personal medicine wheels are a reflection of our relationship to the natural circular evolvement of the world." [1]

In contrast to the Indigenous knowledge framework, the Western science approach treats each aspect of the Medicine Wheel as a separate problem in search of a solution. If the holistic approach of the Indigenous knowledge framework is not used, the "symptom" of a problem will most likely be identified and not the "disease" causing the problem. An analyst may be unaware that they are

---

[1] http://www.dancingtoeaglespiritsociety.org/medwheel.php; Photo Source: U.S. Forest Service Photo — http://www.fs.fed.us/r2/bighorn/, Public Domain, https://commons.wikimedia.org/w/index.php?curid=3358774

resolving a symptom of a problem when statistical tests produce significant relationships among the studied variables. Thus, solutions developed address the symptom and not the "real problem". One of us remembers being at a meeting where a well-known scientist presented research results of significant ecological relationships but was unaware that the highest correlation occurred between their variables and the sample numbers. What kind of solution is derived from knowing the sample number! But it was statistically significant!

Another strength of the Indigenous knowledge framework is its ability to address **scarcity of knowledge**. Since the scarcity of knowledge is local, the focus of the Western science knowledge framework on "airplane level science" will define the problem at the larger landscape scale and not the local scale. The decision process will most likely identify the "symptom" of an environmental problem and not the "disease" causing it. When a bandage is put on an environmental problem (e.g., decision process treats the symptom not its causes), the impacts of the disease (or its cause) could persist for decades and cost societies millions of dollars in mitigation efforts. This has been the problem the world has faced when researching solutions for complex environmental problems.

If we do not treat Indigenous knowledge as credible as sources of information, global environmental problems will probably continue to persist for many decades. **While consuming resources, industrialized societies damage the interconnections in global carbon and nutrient cycles. In contrast, Tribes and their ancestors live resiliently in nature and leave small footprints on the land. Tribes were not passive consumers of environmental resources but sustainably managed nature for thousands of years, often with comparable human population sizes as found today.** We have converted what once used to be tribally managed lands to urban landscapes that support little of nature (*see* photo collage at the front of the Preface). We are not going to eliminate urban landscapes but there is a need to re-establish the balance of nature to the urban landscapes for both human and ecosystem health.

In the Indigenous knowledge framework, the foundation of people's ethical behavior towards the environment is rooted in the knowledge gained through inter-generational transfer of holistic information that interconnects local and landscape scales. This knowledge provides an approach to address an environmental problem. But it does not lead you to adopt a pre-determined endpoint or solution. This process allows Native American Tribes to prosper despite environmental/economic and social boom-and-bust cycles. It is a more environmentally ethical approach to distribute the bounty of nature. The process recognizes and includes multiple stakeholders, along a continuum of space and time, who all have a stake in the solutions. The Salmon Chief exemplifies this ethical behavior of an Indigenous knowledge framework. The Salmon Chief was selected by multiple Tribes to make decisions so all the people living up river and down river received an equitable amount of the salmon that were swimming up the river, even if some communities were enemies and fought battles against one another. This approach precludes special interest groups from controlling which solutions are adopted or what regulations are implemented.

A holistic knowledge-forming process is more likely to identify the "disease" and not the "symptom" of an environmental problem. To treat a "disease", we need different problem-solving skills and we should not be bound by disciplinary-based science experts. We have all talked to scientists who *"can't see the forest for the trees"* because they are overwhelmed by the minutia of massive databases in their discipline. Further, data that does not fit the commonly accepted facts are frequently ignored or thrown out [7]. It took Alexander Fleming to notice the antibacterial substances being secreted by the *Penicillium* mold growing on petri plates. He could have just thrown the petri plates out after noticing they were contaminated as had many others before him. But he looked at them more closely even though they did not fit the "normal" image of what science said he should see. He is credited with developing an important medicine, i.e., penicillin. **We don't want the Indigenous knowledge framework to become a contaminated petri plate that is throw out because it doesn't fit the knowledge-forming process of Western science.** We contend that an ethical environmental and social future depends on us acting more like Alexander Fleming.

The Medicine Wheel is described by Dr. Mike Marchand and supports why it's our book title.

1.1 Bighorn Medicine Wheel Story*

This Bighorn Medicine Wheel, located in Wyoming, is an intersection of human thought on many levels. It is an intersection between the physical world and

Dr. Mike Marchand.
Photo by Benjamin Drummond.

---

* By Dr. Mike Marchand.

the sacred world; an intersection between the physical universe and the human mind; an intersection between the ancient times and today. It crosses time, no one knows for sure when it was built. Some of the local Tribes say it was there before them and they have been there for thousands of years. The local national park sign said it is only a few hundred years old, based on carbon dating of a fire pit. Rock formations are difficult to date. It seems likely to me that the fire charcoal they dated was built long after the wheel was created. It is a sacred site to the Native Americans. The circle is a sacred symbol.

Likewise, science has many cycles: Hydrologic cycles, biological cycles, astronomical cycles such as months and years. Native Americans tended to like to live in circular homes, with round tepees and round lodges of all sorts. There are 28 spokes, the same as the lunar calendar and that of the female cycle. It is located high on top of the Bighorn Mountains and connects the sky to the earth. The wheel is constructed of stones laid out in a circular pattern about 80 feet in diameter. It is located on top of the mountains on a northwest facing ridge top. There are stones aligned with certain stars in the sky, the moon, and the sun. It is constructed based on observations of natural phenomena, which is the basis for much of modern science. The site was used for ceremonies and children were sent there for their personal vision quests, a common Native American practice for understanding themselves and the world around them. It was at a crossroads for many Native American Tribes who once crisscrossed the North American continent. There are other medicine wheels in North America, but this must be the most spectacular one to visit.

Jutting straight up 12,000 feet out of the Northern Plains are the Bighorn Mountains in Wyoming. They stand like sentinels over the lowlands. One can see for miles and miles from the top. From the high perch of the Bighorn Mountains, the earth looks like a hoop. The sacred hoop is an important Native American symbol.

As a newly married young man years ago, my wife and I were driving cross country. We had watched the daily rodeo in Cody, Wyoming and planned to drive over the Bighorn Mountains on our way to South Dakota for more sightseeing. As we climbed the windy road full of switchbacks to the top, we noticed deer heading downhill. The sky was black, there was a storm on the top. Apparently, they were heading down to get out of a storm. As we neared the top we entered a big thunderstorm. It was very windy, very dark. At the crest of the climb there were two-four-horse trailers blown upside down by the wind. Most of the horses had been thrown out of the trailers and were running around in sheer panic. One horse stayed in his upside-down trailer and was visible under the interior light of the trailer. Some cowboys were running around trying to catch the horses. A couple of cowboys were on the road to warn traffic. It was a bizarre sight. Chaos. The horses looked slightly injured but were mostly scared. The power and fury of nature was clearly visible. Wind. Lightning. Humans and animals and metal vehicles tossed around like toys. Insignificant pawns to the power of Mother Earth and the Creator whom my people call Quillenchooten. That night left a lasting impression on my mind. The power of humans versus the forces of nature and the creator. We are pretty small creatures on that chessboard.

1.2 The Medicine Wheel: Non-linear Knowledge-forming Process

A fundamental difference between the Native American and Western science view of the world is apparent from the shape of a home you prefer to live in. The Western world prefers squares shaped homes while the Native American prefers circular shaped homes. This quote published by Jonathan Hook in 1997 in his book entitled "The Alabama-Coushatta Indians" [8] describes the American Indian preference for circles (Fig. 1.1) and White man's preferences for squares:

Fig. 1.1 Colville reed tepee, Colville Indian Reservation, ca. 1903.
Photographer — Edward H Lathan. University of Washington Libraries. Special Collections NA964.[1]

---

1 Accessed on June 5, 2018 at http://digitalcollections.lib.washington.edu/cdm/singleitem/collection/loc/id/1529.

"You have noticed that everything an Indian does is in a circle, and that is because the Power of the World always works in circles, and everything tries to be round... Everything the Power of the World does is done in a circle. The sky is round, and I have heard that the earth is round like a ball, and so are all the stars. ... The life of a man is a circle from childhood to childhood, and so it is in everything where power moves."

In LAME DEER, SEEKER OF VISIONS, Richard Erdoes describes the life of a Lakota medicine man, John Lame Deer [9]. He writes that Western society is rooted in a dramatically different world view:

"The white man's symbol is the square. Square is his house, his office building with walls that separate people from one another. Square is the door which keeps strangers out, the dollar bill, the jail. Square are the white man's gadgets—boxes, boxes, boxes and more boxes—TV sets, radios, washing machines, computers, cars. These all have corners and sharp edges—points in time, white man's time, with appointments, time clocks and rush hours—that's what the corners mean to me. You become a prisoner inside all these boxes."

The differences in linear versus circular thinking are important to recognize because it determines how each person approaches problems or forms knowledge. Indian Country Today summarized this well [10]:

"Linear thinking versus circular thinking are explored in the program, as is the hierarchy inherent in American business and life. With linear thinking, we rely on logic, institutions, and others to try to protect ourselves. Circular thinking originates from the earth, the universe, and the Creator; we are all connected and all safe. Two Bulls explained at a recent conference. The Medicine Wheel Model is about social change; it's about having to look internally before you can work externally."

Astronomy depicts and favors circles when describing space and time. The Medicine Wheel depicts several circles. Dr. Mike talks about his grandmother wanting to live in a round shaped house — her tepee. She didn't want to live in a square house that the White people built for her to live in. Dr. Mike Marchand said that she finally put her animals in the square house while she lived in the circular tepee. Similar preferences for round shaped houses are found in Navajo country (see Box 2).

> **Box 2: Description of a Traditional Navajo Home**
> (Dr. Nancy Maryboy, Dr. David Begay)
>
> *Hooghan*: A traditional Navajo home, is a circular dwelling with a dome shaped top, a replication of Mother Earth and Father Sky, the cosmos, as viewed from earth. The fire is in the center of the hogan and is associated with the North Star, according to traditional star knowledge holders. The term *"hooghan"* is sometimes spelled "hogan" in English.

In the Western approach to science, the linear way of thinking is based on using the scientific method. Linear thinking in science is tested using experiments. If it cannot be tested, it will not be accepted as fact. For example, Einstein's Theory of Relativity needed 98 years before an experiment was designed capable of proving what he had hypothesized[1].

The Medicine Wheel is circular and not linear. Dr. Nancy Maryboy et al. [11] describe well the Native definition for science and why it is circular:

> *"For many Native people, science is better understood in the context of place-based knowledge, knowledge that comes through generations of experience, normally tied to the geography of the land on which the people have prehistorically and historically lived. In native cultures, place-based knowledge is descriptive; it is often tied to specific and interrelated empirical ways of knowing."*

In "The Cosmic Serpent", Dr. Maryboy et al. [11] describe the cosmic serpent as a global symbol that highlights the interconnected nature of fundamental concepts for earth, space, life and environmental sciences" (Fig. 1.2). Their cover of their book depicts a double headed serpent representing Tribes around the world and a process for engaging both Western science and Native people's ways of forming knowledge [11]. Both ways of forming knowledge cannot be merged to form a more robust knowledge-forming process since each loses the basic elements that make them unique.

Dr. Maryboy et al. [11] further write that when Native people interact with scientists they have to translate their science using "scientific practices" or what is typically called hypothesis testing and linear design approaches. Further, for Native people *"transferring holistic knowledge, a world of interrelationships, to a reductionistic method of thinking"* loses what fundamentally forms Indigenous knowledge [11]. Therefore Indigenous knowledge converted to Western science no longer accurately represents Indigenous ways of forming knowledge. Convert-

---

1 Accessed on February 23, 2018 at https://www.space.com/37018-solar-eclipse-proved-einstein-relativity-right.html.

Fig. 1.2 Cosmic serpent.
Source: Maryboy et al. [11].

ing the Native people's web of science, that is a "spider web" of network of relationships, [where] everything is interrelated is not practical. They further write:

> "In both Western science and Indigenous knowledge, we see the use of observation, prediction, cycles, and processes, to provide understandings of nature and the universe in systems worldwide".

Circular thinking is more difficult to test using the Western scientific method. It is not amenable to testing a holistic form of knowledge that includes spirituality as well as art, plants, animals, geology and physics. Further, the end-point parameter varies in the circular thinking framework, but the Western scientific method needs a parameter that doesn't change or you can't test the hypothesis.

The traditional Western world environmental economic model is linear (Fig. 1.3) while the Indigenous business model is circular with no consistent starting point (Fig. 1.4). A quick glance of the Western environmental economic model looks like it is also circular, but it has a starting point and end point that is repeated in each model run. The fluxes or inputs into the state variables are selected from a suite of parameters that feed into the model and "unknown variables" are not generated by the model run that could modify the order in which inputs are assessed. All parameters are selected up front and the range or threshold of each variable is defined before the model run. The modeler evaluates indicators (fluxes) going into or out of each state variable (economy,

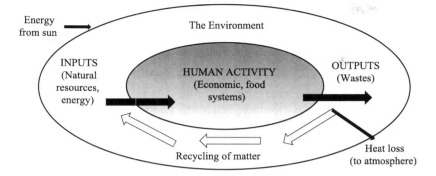

Fig. 1.3 Adapted from a figure depicting a Western environmental economic model. All human activity is embedded in the environment in the environmental economic model[1].

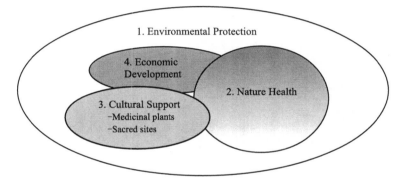

Fig. 1.4 This characterizes an Indigenous business model and priorities of Native people for their forestlands (*see* Section 5.3.1).
NOTE: Economic Development has a fourth priority for tribes from their lands.

environment, society) that are based on consistent relationships formed over decadal time scales. Therefore this is a very linear model with fixed starting and end points that follows a predetermined path in a model run.

A Western environmental economic model has inputs, outputs that sustain human activity. Outputs or wastes are not lost from the model but are calculated as part of the model predictions. Santone [12] describes this as:

> "Reflecting these ideas... shows the true relationship between natural systems and human systems (the economy). Here, the environment is not merely a factor of production (as it is portrayed in the conventional 'circular flow' economic model), but rather the containing system for the economy. The diagram shows that environment is the source of all

---

[1] Accessed on March 19, 2018 at http://www.pelicanweb.org/solisustv07n05page8.html

materials humans use and the 'sink' into which all wastes go; moreover, wastes stay in the system and do not go away."

Western world economic models are still researching how to include people's behavior in making economic decisions. So, the emphasis on understanding how culture explains economic and financial decisions is just emerging in economics. In 2017, Richard Thaler received a Nobel Prize in Economics for his research that included behavioral economics where humans are irrational players in making decisions[1]. In giving the prize to Thaler, the Swedish Academy of Sciences wrote:

Limited rationality: *Thaler developed the theory of mental accounting, explaining how people simplify financial decision-making by creating separate accounts in their minds, focusing on the narrow impact of each individual decision rather than its overall effect. He also showed how aversion to losses can explain why people value the same item more highly when they own it than when they don't, a phenomenon called the endowment effect. Thaler was one of the founders of the field of behavioural finance, which studies how cognitive limitations influence financial markets."*

Before that, economics had considered humans as being rational players when making economic decisions. Thaler showed that people make bad decisions in economic markets because they are irrational. His idea was that people's ownership of something increases their view of the value of the owned asset, where they would charge more money for it. Thaler's perspective reduces the economic discussions to a general focus on ownership and how ownership changes the value a person assigns to something.

The concept introduced by Thaler is contrary to the perspective held by Tribes who do not own land but use resources from that land while living on it [13]. Dr. Mike wrote about how the most important role for a leader was to give things away (*see* Section 2.2). This meant that the leader had the shabbiest house and living conditions compared to the rest of the tribe. Tribes make economic decisions based on their culture and not whether human behavior is rationale or not.

Thaler's economic theory explains the irrational behavior that occurred in 2018 when Cooke Aquaculture's operation collapsed in the San Juan Islands and released more than 200,000 non-native Atlantic salmon into the waters of the Northwest United States. They valued the non-native species because they owned them. No economic theory would have been able to pro-actively respond to the potential disaster that unfolded. It took 21 Tribes living in the Puget

---

[1] Accessed on June 21, 2018 at https://www.nobelprize.org/nobel_prizes/economic-sciences/laureates/

Sound region to respond to this event by sending a letter to the Washington State legislators. But it was too late to prevent the impacts that occurred. So, the business model was not designed to address these types of problems or environmental externalities of the business decisions. The Ecotrust BLOG written by Lisa Watt summarized the situation as [14]:

"Our investigative team doggedly pursued the truth," said Maia Bellon, director of the Washington Department of Ecology. "Cooke Aquaculture was negligent, and Cooke's negligence led to the net-pen failure. What's even worse was Cooke knew they had a problem and did not deal with the issue. They could have and should have prevented this." Cooke disputed the findings.

The Indigenous business model is more like a Medicine Wheel that includes more variables and no fixed variable end points or expected outcomes (Fig. 1.4). It can be difficult to identify the starting point of the model or where the model ends. The importance of input variables, e.g. outputs or wastes, are not included since they are not an important output for Tribes. In the Indigenous business model, life style reduces wastes sufficiently so that they become a minor part of the output.

Indigenous business model is better depicted as a web where cultural and environmental protection are more important (Fig. 1.4, *see* Section 3.2). This model makes it hard for you to fix on one point as being the most important input variable to the model. Some people may not like this type of model form since you have to think of more connections and links then is possible if you focus on inputs and outputs. It forces you to explore more options and tradeoffs instead of defining up front the pools and fluxes ahead for each problem. Lane [15] describes this an "Indigenous view of Sustainable Development. ... The closest equivalent that the Cuna Tribe of Panama has for the term 'sustainable' is the word *harmonious*". Sustainable development has very different interpretations in the Western environmental economic models. It is interchangeable with the business model, in contrast to the tribal business model. In the environmental business model, economic opportunities are more important than in the tribal business model. Dr. Nancy Maryboy calls this "balance".

The Medicine Wheel concept is very similar to the design used by spiders to weave their web (Fig. 1.5). The spiderid.com website describes spider webs as:

"Spider webs can be quite delicate, or exceptionally strong, depending on the species and age of the spider. Webs of black widows, for example, are expansive (usually about a cubic foot) and incredibly elastic. You can pluck the threads like guitar strings without breaking them. The webs of Nephila *orb weavers from tropical regions are so strong that*

Fig. 1.5 Photo of a spider web[1].

> *native peoples in Papua New Guinea use them as handheld fishing nets. In 2010, it was found that a species of orb weaver from Madagascar, <u>Caerostris darwini</u>, produces the world's toughest biological material. Spider silk, in general, is widely regarded as the strongest natural fabric known, at least half as strong as a steel thread of the same thickness, and much more elastic."*

Spider webs are similar in structure to the Medicine Wheel which is also incredibly elastic. You only have to walk through a wet tropical forest to experience how strong spider webs are. When you accidentally walk into a web, it is a struggle to get rid of the web. At the Puerto Rico NSF LTER site, Drs. Daniel and Kristiina Vogt frequently ran into spider webs in these forests. The spiders lived in condos with multiple individually built webs interlinked together. Can you image running into one of these condos and then trying to detach yourself from very strong and sticky web filaments? What we experienced in the forests is this elasticity and strength of connections. It describes the strength of connections between Indigenous cultures and how culture provides the elasticity to a community. The lack of cultural strength in Western economic models make them brittle and easily broken. It suggests the importance and strength of circular structures compared to linear models where one repeatedly ends up with the same end point each time an analysis is conducted. The Medicine Wheel approach to a business model does not produce a predictable end point which makes the model results elastic and not brittle.

The Indigenous business model is easier to understand by looking at the iceberg metaphor (Fig. 1.6). In the Western environmental economics model, most of the decisions play out in the exposed part of the iceberg. In contrast in the Indigenous business model, decisions play out in the deep culture which is not

---

1 Accessed on March 19, 2018 at https://www.pexels.com/photo/spider-web-34225/

Chapter I Indigenous Knowledge Framework and the Medicine Wheel

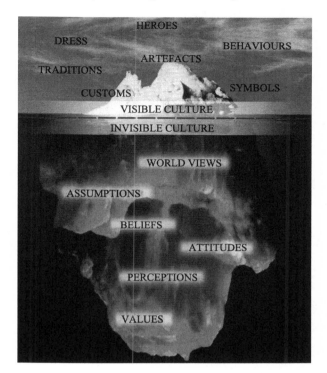

Fig. 1.6 A metaphor for surface (VISIBLE CULTURE) and deep (INVISIBLE CULTURE) culture with the Visible Culture including "Language, Folklore, Laws, Music, Customs, Food" and the Invisible or Primary Culture including "Patterned implicit rules of behavior, hidden cultural grammar"[1].

The iceberg image was produced in 1999 by Ralph A. Clevenger, faculty of the Brooks Institute of Photography, Santa Barbara, California.

visible or recognized by the casual observer (*see* Section 3.2). Lane [15] equates the Medicine Wheel Model as a powerful mapping tool and further describes the hidden culture as:

> "A set of unspoken, implicit rules of behavior and thought that controls everything we do. This hidden cultural grammar defines the way in which people view the world, determines their values, and establishes the basic tempo and rhythms of life. Most of us are either totally unaware or else only peripherally aware of this."

---

1 Accessed on June 3, 2017 at http://civet.dedi.velay.greta.fr/book/export/html/36.

Chapter II
What Is Needed to be a "Leader without Borders"?

Iceberg image source: http://www.freedomisknowledge.com/otw/stuff/tipoftheberg.htm

Native people have very different models of leadership compared to the Western world. The Western world views of nature are short-term, and the scientific method determines what factors need to be included as driving the changes that are observed. Nature knowledge is not informed by intergenerational knowledge passed down through the generations. Many of these differences are highlighted in the following piece written by Dr. Mike Marchand.

2.1 My People's 9,400 Year Ancestral History*

My people are Native Americans in the Pacific Northwest. I live on the Colville Indian Reservation, established in 1872. About 20 years ago an intact skeleton was discovered in Southeastern Washington State and he became known as the Kennewick Man. Carbon dating determined he had lived 9,400 years ago. This sparked a controversy between Native knowledge and scientific knowledge.

The Native American Graves Protection and Repatriation Act (NAGPRA), dictated that skeletal remains such as this be repatriated to the local Tribal descendants. But the science world balked in this instance. We saw in the

Dr. Mike Marchand.
Source: Mike Marchand.

---

* By Dr. Mike Marchand.

newspaper that the Smithsonian Museum was intervening and was advocating the transfer of this skeleton to their museum in the interest of science. I along with the Tribal Police Chief and Tribal Historian took a jet to see the Smithsonian. The Smithsonian Museum was the leading proponent to break the law and said there was no scientific proof that the Kennewick Man was in fact a Native American. In fact, they believed him to be European. There were many wild theories flying around ranging from him being a Viking to a look-alike of the Starship Captain of the movies. The Smithsonian paleontologist acted superior and matter-of-factly, stating that we had no legal claim on the remains and the law did not apply. Further, they said that we could not prove he was our ancestor. I said that in our traditions we know that we were here for thousands of years and that he was in fact our ancestor. He repeatedly said we cannot prove that. That is how the meeting ended.

It ignited over 20 years of conflict and litigation, involving many Northwest Indian Tribes versus the federal government and science forces such as the Smithsonian Institute. Eventually the science of DNA progressed to the point where a scientist in Denmark took DNA samples from me and 21 other Colville tribal members. He established a DNA connection between my tribal members and the Kennewick Man from 9,400 years ago. He was in fact our biological grandfather.

It still took an act of Congress to force the transfer to the Tribes. I was present that day at the Burke Museum in Seattle, Washington. Kennewick Man was transferred from museum boxes and plastic bags onto a buckskin robe. Spiritual leaders sang sacred spiritual songs following ancient protocols. The Chairman leaders from over 30 northwest Tribes were standing witness to the ceremony for hours. Finally, during a break we looked out from the back of the museum on the loading docks to get some fresh air. Someone said look up! There were two eagles spiraling over the Burke Museum during the transfer of the Kennewick Man to his descendants that day. As we watched the two eagles, they started to climb up into the sky straight up, spiraling out of sight.

2.2 Becoming a "Leader without Borders": Interview of Dr. Mike Marchand[*]

Kristiina: I want to get an introduction to you as a person and what impacted you the most as you grew up and still influences your thinking today? We want to see what makes you a "leader without borders." Before we start our interview, it is important to mention how we first met. You applied to work with me as your PhD advisor under the auspices of an NSF IGERT (The National Sci-

---

[*] By Alexa Schreier and Dr. Kristiina Vogt

Photo Dr. Mike Marchand (left), Alexa Schreier (middle), Dr. Kristiina Vogt (right).
Sources: Mike Marchand, Alexa Schreier, Kristiina Vogt.

ence Foundation's Integrative Graduate Education and Research Traineeship) program focusing on energy production from forests and Tribes. This program was jointly managed through the College of Forest Resources (my home program) and the College of Engineering. You blew everyone away before you even showed up at the University of Washington because of how well you did on the Graduate Record Examination (GRE). You did not study for the GRE exam but got almost a perfect score of 100%. This is an unheard-of result since even the smartest people don't obtain such results. It especially made the engineering faculty take notice of you since you clearly are smart. This is something they respected. The other PhD students in engineering also respected your knowledge and would look to you to understand Tribes and their practices. I heard them quoting you on several occasions even though they never quoted anyone else from the College of Forest Resources. We should also note that you're the former Chair of a very large tribe (Confederated Tribe of the Colville) — a position that you have held before and you have also been voted in several times by your tribe to serve on the Colville Tribal Council. You also received a PhD degree from U.W. which makes you a HUSKIE!! So, let's have you introduce yourself so we can begin to understand what makes you Dr. Mike, who we are calling a "leader without borders".

> **Mike:** *Well I was born on the reservation in a tribal hospital, which there are not a lot of tribal hospitals. I was raised on the reservation. I was raised in an unusual family, my family were leaders, tribal leaders, and elected leaders, and back in time they were hereditary chiefs. It's been like my family business. One common denominator with Tribes is that everyone in a tribe has a role, whether it's as a leader or as a worker. Maybe some people are good with their hands or art or something, but ideally everyone has a role in the tribal society. Now we aren't always ideal either, but that's the general goal and philosophy. Just because of your role, it doesn't mean you're more important than other people. They value you by different criteria than what are used in*

Dr. Mike Marchand (middle) after defending for his PhD at the University of Washington in Seattle. Left: Dr. Daniel Vogt (left), Dr. Kristiina Vogt (right).
Photo Source: Daniel Vogt.

modern cities. So, in the United States "What's important?" Money's important. Trump's the President, a billionaire. In a lot of tribal societies, its different values that determines whether you are important. Often you see how generous a person is by how much they give away. This is valued more. Or, someone that helps other people is valued more. Money is not really bad, but it's not the main thing that they judge you by.

Kristiina: *So, these are the types of things you saw as you were growing up?*

Mike: *Yes, I think Tribes, at least my tribe, assesses every individual and tries to encourage them to be good at whatever they're born with or as far as their talent will allow them to achieve. I had four younger brothers but none of them were treated like me. They were all treated different, I'm not saying I was treated better, they were just treated differently. They plucked me out very early on. They were kind of training me to be a leader. I didn't know that at the time, I just thought it was normal. But each one of my brothers was treated a little bit different, they were treated as individuals, whatever talent they had, they tried to make that blossom. I think my parents were pretty good at that, they're very positive, my grandparents were the same. We're very positive with providing positive reinforcement, we didn't beat people or coerce them. They hardly ever got mad at you. Many times, I thought I'd be in trouble and no one did anything, they just kind of said "Be careful."*

We were always taught that traditions are important, the past is important, the community values are important, helping out your tribe is important. I was always taught that from the time I was a baby. For example, we're still a hunting culture, we're still a fishing culture, and traditional foods are valued. You can go get food at a Safeway or QFC and that's okay, but higher up in the hierarchy is traditional foods, natural things like deer meat, salmon, roots, or berries. Those have a higher value in our traditional system. Plus, those are healthier for you anyway, they're not processed or anything. So, traditional foods are important, water is important.

When they start to eat a traditional meal, they have a protocol where they start with certain ceremonies. They line up food in a certain way. But water is the number one thing. That's very common in tribal societies. Like this summer, when there was a big protest at Standing Rock, it was all about the water. It's pretty universal among the Tribes, water is life, people almost make fun of that, but it's true. So, as a child, we'd have family — I'd call them a clan, extended family — or clan dinners and get together like for Thanksgiving. These were celebrated just like normal American holidays except it was a bigger feast. We would have extended families and cousins show up from all over the country. A lot of them lived on the reservation but they had things like traditional turkey, like pilgrim dinners, but they also served deer meat and wild game.

When I was a young boy, if I shot a deer as a child, my grandmother would make a big point to say "Look at this meat, Mike shot this, he's providing for the family." Whoever shot the game were treasured people that supported the family or the community. Well, the family first and then the community. You're treated like royalty if you brought food to the table such as deer or salmon or even huckleberries. You were treated as special for helping out.

Kristiina: So they never really had to say to you, this is how you should behave?

Mike: *No, it's just positive reinforcement, you see everyone eating and they're happy and having a good time and you get credit for it. They'll say, "Mike went out and shot this deer that we're all eating." Maybe I was 10 years old and to a young child that's a big thing. Everyone's patting you on the back, so you're praised whenever you do something good. Usually it's associated with you giving to the family. So, it was food or just chores. When I was a young child, a lot of places didn't even have electricity or running water. So children were expected to pump water and bring it into the house with buckets. People cooked* with wood. They didn't have electric ranges, so you were chopping wood and hauling wood. These are just menial chores that children would do. But they praise you for doing that, reward you, and pat you on the back. All positive reinforcements. So you're taught when you are a small child that you're helping, you're doing something important. Everyone's important.

Kristiina: That really seems different from how children grow up in today's non-tribal communities in the US. You don't think about whether your child should experience unique activities tailored for them to become a future leader. Parents plan mostly to expose their children to popular team sports or the arts, but every child is given similar experiences. Parents complain about having everything planned but still take their children for their music lessons followed by their sports activity. As a child somebody else is planning everything you do and how you do it and there are no tailoring activities for each child. Children are disconnected from the world. What are your thoughts on this?

Mike: We see both worlds as an Indian child. Although when I was young, I had no idea why they selected me. They must have thought I had the potential or something. My grandfather was the Chairman of the tribe. To me he wasn't the Chairman of the tribe, he was just my grandfather. Looking back at pictures, he was wearing a suit, he was wearing a fedora. He was all dressed up, even by today's standards (Fig. 2.1–2.2). He was a businessman, but he was also a rancher, so he'd take those clothes off and put on his cowboy clothes. But to me he was just my grandpa.

Fig. 2.1 Photo of Dr. Mike in the middle, his mother on the left and grandfather on the right.
Source: Mike Marchand.

Fig. 2.2 Dr. Mike Marchand's grandfather.
Source: Mike Marchand.

He would drive me around and he would say, here's a site where villages used to be, or here's a site where the Tribes did this, or here's a sacred site. To tell the truth, I was getting bored of that, he'd drag me around and I'd have to hear all these stories. We pulled up to this one site that was like a picnic ground, a state picnic ground and I was thinking what sacred site grandpa's showing me today. He said there were these caves near Lake Chelan, and they're ice caves. Then I asked, "Well what'd they do there?" I was just kind of being ornery and sarcastic. He said, "Oh that's where we used to put our beer." Well I was only in like the first grade or so, so I barely knew what beer was. But it sounded funny though because I wasn't expecting that kind of answer. I was expecting some big Indian traditional story. But he was kind of joking with me.

We went a little bit further down the road and the next stop down the road after Chelan was Entiat. This happens to be where my mother's people came from on little Entiat River, north of Wenatchee. They had a little village there and were called the Entiat People. Traditionally there was a little river there. My grandpa stopped by the road where there used to be a bridge across the river. But the Rocky Reach Dam was built in the 1960s and flooded all the banks resulting in a big reservoir in the Columbia River. My grandpa said right under the water here, it's all flooded now, there was a village. That was our people's village, the Entiat village. He said there used to be a lot of tepees in that village down there. Then he pointed to a spot and said that's where the Chief lived, that was your great-grandpa and his name was **Silcosasket** *that means "standing cloud" in English. He lived a long life and he had lots of children, he had several wives. It was dangerous being a woman in those days, lots of women died during childbirth. So, he had three wives, not at the same time, but one right after another. He had a lot of children. In that tiny little village, all his kids had big families. Like half of the Colville tribe today are his descendants. It was this little tiny tribe and now we have 12 Tribes today and a lot of them descended from this chief, he's not just my great-grandpa, he's a lot of people's great-grandpa.*

Anyway, back to the story, my grandpa said, "That's where the chief lived." I asked him "What did he do for a living?" I was trying to imagine what they did, what kind of jobs they had. He said, "Well he was the chief". And I said, "What did he do for food?" He thought about it and then said, "He didn't have to worry about that, everyone brought him food." I said, "Well that's pretty good!" Then he asked, and now he's trying to teach me something, "What do you think his tepee looked like?" I'm thinking, well he's the chief and they brought him food, so he must have been like the emperor or something, so he must've had the best tepee in town. My grandpa just laughed and he said, "No he didn't have the best tepee, in fact he had the worst tepee." He had an old, holey, falling down tepee. He looked very poor, he didn't have fancy clothes or anything, because he gave everything away. He gave everything to the orphans, to the widows, to everyone that needed help. He gave everything away. When he was younger, he was a warrior, so he was willing to give his life away, fighting wars for our people. But that's the idea, he's the most respected person, he gave everything he had to his people. They took care of him, they would feed him and invite him to their tepee. He was the VIP, they treated him well, but he gave them everything too. So, there was an exchange, a kind of reciprocity there, where when you're good, you give, and the tribe takes care of you.

Kristiina: Was your grandpa the most important person impacting you when you were growing up or did you absorb things from lots of people? Did he make you think about things?

Mike: *It was probably my mother and my grandpa, because my mother was a Council Member too and probably when I was younger it was both my grandparents. That was kind of traditional. Grandparents kind of raised the kids, because back then, a lot of times parents would be working. In traditional times, parents would also be working, but they* *might be hunting or fishing, or gathering roots or berries. So even thousands of years ago, the grandparents played a big parental role, and uncles and aunts, along with your extended family. So that's been common for thousands of years and it's a little different now, but that's how I was raised. A big parental role was played by a large extended family. It is changing now. But this was how I was raised.*

When I was a very young child my grandparents had a big influence on me; the one I was talking about was my maternal grandparents, he was a tribal chairman. When he was young, he was one of the first people educated in English. He went to a Catholic school. It was one of the first schools in the Northwest. It was called St. Mary's Mission and it was run by Jesuit priests and nuns from Austria (Fig. 2.3). They taught a classical Catholic education, which is math, geometry, Latin, and Greek. It was a universal Catholic education taught around the world. They taught everyone to read the same classical literature, and even though it was in a frontier reservation, that's what they taught. That is probably the same education you got taught in Vermont or Boston or even in foreign countries. It's actually a pretty high level of education. My grandpa stayed there until the eighth grade. My grandpa

Fig. 2.3 Colville Indian Reservation school, Washington, ca. 1908, Palmer, Frank, 1864–1920. Twelve schoolchildren and their teacher pose outside their one room schoolhouse on the Colville Indian Reservation. The school is a log cabin construction.[1]

---

1 U.W. Digital Collection, Negative Number L84-327.1103.
https://digitalcollections.lib.washington.edu/digital/collection/loc/id/1094

was raised in the rancher's family and his dad said, "No one needs this much school, you need to get out of school and get to work." It was a small school, so he was probably doing college level work in eighth grade. He was offered a scholarship to Oxford University in England, so he was a gifted guy. His friend was named Pastel Sherman. He got educated too and became one of the early tribal lawyers. He went back east to D.C. to practice law and he was pretty influential in Indian politics too.

Kristiina: So, a lot of the education was based on learning from books and you probably needed to even memorize the Bible, right?

Mike: I'm sure he did. He also had to learn the Indian side too. He had both a western education and a Native American education. So, it's like two educations. My grandpa's talent was for languages, he could pick up languages fast. A big problem with the Tribes in those days was that none of the Tribes could speak English very well, so he was the interpreter type guy, who could speak for Tribes and communicate with the United States. He picked up English fast and other tribal languages too. He could speak almost any language in the Northwest and he could speak world languages like Greek, so he was a gifted guy.

Tribal leaders in those days didn't have money to hire lawyers for the Tribes. So, they'd have to be the lawyer for the tribe, and they would have to communicate to Congress and congressional lawyers. He was pretty self-educated in law. He knew Congressmen and Governors.

Kristiina: You have talked about in the past where tribal members went to Washington D.C. to deal with tribal issues and did not just stay on the reservation. They would get on a train that took a while to travel to D.C., so it was not necessarily convenient. It seems that the tribal responses have always been to never isolate themselves. It would have been very easy to close the doors and hide out since they had not been treated very well by the westerners and there was not a lot of trust between them. They have been open to other cultures. Tribal people recognized that they needed to be heard and D.C. people were going to influence what they would be able to do. Can you comment on this?

Mike: I think that varies between Tribes. Our tribe has always been kind of aggressive and open and taking care of their rights. Other Tribes maybe did try to stay out of sight, out of mind, and that's just how they looked at the world. I don't know. Maybe they were too small to have resources. There are some Tribes nearby us who do kind of hide out on the reservation, even now. My tribe has always kind of reached out, to *protect their rights, go to Washington D.C., take the train to D.C. In those days, that was the only way to get there, there were no airplanes.*

That's partly U.S. tribal history too. There's always been a link between Tribes and Washington D.C. since before there was a country. It goes back to the British rule. What happened was as settlers were coming into North America, they tried to make deals with Tribes, one on one. Some guys would get off the boat, like the Mayflower, and try to go buy land from an Indian. Well the Indians didn't really have the same concept of land, so it'd be like coming to Seattle and saying, "I'd like to buy some air, go down to Pike Place Market and buy some air." Someone would say this guy is crazy, so they'll sell you air. Give me $20 or something for the land. Then the next white guy would come and they'd sell him the same piece of land, which would cause fights. The Indians thought well this is silly, they're paying money for something you can't even buy and sell.

What happened was the British Empire said this has to stop. The only ones that can negotiate with the Indians are the official convoys of the King which was King George at the time. So, only King George could buy land from the Indians. So, if you were just a regular pilgrim, you couldn't deal with the Indians. It had to go through the King, because otherwise it just causes chaos and conflict, and everyone gets mad and kills each other. So that's how it was in the 1500s, 1600s. Then when the United States started in the 1700s, they kind of took over that role of King George, but instead of King George, you dealt with the President. So, there's always been a link between Tribes and the President and Congress and they kind of bypass the states and that's just kind of the history in the United States.

Kristiina: It seems like the form of tribal rights is very unique to the U.S. or the Americas. So many other places, I think of Indonesia, the Indigenous people have absolutely no rights, they have no treaties, and nothing was ever written down on what the people were entitled to.

Mike: *I think that's because North America is so big, and the original inhabitants from Europe were relatively small and there were large numbers of Native Americans. They couldn't just dictate all of North America, they had to make deals, bargain, and negotiate with Indians. By the time they took over the United States, western powers had gotten so powerful they could dictate to those countries. They had more guns, more armies after they became powerful. But when they* first started getting foot holds in North America, they were relatively weak, and Tribes were relatively strong. The Northwest, even in the 1850s there were battles, 1870s there were battles, so they wouldn't have made treaties unless they had to. So, treaties were written. Although my tribe doesn't have a treaty, most Tribes have treaties. The idea behind treaties was that they wanted Tribes to cede large tracts of land to the white people and in return they got promises for the reservation, promises for services, for schools or different things. Today, we call it trust responsibilities, but it goes back to the treaty idea. In Washington State all you needed was land, they made treaties, Indians took these reduced land bases on reservations and in return the United States is supposed to return stuff, like a grant almost (see Box 3).

Box 3: Shrinking Land Base of the Colville Tribes
(Dr. Mike Marchand)

Tribes comprising the Colville Tribes once roamed a vast landscape from the Columbia River in Washington State east to the present states of Oregon, Wyoming, Idaho, Montana, North Dakota, South Dakota, and into Canada. They were both salmon fishermen and also buffalo hunters.

But by Executive Order of April 9, 1872, the tribes were confined to an area of eastern Washington. The western border was the crest of the Cascade Mountains and extended east to the Idaho border. This existed less than three months. Then the Executive Order of July 2, 1872 shrank the reservation size down to 2,900,000 acres in north central Washington State. This was again reduced down after the discovery of gold to 1,500,000 on July 1, 1892. Additional lands were deemed surplus and made available to non-Indians for agriculture and mineral purposes, reducing the reservation to 1.4 million acres.

(Source: The Year of the Coyote: A Centennial Celebration Publication, July 2, 1972, Confederated Tribes of the Colville Indian Reservation)

Kristiina: Can we go back to when you got your other name and when did that happen? In Finland, our family name is similar to your other name. My family name is "Virtanen" which means "a family living near a water". Most people do not know that Virtanen means anything or that it describes where the family lived in nature and not a profession like in the English-speaking European countries, e.g. Smith being a Blacksmith. Your other name is unique because it is different from your family name. When do you get your other name and how important is it for you?

Mike: *It varies depending on each family. It wasn't that important in my family, at least with my immediate family. They didn't give me a different name. It makes me think of my grandpa when he went to this mission school. A big objective of these schools was to erase traditions, erase history, and erase tribal culture. So anyone who went to that mission school in the 1800s, the first thing they did was cut all their hair off, they took all their clothes away, their Indian clothes, and gave them western clothes. They were forbidden from speaking their language, if they were caught, they were tortured. They were literally tortured. My father-in-law was caught speaking Indian with one of his friends. It was in the winter, it gets cold in Omak, Washington, so it's like 20 below zero. They stripped him of all his clothes, threw him in the basement of this church. It's below zero and he's still a kid, but they just routinely tortured kids like that. If you weren't going to shuck off your old Indian-ness, they were tortured.*

Kristiina: So, they were used almost as an example, because then any other kid would say *"don't do that"* because they knew what would happen.

Mike: *Yeah, so you have generations of Indians all across the country that are actually ashamed of their culture, of their traditions, their language. You really saw that in 1950s, 1960s. I was born in 1953. My dad could speak his language, but he didn't teach it to me. He just thought it was a waste of time, or it would be one extra thing you don't need in your life. I wish he had, but he didn't. So, in my family, they didn't give me an Indian name, but traditionally they'd give you an Indian name when you're a young child.*

You have a vision quest practiced to a place that varies from tribe to tribe. But in general, they take a young child to the mountains, to a sacred spot, and basically just leave them there by himself for a few days, maybe 1 or 4 days. Then they would just have time to think, and each one's experience was different but often they'd come out with a name from that quest. It can vary, but sometimes they would have a vision, or sometimes they would have a dream, or sometimes they might think of something.

I didn't really go through that myself until I got older (see Section 3.9). I got married, and my wife's family is more traditional than my family. They said, "You need a name." I had a mother-in-law who was very old and an aunt who was very old, they were very traditional. I was talking to them one day. We weren't even talking about names, but I told them one time I was by this creek, they knew where it was, when I was a boy and I saw a Wolverine. This is a very rare animal, especially in Washington State. I was looking for a fish, we'd catch these little fish in creeks. We'd call them crick-fish, little kids, fish for them and we can catch like 20 or 30 of them pretty easy. I was just this little kid walking by the creek and saw this snake sticking out of the creek about 5 feet high. I hear a big crash in the water and a Wolverine jumped on top of this whole snake and he's like eye level from me. He just looked right at me and then I don't know if he was actually communicating or if I was just making this up, I honestly don't know, but in my mind I got the impression that he was telling me that I should live there. And that if I did live there that his spirit would protect me. So, I don't know if it was ESP or if I dreamt it up. I wasn't that traditional at that time, but it stuck in my head, like how often do you see a Wolverine like that? I mean, it definitely wasn't scared of me and it's the only one I've seen in real life. Really. So, I didn't think too much of it, but 20 years or so later, I was telling my aunt and she was telling me, "That was a big Indian event in your life, that should be your name." It has all kinds of meanings and she asked, "Did you move there?" And I said, "Yeah, I did. I actually lived there. It's a pretty place." I guess it's worked, because I got everything I wanted. The other funny thing too is that Wolverines are known for not being scared of anything, they like to fight, and they take whatever they want. It is funny when you look at YouTube, you see wolverines fighting polar bears, or fighting grizzly bears or a pack of wolves.*

Kristiina: What's interesting though, Mike, is how you behave. It's not like you're fighting like a wolverine. You aren't going to physically attack somebody. You have a different way of getting attention, like what you did when you were in the PhD program. You had a way about you, which I don't even know what it is, but you really got the respect of the engineering PhD students. This was interesting to me since they didn't respect the students in the College of Forest Resources since our research was considered too practical. You always had the feeling that they thought we weren't really *real* scientists. But you had a way of interacting with them and they respected you. They even quoted you, which I never saw occurring with any of our other PhD students.

Mike: *There's different ways of fighting and probably the closest thing that fits my way of thinking is the Chinese book,* The Art of War *(written by Sun Tzu) [16], which is a small book and it's vague enough that you could probably interpret it to mean almost anything you want. But this fits my brain perfectly and a lot of it is about understanding who the enemy is, or understanding the environment, or what you really* want to do. So, if you're a very, very smart, good general, you get everything you want without even going to war or firing a shot. But it takes deep understanding of who you are, what do you want. It takes a deep understanding of how your enemy thinks, what do they want, and the world around it. At least that's what I get out of that book. It is a very small book.

Kristiina: I have seen how you assess the environment around you at several meetings where you were going to talk. I remember one Associated Tribes of Northwest Indians (ATNI) meeting where you were going to present. I watched you sit and listen to everything that was going on. During that process you were digesting, you were assessing the whole audience and you had figured out where they were coming from and what they were going to do. In a matter of minutes, you had fine-tuned and re-packaged your message and adjusted your message for that particular meeting. You have an ability to very quickly think about things and change what you are going to say.

Mike: *That's very much* The Art of War *thinking. You walk into any place, it could be a classroom, and think, okay who's in here. You kind of assess what's each person is thinking, what are they about. You don't always know for sure but usually if they say a couple of words, you can figure them out pretty fast. That's just how I think about everything. If people are kind of flowing the way which I want them to go, I don't even say anything, cause I'm winning. But then sometimes the tide's going this way and you've got to reverse it and that's a little harder. It's just about assessing each situation, assessing each individual. So, Sun Tzu's* Art of War, *that's my bible. It's Chinese, but I can interpret it for my life.*

Kristiina: How did you even find it? You really read a lot, right?

Mike: *It was just kind of random. But I like military books and I like how generals think, so all the old epic tales like Alexander the Great, I like all that kind of stuff. I went through a period where I was a young child and my grandpa kind of raised me. He was the leader. He used to meet with tribal leaders in this farm house and I got to observe all that. In the old days, their whole house would be as big as this room almost, but there would be the woodstove in the corner with an old coffee pot. The tribal leaders would come into the room, in those days they all smoked cigars and pipes, or cigarettes, so there was a big cloud of smoke in the room. It was all men, all the women would go into another room because they thought politics were just boring as hell. I thought it was great, so I used to fall asleep in my grandpa's* lap. They'd be talking about legislation, national policy, way out on the farm lands in the middle of nowhere. I was exposed to that at a young age. They'd be talking about the Omnibus bills or this bill about such and such and I thought that was just great stuff. I also think I noticed they all kind of looked up to my grandpa too, so I thought that was pretty neat. I wanted to be like my grandpa. He was Chairman. I was maybe 5 or 6 years old and I'd fall asleep on his lap, and I just said "I want to be the Chairman."

And now I'm the Chairman. So, I'm lucky, at 5 years old, I knew what I wanted to do, I knew who I was going to be, I didn't have to search for that, I knew what I wanted. But I guess the other side of the coin is that as part of education you need to know your people, what their culture is, what their traditions are, but then you need to interact with the outside world. So, part of education is learning how the outside world works too.

I was always encouraged to read and write and learn as much as I could. My mother kind of became important in that, she went and studied education. She wanted to be a teacher, but mainly she wanted to teach her own kids. She had five boys and she was a certified teacher, but she never did anything with it. Probably by the time I was in the 3rd or 4th grade, I'd read every book in the Omak library. I just liked to read. My brothers didn't, they didn't like to read. I had all this stuff in my head, I don't know if I understood it all, but it was like I was born for standardized tests. In grade school they give you all these standardized tests, where you check the box, and I could do those in my sleep.

Kristiina: But there's a big difference Mike. There were some people at Yale University that would come in and everybody would say they're really smart. But they had just memorized facts, and they weren't connecting the facts. What you do is you're connecting the facts. You might end up with something that's totally different. You might think it's not even relevant, but somehow it helps you to think through and you're making those connections. So that means that you come up with very different decisions or conclusions than somebody who has just memorized a lot of facts but doesn't really know how to put them together.

Mike: Well, for academic stuff, I was just really a gifted child especially related to standardized tests. I actually broke into the school one time, I just wanted to see if I could do it. This was like a grade school. I was sitting there with my friends and I said, "I bet I could get in there without anybody knowing I was in there." I was just kind of joking around, but I did it. I got through the locks, went through the alarm system, and who does stuff like that? But I did, I went in there and I thought okay, I'm in, now what'd I come in here for? I don't steal, I'm not a thief, but I was curious what my file said. I opened up the file cabinet, looked at my file and there were all kinds of stuff in there that I never knew about. There were letters from educators saying, "Michael is the most gifted child we've had this county, ever" or there'd be test scores, which they don't show you. There were tests that said his IQ is like 180 and stuff like that. I just closed everything up, like I was never there. I unbroke my way out of the school. That's the only time I ever did that. Then after I got out I was thinking that was really stupid, I could've gotten caught. But it was just like a puzzle, a mental challenge.

Kristiina: But even seeing that information, it's not like it affected you. Some people would all of a sudden be like "I'm smart" and you can't forget it. It sort of takes over your life, we saw several of people at Yale that were like that. Where they assumed that since "I'm smart, so of course you've got to listen to me." They really didn't listen to anyone else because they thought they had all the answers.

Mike: *I guess, I'm aware of that. I went to MIT and I thought, yeah, these guys are smart, but I know I'm smarter. But I don't feel like I have to flaunt it or anything. I know inside myself.*

Also, I don't think I'm as good at those tests now as I used to be. It used to get me excited to get like 99th percentile on these tests. I used to enjoy that, but now it's like, I've been there, already done that.

Kristiina: But you still did amazing on the GREs and the engineers couldn't believe it. I know their GREs were not even close to what you got. And they were impressed, that's all it took.

Mike: *I didn't go to Kindergarten, but by the 8th grade, I just read and absorbed everything in the world I could. As for the Omak library, now it's this tiny town that, they didn't even have internet then. But I was way ahead of the curve at that age and then kind of lost interest in it, so I doubt if I could even get those scores today.*

Anyway, the Bureau of Indian Affairs had a program where they scanned all the Indian's standardized test scores and they'd pick out the top 1% and ask them if they're interested in going to Prep Schools. They actually had a program that did that back then. They picked my name, kind of out of the hat, since I scored high (Fig. 2.4). They asked, "Are you interested?" and I thought, "Well, tell me more." I just told them, "Well send me to the best school in the world." I was just being a smart ass. They came back about a year later with tickets, saying "Are you ready to go?" So, I said, "Yeah, I'll go," figuring if I didn't like it I could come home. They didn't force me. So I went to New Hampshire. I think I was about 15 maybe, plus my parents were driving me crazy so I wanted to get out of the house. When I was 15 I couldn't stand my dad, he was driving me crazy. Later, we got along fine, but there was like period when I was 15–20 years of age that I just needed to be by myself and somewhere else. The school in New Hampshire was a great school.

Kristiina: Back on the reservation, did people react negatively to what you were doing? Did you ever have any problems because you were clearly smart, and being identified as being smart by others?

Fig. 2.4 Newspaper article on Dr. Mike Marchand going to a famous New England preparatory academy — Exeter Academy.
Source: Mike Marchand.

Mike: *I think there was resentment by some. There was some racism. You know, it's a border. It's kind of half way on the reservation, half way off. These tend to be racist places. There was racism and even amongst my people too. It's kind of like in those days, if you had a smart girl, she'd actually act dumb, because she didn't want people to think she was too smart. Or tall — tall people would stoop down a bit. But to me, I was kind of shy. I just kind of stayed by myself, I still am kind of shy. But it's kind of funny, now I can make myself go talk at big conferences, or I can do public speaking. I can do all that stuff. But it's not natural. I had to make myself do it.*

Kristiina: Really? Because it seems natural. When you're doing it, it just sort of comes out in cohesive sentences. It's not like you stutter or anything.

Mike: *It's like the same thing when you're in a big class, asking where knowledge is at, what kind of things are they interested in. You try to tune into that class a little bit. But you can't keep everybody happy.*

Kristiina: So other than *The Art of War* [16], what other book or books had an impact on your thinking or how you put ideas together? Who else has impacted you?

Mike: *Probably a lot of the American literature, the revolutionary literature. I like a lot of revolutionary literature. In the 1960s, it was a popular genre of literature, the revolutionary literature with Che Guevara, Cuba, and communism. There were real questions being asked during that time about whether the United States even works. There were protests in the street, they were blowing up things, and people were getting shot. The Vietnam War caused a lot of unrest. Things were kind of crumbling a little bit. There was a lot of literature that came out of that time period. It probably wasn't real great literature, but there was a lot of stuff coming out. People thinking about change in the United States, change in philosophies, I guess. I was really interested in things like that.... A lot of it was not intellectual. They were just blowing stuff up. I never did do that.*

Kristiina: I think that was a very common theme in those days. I was a young kid living in Europe for some of this time and people were really thinking about what type of government works best and what needed to happen to deliver justice to society. The Cold War was big then. It was a very common theme.

Mike: *All these different groups went through civil unrest. Native Americans were one of the last groups to get involved. My people got* *involved in it around 1969, 1970. I happened to be in Prep School when the Native American Movement was kind of getting started. I went to their first big protest and that had a big impact on me too. I went to Prep School, by then called Philips Exeter Academy. It was always ranked in the top, in the top five in the world in schools. At that time it was an all boy's school and tended to be mostly all rich kid. They were 99% rich kids, all smart kids, as far as book learning goes. Anyway, I was amongst a small group of scholarship kids. We were the scholarship, low-income kids. All my classmates would fly home on the holidays, but the scholarship kids. We didn't have any money, so we were all stuck there. We thought, "Well what are we gonna do for five days?" This lady who worked for the BIA came by. At the time I was thinking she was old, but she was probably only like 20. I was like 15, so when you're 15, 20 seems kind of old. So, she had a station wagon, this government station wagon and she had four guys in there about my age, maybe a little older and she said "Come on, do you want to come with us?" I said, "Sure, I'm not doing anything here."*

She looked official. I said "Where are we going?" She said "Boston." Boston was only like 30 miles away. I had been there before. It was close by the school, so I said, "Yeah, I'm not doing anything here." There's no one left on campus. I went with them. We pulled in at the Cambridge Hilton where they checked us in. It turns out the American Indian Movement got some church group to sponsor them to be at the Cambridge Hilton. The Cambridge Hilton was just full of Indian kids. But I had never heard of the American Indian movement, I don't think. We all checked into this room and they put about maybe 4 kids in each room. I met three new Indian guys and they were from different places.

The leaders of the American Indian Movement became famous later. Maybe they were famous then, I don't know. But Russell Means was one. Dennis Banks was another and he just died recently. There was John Trudell, who was kind of like a poet, speaker type guy, he was there. But I had never really heard of any of these people. Anyway, we were there, and they were having protests in the streets, they were marching down the street. The main street in Boston is called Boylston Street. A big rag tag group of Indians was marching down the street.

I was a weirdo since I was dressed up all preppy style. Most of them were radical kids with long hair and looking pretty wild. A lot of them came from Alcatraz. There were protests in Alcatraz, they took over Alcatraz Island for many days. They all had these homemade jackets that said Alcatraz on the back, kind of like a letterman's jacket. I thought they were so cool, and I wished I had a jacket like that. But then we were all just kids. Everyone was nice to each other, that was kind of different then. I think now things are more cliquey and there are different gangs, but then there wasn't. We all got along, they were all nice to me. We went to this protest. They had this protest planned at Plymouth Rock on Thanksgiving Day in 1970. One more thing about Boylston, everyone was scared of us. These stores would close their doors and shut their windows and run, they literally ran away from us. I guess they thought we were gonna kill them. Then these big policemen - Boston is famous for having these big Irish policemen all lined up like guards. I was watching those policemen, they were cocking their guns, getting ready to shoot, I guess. That's when it gets serious. I was raised around guns hunting, so hearing a gun cock, I know that sound. I thought, this is getting kind of serious.

Some of my roommates were crazy, they just wanted to cause trouble. They wanted to get arrested, and they wanted to be famous. I'm thinking like maybe 10% of the kids here are crazy and want to get shot, they do. But I didn't want to get shot. I wanted to protest but I didn't want to get shot. I was just hoping my roommates didn't get us shot. Anyway, we didn't get shot the first day, but the guns were getting cocked and when you get all those people together, you don't really know what's going to happen, you know? It could blow up. You can't control everything. Just think, if a policeman had shot a kid? It would've caused a big riot. But nothing happened the first day.

Then they went to Plymouth Rock. Like a tourist thing, the pilgrims had this big Thanksgiving spread all laid out on these long tables and they were all dressed up like pilgrims and pioneer ladies. There was food spread out over every table and all these kids were lined up eating and I thought this is kind of neat, you know I wasn't expecting this. I'm a kid, I'm 15 years old, and I can eat a lot. I was into sports, you know we could eat like a million calories a day and not get fat. I was eating away and then one of my roommates who I had just met the day before, he was a couple tables over, he said "Hey Mike" and then he grabbed some kind of food. He threw it right at my head. I ducked but it hit somebody behind me. That started a food fight like in those comedy movies. Food's flying all over. Then some of the radical kids, they were just trying to ignite it and they were saying like "······ We're not going to eat this stuff." Then they started knocking stuff off the tables, they started flipping tables over.

It was this real nice banquet room and then our hosts, the pilgrim people, they were going into shock. The leaders were kind of wondering what was going on too, they didn't plan this. The kids just went kind of crazy. The room turns into a riot and then they start spilling out of this fake fort. I don't know if it's still there or not, but they start spilling outside and there's stockade walls around there and these kids are climbing over the wall, like in the movies. But I'm looking at the wall and there's a door right there. Then the riot starts rolling down the hill and I'm thinking I'm just going to go with the flow and hope we don't get shot. I was just trying not to get shot (Fig. 2.5).

Down the hill there was a Mayflower replica ship, and the Plymouth Rock. They go down that way, it's a tourist attraction. There's a little ticket booth there and this guy in the ticket booth. His eyes are getting big. All these Indian kids are coming, and he just says, "You guys can all come in for free today." He kind of just runs away, abandons his post. These kids are just looking for trouble, they knocked over his little stand and they took over the Mayflower. Me and my roommate went and got a good seat, he got up in the crow's nest and I sat in the neck, those nets in the front. I was just sitting there looking around and all these kids are having a riot and then all these photographers started coming. There's huge flash bulbs going off all over, it's back when they had flash bulbs. Then another group went over to the Plymouth Rock, which was right next door. Trudell got red paint somewhere, he must have had this planned out I guess, he poured red paint on the rocks. I wasn't really paying attention to that, I didn't know who Trudell was back then. Turns out he's a real eloquent poetic guy, he had a long

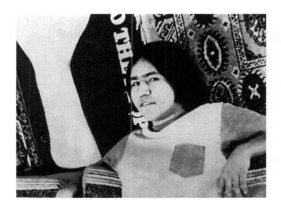

Fig. 2.5 Photo of Dr. Mike Marchand around the time of Plymouth Rock incident. Photo Source: Mike Marchand.

career following this riot. But at the time the kids were kind of out of control, the leaders weren't really planning any of this. All these kids took over the boat and they're all sitting on the ship. So then as a step 2, the leaders jump up onto the ship. They kind of catch up to the riot and they start speaking. They were good speakers. Russell Means is a great speaker. Dennis Banks was speaking. Later I was reading Russell Means' biography where he talks about that day and says it was the best day of his life. He must've been in like 10,000 protests.

Kristiina: But it impacted you though, right?

Mike: Yeah, it impacted me because they were talking about how the government is bad. Sovereignty is good. Indian nations need to rise up and we need to build our own nations up. Then a second light bulb went on and I thought, "Who's going to build these nations, who's going to build this society?" I thought, that's me, that's going to be me. I got interested in planning. I wanted to learn how to plan communities, businesses, institutions. I got a master's degree in planning and it kind of comes back to that day.

Kristiina: It stimulated you to think about how you wanted to do it?

Mike: Yeah, because we had a whole army of kids who wanted to tear everything down. We had leaders who wanted to tear everything down. I thought they were going to do that. I thought they were going to tear everything down from an institutional standpoint. But then when everything gets torn down, who's going to build it back up? I'm thinking that's going to be the hard part and that's what I wanted to focus on. Like, okay, I think we've got enough armies to tear it all down. We're going to need an army to build it back up and I want to be in that crew. So, that's what I did. That was in 1970. From 1970 on, I wanted to be one of the planners who rebuilds Indian Tribes. That's what I've been doing since 1970.

Kristiina: Do you think that a planning degree would give you an approach or way of thinking through what's going on?

Mike: Yeah, I think I was lucky. I got lucky at Eastern Washington University. I had a professor who had been a planner at Fort McDowell in the southwest and he had been a planner at Dartmouth. He was not an Indian, but he went to school at ASU. But he knew a little bit about Indians, he knew a little bit about sovereignty. But the main thing he knew was the planning processes. Other disciplines are taught *planning using the problem-solving model that defines the problem, develops alternative solutions to the problems, and then implements it. It is kind of simple. But planning has to include different variations. In planning you're problem solvers. Where do you want to be in the future? Where are you now and figure out how to get there? So that's been a simple model I've been able to apply to everything. Probably a lot of it is coming out of your own head too, like the Art of War, I don't know if the book actually says that much, but it's useful for me because it jogs my brain. I don't know if other people get anything out of it. It's probably like talking to the Wolverine. Other people would see the Wolverine and say there's a Wolverine, but I got to bounce these ideas off him. I don't know where my brain starts and where the world starts, I don't know where the line is.*

Kristiina: This gets into an interesting question about how you sort different sources and forms of knowledge. You have knowledge from Eastern Washington University where you studied planning, but you learned a lot as a child when you read all the thousands of books. There's a lot of discussion in the literature and scientific community about how we're going blend the Western science with indigenous ways of forming knowledge. The Western science thinks that there's going to be an amalgamation of these different forms of knowledge, which I really don't think is a good model. Can you talk about that a little bit?

Mike: *Maybe that's why Winchell was useful for me. He could kind of see the old traditional model versus the Native American Indian model. The old traditional European U.S. planning model is really kind of like the engineering model. It's very linear with very linear problem solving. It's like when you start a plan in January that ends in December. You could plan in the winter, start building in the spring, and be moved in by fall. To me that kind of plan is kind of like the engineering plan or site plan. In terms of Native Americans, we do that too, but we think about it a little differently. When Eastern Washington University teaches planning, they say, we'll start planning in January. Then they ask what do we want to do? We want to meet with the community. We meet with neighborhoods, to get input from these neighbors, people in the community. They kind of set the goals for the project, and they kind of set the direction we're going to go. Then we can include technical people like engineers or scientists if we need to implement the vision that the community wants.*

That's kind of okay in general terms, but it kind of depends on who makes up that community. If Seattle is the community, people in Seat-

tle are diverse but do have some kind of commonalities. I think the difference in the Indian communities are quite a bit different than Seattle. There's some overlap with how Seattle people think, but it's probably more different.

In my community, I think a good example is my office. We have this modern glass steel office we just built, it cost 60 million dollars. Its 3 stories high and it's like a bank or something. So upstairs, I get the best office in the reservation, with a view of the mountains. I have 2,000 employees. If I have an idea, I tell those guys to get to work on my idea. But you go in our chamber room where we meet, it's a big room. Around that room are pictures of Chiefs (Fig. 2.6). I didn't plan that room, but I like the room. So, when any kind of issue comes in, I think, "What would these Chiefs think about this issue?" I think that's different. Like if you're doing a plan for Seattle, I don't think the topic would be, what would Chief Seattle think about this? Even though he came from here, his name probably wouldn't come up. But in our minds, traditional things are good, they're kind of the definition of good, that's kind of an important building block. Traditions are good, at least some of it is, most of it probably. We start from there and say do these ideas bring us back to the good parts of traditions or not? So superficially, it's kind of the same process, except I think what our ancestors think is more important, our culture is more important, so that's probably the main difference.

Fig. 2.6 Colville Council Room with photos of three former Chiefs including Dr. Mike Marchand.

Photo Source: Mike Marchand.

Kristiina: But what it really means is that the whole Western science approach uses bits and pieces of the facts. But you aren't going to use their framework. You aren't going to use their process for thinking through a problem because it really doesn't work for you?

Mike: *I think there's also a tendency to think that the more ancient something is, the less important it is to you. If something happened last year it's kind of still on people's minds, maybe. But if it happened in the 1800s, people might think, that's before my lifetime. Who cares? You had students in your class who said, that happened 20 years ago, and I wasn't even born then. So, who cares?*

Whereas in my tribe, we have descendants of Chief Joseph in my tribe. In fact, my grandson is one of them, so you can't escape the past. It's important. I had a grandmother; her dad was the Chief of the people called the Arrow Lakes people. They had the best fishing site in the world, salmon fishing site. They built a big basket under the falls, fish would try to jump over the falls. They'd fall into the basket. They'd pull up 2,000 fish a day out of this basket. It was like the Garden of Eden. For 10,000 years they were living in the Garden of Eden.

My grandmother would tell me these stories. I was born in 1953 so I was maybe like 5 years old when she was telling them to me, so maybe in like 1958. But she'd tell me all these stories like this happened yesterday. But that place was flooded in 1942, so it was gone before I was born. But she'd tell me these stories like it just happened or like it was still there. To her the past is still important. I think that's a big jump off difference that time scale is different and also the fact that some of the ideals of the past are still important. A lot of our people think that the problems we have today are because we've kind of deviated from the past, or we've forgotten the past and that maybe part of the solution is trying to figure out how to bring that past forward again. Maybe that's not always true but we kind of tend to think it is.

Kristiina: If you were in Seattle, and some parents approached you and say, "*We want you to take our kids*", let's say they have 100 kids, and they want you to "*train them so that they're better at making more ethical decisions.*" What would you do? How do you take people who are growing up in a very consumer-oriented society that are trained in a particular way? They don't really have family, because many family members live far from other members of the family? What would you do based on what you've experienced to produce leaders that are going to make a difference in making more environmental, and better societal decisions? What would be your model?

Mike: *Well I think I'd try to do similar sorts of things, but I think it's probably easier on the reservation because we have a fixed land base. It's roughly 50 miles by 50 miles. We can kind of teach the kids that this is your homeland and you need to take care of it. You can mess it up, but we still got to live here. Maybe you can mess it up and not deal with the problems but then your kids or your grandkids face the mess you created. From that standpoint, this is your home, you can't mess it up, you've got to keep it livable, and preferably make it even better because it's kind of messed up now. I think you could do the same thing.*

Now a days, on the global scale, it would take some education. We know all these things have global impacts. Climate is a global concern. All these natural cycles are global like water. The water that's coming out of Fukushima but the whole Pacific Ocean is one big bathtub really. Technologically we could probably show that now. We have a fish hatchery at Colville, 200 miles up the Columbia River. We let out a little baby salmon three years ago and it's at Fukushima. He shouldn't have swum there, but the world is kind of shrunk now with internet and technology. It's like the world is more like one place. I think you could teach that now. A lot of these environmental things, it's not that hard to show their worldwide impacts. Like Carbon goes around the planet, water goes from the ocean to the clouds to the mountains to the rivers, there's a cycle. Most things are a cycle. Then there's visible things like garbage, throw out a plastic jug in a river here and it's likely to end up in a big cesspool in the Pacific Ocean somewhere. There's big giant islands of garbage out there now. They find big dead zones in the ocean where there's no oxygen. The technology's there to monitor that kind of stuff now. But 50 years, it didn't exist, but now it does.

Kristiina: It's almost like you have to get people to think in a different way. You're saying that if you're going to teach non-Tribal people, they need to start thinking about the fact that there is a defined land. You're like a Tribe, on a piece of land, and you're really being impacted by the decisions that you're making. Maybe that's like with the stories you've had, you've heard the impacts of certain decisions you might make. When I think about people coming to university, that's when they hear about science for the first time. They don't hear about the holistic science that has implications on their decisions.

Mike: *The Grand Coulee Dam was built in 1942, the biggest structure ever, that cut off all the fish in the Columbia River. At that time there were no laws requiring that they needed to know what the impact was of building the dam. Nobody cared. They just killed the fish off. There were* no fish ladders. The thousands of miles of habitat above the dam was just cut off. These fish are dead. But the fact is that all those salmon that used to be produced up there, they used to come down the river. They used to go into the Pacific Ocean. Probably a big bulk of those fish used to actually be caught in Alaska or California. The ones that were impacted by the Grand Coulee Dam, they were us. But they were Alaskans, people in California, probably clear out to Japan. Because of the ignorance of science and technology in 1942, no one cared, no one even cared what was going on. No one even knew where those salmon were going. Today it'd be easy to show that, I think.

Kristiina: What you're really saying in a way is that by having the knowledge that you have of the past, we look at a site and say this is where the salmon are. We have a dam, we know they aren't going past the dam. But what you're saying is that they should be going past the dam. You have an ability to say where in the environment salmon should be found. If you really want to deal with an issue you need to figure out how we get them back to all the different place they used to be found.

Mike: What's happening for the last two decades is that around 15 Tribes along the Columbia River have been suing the United States government to fix the dams to restore salmon runs. These Tribes have been suing them, negotiating with them. About ten years ago, a settlement was reached for about a billion dollars from the Bonneville Power to fix the fisheries. The dam structures are being modified in response to Tribes. They're modifying how dams are managed.

They're restoring habitat that was lost over time, which kind of feeds on itself too. Like orchards. Okanagan Country is famous for apple orchards. They grow apple orchards, they dewater the streams to irrigate the orchards so there's no water in the streams. This means that there's no salmon. But the Tribes can come in and say we want salmon back because it's important to us. It's important to our history and our culture and to who we are. So even though we have limited resources, we put those resources — lawyers and scientists — and we're fighting Bonneville to give us money to fix it, and we're winning. So far, we're winning. We got nearly a billion dollars and with that money we mitigated some of the problems. For example, the Okanogan County irrigation systems are inefficient, they're just soil lined ditches. Half the water just sinks into the ground and goes away. We're lining some of those ditches, so now there's water for the trees, there's also water for the streams, and for the salmon.

> *One of the streams was shut down for logging probably in the 1930s. They built a railroad. They blasted rocks in the creek which stopped migrating fish. We took those big boulders out of the creek so salmon can get back up the stream. Case by case we're fixing everything and it's a matter of just physically fixing them and finding money to fix them. To restore the fisheries, we built the hatchery. It takes people. You've got to have scientists, hatchery people to run it, people to monitor it, people to put nets on the river to catch the fish.*

Kristiina: I don't think if the Tribes weren't actively pursuing changes in the fisheries management that anything would change. It's almost like non-Tribal people are not very good at identifying what the problem is or figuring out what changes will improve fish stocks.

> **Mike:** *We're fighting people all the time. A lot of the agriculture people and just conservative people in general don't want money going into bringing the fish back. They ask, "Why do you want the fish back?" We don't understand this perspective. But we're fighting to bring the fish back. We've got plans to bring the fish above the Grand Coulee dam again. We don't know when it's going to happen, but that's the goal and we're moving that way. We think we're going to win, but it's just a matter of time.*

Kristiina: I think this identifies an important point in how you define what the problem is. This differs from the Western science approach which is more linear and usually in response to a narrow problem. It's more reactive than proactive so it focuses on what is identified as the current problem.

> **Mike:** *That's one side of the coin and the other side is that we have a vision of what we want. Basically, we want to return what was there in the first place. So, we're determined to bring salmon back up into Canada. We've actually had some success in that too. The Okanogan River is a tributary of the Columbia River. Canada said there's no fish in the Okanogan River, they don't want American fish up there. But we work with the Tribes up in Canada and we said, "We have the expertise to get salmon up to your tribe, do you want us to work with you?" They said, "Yeah, they want salmon back." So, we've been helping them get salmon back.*
>
> *The ministers of Canada were in denial, they said "You can't get salmon up this river, there's no salmon in that river." Their official position is that they're extinct. So, it's almost comical, they had a*

conference on the Okanogan watershed up in Canada. The national ministers of fish were there, and they took them out on a bus on the upper Okanogan River where there were thousands of salmon spawning. We were all looking at that guy to see what he'd do, and he just laughed and said, "I guess there's salmon out here." But that's just a vision that the Tribes had, the Tribes wanted salmon there. But now, the Columbia's tougher problem is probably bigger dams and bigger barriers. But it's solvable we think. Some of the solutions are kind of funny too. They have water cannons that allows us to shoot fish up the river. We're working on that, we're spending money on that. I don't know what the final solution will be, but we're looking at options like that.

Chapter III
How Do You Become "Cultured"?

> "......Everything on the earth has a purpose, every disease an herb to cure it, and every person a mission. This is the Indian theory of existence."
>
> ~ Christine Quintasket (1888—1936), Mourning Dove, Salish ~

Our need to understand what it means to have culture or be cultured transcends culture as understood by the popular media. If we don't understand what culture means and only look at it as the surface characteristics of a group of people, it will be impossible for those people to understand why they should consider developing a "Tribal" culture based on a holistic and interconnected view of nature. Most cultures are not nature linked but hold more egocentric views of nature. Such an approach doesn't force any person into taking responsibility for their actions related to the environment. It classifies a person as being cultured when you attend a concert, symphony or opera. It doesn't move you away from thinking about each person's behavior as a solitaire decision that doesn't impact the environment. Therefore, it is important to think about this since our decisions should be made as a group and not as an individual or one family.

An individual perspective on nature is not balanced. You can use it to decrease your own individual carbon footprint, but this is not measurable at the community level. There is value in recognizing how your individual behavior impacts your carbon footprint because it forces each person to recognize that their actions have repercussions on nature. But most individuals do not make the connection between their actions and what is happening around the world. This has made it easier to focus on economic activities as the primary tool to evaluate the important elements of your culture at the community level. This is not a balanced and unbiased approach. This means that nothing changes related to nature — this has been the Western European views of nature over the last 40-50 years. Respect for nature has to become a community response since this is when behavior changes based on community rules. Further, all community members will have knowledge of nature and how it is impacted by our activities.

Tribes are not egocentric towards nature but make decisions after developing consensus as a group. This is why it is important to think about what a group-level culture really means. We need to understand how individual cultural attributes are defined by the superficial appearance of our decisions. Individual cultural attributes have to scale to assess how our decisions link to the rest of the ecosystem. Western European culture tends to be individualistic while the indigenous culture reflects the group.

Champagne [18] expresses well the conflicts in Western and Indigenous traditions. The Western tradition is more egocentric while the Indigenous is more community based. As Champagne wrote:

> *"The goals and values of Indigenous Peoples stress that the world is full of give and take. Life is a great gift. One has a role to play in community and society. Individuals seek to find that role or life purpose and to fulfill one's given task. Ceremonies are often about seeking personal and tribal understandings and directions. The world is full of meaning and purpose, although people are not gifted with a complete understanding of the future or present. Tribal knowledge is made up of ceremonial interpretations and human experiences. Elders collect knowledge on their long life journeys and pass information onto other generations.*
>
> *A cultural theme within contemporary modernism is the increasing rationality of the world. Markets are favored, in part, because markets are efficient, productive, and profitable ways to distribute goods. Science brings greater understanding of the organization and activities*

> within nature. Science dominates over religion and culture. Culture and being are subordinated to the requirements of efficiency. Religion and culture are preferably separated from government and economic decision-making.
>
> In Western tradition, the earth is made up of raw material waiting for transformation into a product useful to humans. A major purpose is the transformation and control of the world for political and economic domination. The earth, full of wild and useless beings, needs to be transformed into objects that serve the goals and purposes of humans and nations. Making heaven on earth is a deep underlying cultural goal in Western nations. Heaven, where all human needs and wants are satisfied, is a central goal and purpose for people and of history. History marks the realization of creating heaven on earth. The achievement of utopia, or heaven on earth, will be reward of progress and rationality at the end of history. Humans at the end of history will be the center of the universe and in control of the earth's resources. The heavy emphasis on material goals in life lead to a world bereft of enchantment or cultural interpretation."

This definition is a personal view of culture. It describes how well you have acquired knowledge and how much you are immersed in the arts as well as how you behave. A dictionary defines culture for an individual or group of people but in practice is translated as *"the quality in a person or society that arises from a concern for what is regarded as excellent in arts, letters, manners, scholarly pursuits..."* [1] by Western societies.

If you look up a dictionary definition of culture or what it means to be cultured, you will find several definitions that capture both the Western world as well as the Indigenous peoples' definitions of culture [19]:

> "a: enlightenment and excellence of taste acquired by intellectual and aesthetic training" or "b: acquaintance with and taste in fine arts, humanities, and broad aspects of science as distinguished from vocational and technical skills"

> "a: the integrated pattern of human knowledge, belief, and behavior that depends upon the capacity for learning and transmitting knowledge to succeeding generations" or "b: the customary beliefs, social forms, and material traits of a racial, religious, or social group"

The first two definitions typically characterize what your views are if your ancestry is Western European. *The Daily Mail* summarized a 2015 survey that

---

1 Accessed on July 5, 2016 at http://www.dictionary.com/browse/culture

exemplifies the Western European views of culture [20]. This definition is a personal view of how well you have acquired knowledge and how much you are immersed in the arts as well as how you behave. A dictionary defines culture for an individual or group of people but in practice is translated as *"the quality in a person or society that arises from a concern for what is regarded as excellent in arts, letters, manners, scholarly pursuits, etc."* by Western societies [21]. The last sentence written in *the Daily Mail* was interesting in that most of the 2,000 people interviewed really didn't know what the term meant.

In contrast the definition provided in the second paragraph captures the Indigenous peoples' perspective of culture. What is noticeable about the differences between these definitions is the focus on the individual in the Western world perspective compared to the group or community focus for the Indigenous people. If you follow the Western world definition of culture, there are no rules that define what you should practice in order to become cultured and you can change your culture whenever a new fad becomes fashionable. For the Indigenous people, culture is knowledge that is transmitted through several generations and is retained by a group of people even when they adapt or adopt new fads or technologies.

Next, we will further next explore what culture means for someone of the Western European ancestry and for Native Americans. This will be followed by the story of Native Americans who have successfully lived, studies and worked in both cultures.

3.1 Western European Culture: You Live it, You Wear it and You Eat it

In the Western world, culture is a topic studied by social ecologists, e.g., anthropologists, who are mostly observers of culture. In the Brazilian Amazon, Drs. Kristiina and Dan observed situations where anthropologists were kicked out of Indigenous communities who were tired of having their culture studied by researchers who wanted to learn why they make certain decisions. The typical approach to learn about culture was for researchers to travel from many parts of the world and live with a community or members of a community long enough to observe and interview individuals. This seems innocuous enough and something that anthropologists have been doing for many years.

Studying culture or cultural resources as an outsider is always challenging when you need to keep your distance from the community members and not influence them in any way because your scientific protocols dictate such behavior. Of course, as a scientist you are supposed to be separate from the people you want to study so you do not influence them and jeopardize the results. A scientist must be objective. We were told a story of an Indigenous people's village telling researchers that they commonly eat vulture — which they do not — and "forcing" researchers to eat a stringy and tough bird in front of them. They were laughing the whole time and were increasingly playing tricks on the researchers because of their dislike for being studied by them. They were aware of the fact that the scientists would write papers on their research in the community that would increase the scientists status in the scientific community. Even in Puerto Rico, the Vogt's found local Puerto Ricans were very unhappy about researchers coming in and talking to them, asking a lot of questions and then leaving without giving the people surveyed anything in return. In contrast, we saw Yale students who respected the local knowledge and figured out how they could help improve the livelihood of the people that they wanted to learn from.

Starting in the 1990s, the Vogts observed that many communities were tired of being studied and not having the researchers contribute anything back to the community. We heard about graduate students being kicked out of Brazil because the communities believed that the students were bio-pirating knowledge of medicinal plants because they collected plants for a well-known botanical garden. The community felt that they were collecting plants not for plant identification but to give that information to pharmaceutical companies who would begin to artificially produce compounds isolated from the leaves. In this case, the cultural plants used for medicinal purposes did not result in the co-sharing of knowledge and the financial benefits of identifying plants with potential values. Since most pharmaceutical companies will not continue to need plant materials from the community as soon as they have artificially produced the beneficial compound, the communities rarely benefit from the sharing of their knowledge. The intellectual property of the knowledge rarely is rewarded by financial reimbursement to the communities over longer time scales. In some cases, researchers would try to return something back to the community but had not really determined what the community needed; one community we are aware of had a school house built for the youth in exchange for being allowed to collect plants. Eventually the building was abandoned because it did not really provide the community what they needed. This provides an inkling of what it means to become "cultured" in an Indigenous people framework. It is not just a personal concept where you adopt someone else's values. It is a community concept that touches many aspects of a community from the foods they eat to resources used in ceremonies.

Today, traditional knowledge is mostly translated through a Western science lens or written by people who have not grown up immersed in that culture. It is a foreign lens that is trying to understand the culture of a group of people by observation. It is not unusual for US citizens to think of their culture as derived from their European ancestry. They actualize these cultural attributes by wearing the right clothes or eating the right foods. These are the surface characteristics of a group of people but do not tell you why they behave in a particular way or make the decisions that they make.

People who migrated to live in the US are interested in expressing and maintaining the cultures of their European ancestors. This is not the deep culture since the focus on appearance which tells you nothing about how you should behave if you had a cultural core that drives how and what decisions you will make. How you dress does not directly link you to a functional definition of culture. In ecology we like to think about these as structures and functions, e.g., what we see and those activities that maintain the ecosystem but are not linked to structures [22].

Kristiina Vogt has frequently been approached by Americans whose ancestry goes back to Finland. They feel they are culturally attuned to Finland because they wear Finnish ceremonial costumes during special holidays or eat pancakes and pea soup (in the 1950s to 1960s there was a tradition of eating pea soup and pancakes every Thursday night in each Finnish household). Eating pea soup and pancakes was eaten because there was not enough food to eat; Kristiina didn't eat her first egg until she left Finland. Eating pancakes, even when you add berries like cloudberries collected in the swamps of Finland, does not make you a Finn. Nor does eating potatoes and meatballs make you a Finn. They taste good and are delicious but still don't make you into a Finn. This is also a custom resulting from a lack of food and probably not what something that everyone wants to continue to practice after food availability is abundant.

Kristiina Vogt and her coffee cup.
Source: Kristiina Vogt.

The Finnish characteristic has a name — SISU — which is loosely translated as "Guts, Tenacity of Purpose or Stubbornness". What gives you SISU is not the clothes you wear or what you eat. You could be wearing no clothes and still retain your fundamental Finnish character. It is a character — deep culture — that allows you to survive under harsh and cold winter conditions as well as the low light conditions of the winter months. Many people cannot tolerate such conditions. This explains why so many people in the Nordic countries drink too much or they engage in activities such as cross-country skiing that gets you out in nature, even when there is little sunlight during the winter months. SISU can be traced back more than 500 years and persists today. SISU is a unique characteristic of Finns even though it can be found around the world and describes the mental toughness of a person. Even people who moved from Finland to North America have maintained this mental toughness that is characteristic of Finns. So, to understand culture and how it impacts how you make decisions, we have to search further beyond these visual appearances to see what culture really means and why it persists for hundreds of years. Embedded in four letters — SISU — are important Finnish values and knowledge that persist even in the face of constant change.

It is hard to teach another group of people your deep culture. It is easier to define it using surface cultural characteristics since it is easier to explain it to a foreigner who knows nothing about you. It would take a lot longer to tell someone about your deep culture because they lack a context to understand what you are saying. Most people do not have a frame of reference to understand people with different values and cultural practices. Cultural practices can also not be generalized for all people living on a continent the size of North America. Native American knowledge is not just words but a way of life and philosophy that results in very different decisions being made within each Tribe. European colonialists assumed all Tribal people were the same even when they looked different. They were also not interested in taking the time to learn the different cultures they were exposed to when arriving in the Americas. They were too busy struggling to survive in a very foreign land and fighting to take ownership of the Indigenous peoples' land and resources. The cultures or knowledge learned in Europe did not work well in these new lands. Survival was difficult because they did not recognize the animals and plants that were safe to eat. They were afraid of the forests that were different from the ones they knew back home in some European countries. Forests also covered about half of the pre-colonial US while most European countries had already cut their forests hundreds of years earlier. Expansive areas of forests and animals who lived in them was not in the memory of these early European colonizers.

The European colonizers began to form a modified culture from their ancestry homelands in Europe. Few wanted to understand Indigenous cultures because their view was that these people were not civilized but "savages" [13]. Why would you learn the culture of an uncivilized person? Anyway, Manifest Destiny required them to civilize the conquered people and not adopt their cultures. Over several hundred years and after industrialization, the former European colonizers formed a US culture that characterizes America of today. It tends to be a surface culture that is being extensively studied by researchers and economists today but without much success. Perhaps being a surface culture is the problem since they are struggling to assess the "deep culture" of America. The "deep culture" is very diverse and multi-faceted and difficult to convert into mathematical predictive equations. We tend to read about fads that change every year. Fads are aspects of surface culture since you do not make decisions based on them.

Today, industrialization has left a huge imprint on how culture is defined and practiced by citizens of these countries. These cultures did not adopt elements of Indigenous cultural practices and probably never understand Indigenous culture and why it is important for Tribal people. *The Daily Mail* summarized a 2015 survey that exemplifies the Western European views of culture as,

> *"Owning a library card, watching sub-titled films and being skilled in the use of chopsticks are among the traits which make us cultured, according to a new survey. Drinking "proper" coffee, knowing the difference between cuts of meat and watching the Antiques Roadshow also set you apart from your less refined friends. However, it seems you probably don't have many of those: Seven in 10 consider themselves to be "cultured", although the majority admit to not knowing exactly what the term means."* [20]

European colonialists developed a "myth" view of Native Americans that continues to be fed today by the media and movies — it is not real but is a good excuse to marginalize and to rationalize the need to civilize another group of people. Many movies continue to portray Native Americans as being violent and uncivilized villains. The 2013 movie "Lone Ranger and Tanto" portrayed Native Americans speaking broken English, chanting prayers and worse of all wearing a stuffed crow on your head. You are going to have a hard time finding any American Indian that will wear a stuffed crow on their head. Crows are recognized by several Tribes for their intelligence and are *"used as clan animals in some Native American cultures. Tribes with Crow Clans include the Chippewa (whose Crow Clan and its totem are called Aandeg), the Hopi (whose Crow Clan is called Angwusngyam or Ungwish-wungwa), the Menominee, the Caddo, the Tlingit, and the Pueblo Tribes of New Mexico."* [1]

---

1 Accessed on January 3, 2018 at http://www.native-languages.org/legends-crow.htm

3.2 Culture According to Indigenous People

Western science has struggled, mostly unsuccessfully, to assess and develop solutions for complex, interdisciplinary environmental problems [3]. These tools do not include culture or Indigenous knowledge, and, in some cases, there is no recognition of the validity of other knowledge forming approaches. **The most common Western science methods and economic tools need to be completely restructured and envisioned because they are not designed to solve problems emerging in today's environment. They tend to convert knowledge into discrete numbers that can be statistically compared. This simplification of knowledge misses the holistic, intrinsic, and qualitative aspects of ecological systems that cannot be quantified using linear approaches or easily assigned a dollar value.** What is missing in the Western scientific approach to forming knowledge is culture and an approach to deal with a "lack of knowledge" (*see* Section 4.2). In contrast, Native Americans and Native Alaskans use culturally-based strategies to address the climate change impacts occurring on their lands and water [23]. This raises the prominence of food security as part of assessments since cultural foods are important for Native people and their ceremonies (*see* Section 3.6).

For Tribes, culture has always bound decisions so that not everything has a market value and can be sold. Tribes are risk adverse but at the same time are the "ultimate adapters" and avoid the "tunnel vision" approach to decision-making commonly found in Western science. There is so much we do not know. Our knowledge is fragmented. We need tools like those offered by Native people to address environmental problems and to get creative in decision-making (this is the topic of Edwards 2009 book *"ARTSCIENCE. Creativity in the post-Google Generation"* [24]). Loss of culture/art in science contributes to management problems since it assumes that we know enough to make decisions and can adapt to any new disturbances.

Culture does not have one definition and we should nor fall for the trap of thinking that we have built a better environmental assessment tool by including superficial cultural practices in decision-making for the environment. The iceberg metaphor is an excellent way to visualize the attributes of surface and deep culture. Elizabeth Ferguson wrote about how most non-Tribal people see culture as a surface culture [25]. As Ferguson wrote [25]:

> *"Indigenous people are defined by their culture. Most people outside the culture recognize certain aspects of the Indigenous people. Those aspects are the tip of the iceberg. There is so much deep knowledge embedded in the culture that it does not appear on the surface. This model was developed by the Lower Kuskokwim School District."* (Fig. 3.1) (*see* also the surface and deep culture drawn by Barnhardt [26])

The iceberg was also drawn by James Penstone under a Creative Commons Attribution. It is a powerful approach to connect the visible and invisible aspects of culture (Fig. 3.1). The surface culture — the part of the iceberg that can be seen above the water — typically represents the physical characteristics of a person that are seen by people when they meet someone from a different culture. It also includes part of the folk culture that are seen and reflected in a group's festivals or cultural practices. This contrasts the deep culture — hidden below the water line — which is the values or beliefs of a group of people that impact how they behave or the decisions they make. These tend to be invisible and not known when you understand another person superficially. The deep culture is much more difficult to understand since they can reflect beliefs and values that have been passed down through many generations. Since most interactions with others are superficial, the influence of these values will not be apparent since the connections between behavior and cultural value is not linear. Most of our daily interactions occur at the surface culture stage since people are not interested in learning why decisions were made but what choice someone will make. These

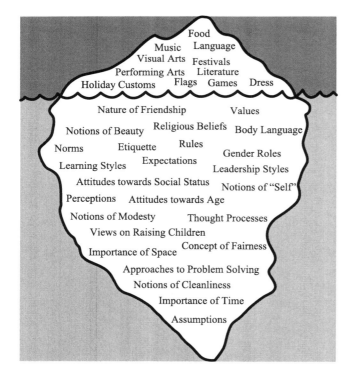

Fig. 3.1 Cultural iceberg drawn by James Penstone (Penstone 2018)[1].

---

1 http://opengecko.com/interculturalism/visualising-the-iceberg-model-of-culture/, accessed January 5, 2018

tend to have linear links between factors.

Culture for Tribal people means they need to walk between two cultures (*see* Section 6.5). This is not a process of assimilation or blending of cultures. This means not losing one's Deep Culture which determines how decisions are really made. In Section 3.3 and 3.5 you will read the stories of two Native Americans who have successfully maintained their cultural roots while also working on PhD degrees at a University.

3.2.1 Cultural Resources Defined by Tribes*

In Washington State "cultural resources" is defined as historical and archaeological sites, artifacts and areas that may be traditionally important to various groups of people and is generally 50 years old or older. Also, identified cultural resources remain the property of the landowner. By federal and Washington State law, when there is a discovery of Native American artifacts the local Tribes must be consulted. Tribal governments believe that areas with significant cultural resources are best co-managed by federal and Tribal government or state and Tribal government. When public land is sold or exchanged the Tribes may lose their ability to manage these resources and their ability to access cultural resource sites.

Many Tribal governments view cultural resources in a much broader sense. To Tribal governments, cultural resources not only include archaeological sites, but can also include places where Native people access traditional use plants, ceremonial areas, and places for hunting and fishing. Importantly, Tribal gov-

Rodney Cawston.
Source: Colleen Cawston.

---

* By Rodney Cawston.

ernments don't view cultural resources as something in their past. Native people are still largely living their culture and have needs of accessing their usual and accustomed areas for cultural resources. In doing so, Tribal elders and cultural leaders pass on their traditional teachings from one generation to the next. Tribes across the United States have a resurgence of teaching their traditional languages and culture to their younger generations.

Many Tribal members continue to gather natural and cultural resources for their own use and also for community use at Tribal ceremonies, dinners and for spiritual purposes. For example, on the Colville Reservation, members of the Nez Perce longhouse each year will hold a first-food feast to honor and respect their traditional foods. Many edible roots, berries, teas, wild game and salmon are gathered for these dinners. Native people have conserved, protected, enhanced, and lived within their environments for thousands of years. Protection of cultural resources and natural environments is paramount for cultural sustainability for future generations.

When Tribes make decisions regarding natural resources both on and off of their reservation boundaries, different standards or metrics are utilized when considering utilization of natural resources that may not be considered with planning processes and decision making models. When planning for the extraction of natural resources from the environment for economic purposes, Tribes make these decisions weighing impacts to cultural resources from their understanding and meaning. However, Tribal governments also need economic development to create jobs for their people and to create revenue to fund many needed programs within their government.

Some Tribes have adopted the term "cultural resource management" to describe their broad definition of preservation. However, the same term is used by many archaeologists and some federal agencies to describe their archaeological programs as:

> "I think that it is important to let Congress know that cultural resources to Tribes are a lot more than what the Anglo society usually regards as historic preservation needs. People in the East think of historic preservation as keeping nice buildings with beautiful facades in place. And that is often as far as people will think about historic preservation. I think it is absolutely critical [to] get the point across that in Tribes, there is a much more holistic view of what cultural resources are [Roger Anyon, Zuni]" [27]

There are twenty-nine federally recognized Tribes in the State of Washington. These are modern forms of Tribal governments each of which have their own unique language(s), culture(s), and socio-legal histories. Some of today's Tribes are confederations of Tribes. When Tribes were moved off of their traditional homelands many different Tribes were moved onto the same reservation. The Confederated Tribes of the Colville Reservation is a confederation of twelve Tribes, including the Chief Joseph Band of Nez Perce whose original homelands were in North Central Oregon. There are also situations that exist where Tribes were separated and now reside on more than one reservation. The Palouse Tribe's members moved onto the Colville, the Yakama, the Nez Perce Tribe of Idaho and the Umatilla reservations. Washington's twenty-nine Tribes were recognized by the federal government through signature of treaties, presidential executive orders, and by congressional legislation. When approaching cultural resources management, one must not only understand Tribal views of cultural resources but one must also understand the unique cultural identities and political histories of each Tribe. This is especially important when contacting a Tribe whose cultural resources may be impacted.

Natural resources include plants, fish, wildlife, and other living organisms as well as nonliving matter, such as minerals, water and air. Often this distinction between living and nonliving resources is made by referring to *renewable resources* and *nonrenewable resources*. At first glance, it would seem that this would be a very easy question to answer. However, searching through the Washington State Department of Natural Resource (DNR) Publications, a definition for "natural resources" could not be found. In a telephone interview with DNR's Policy Director, he commented that there isn't a written definition of natural resources in any of the laws, regulations or publications of this agency. It was his opinion that "natural resources" is fluid and philosophical and can be viewed through three different lenses: through DNR's Trust responsibilities to their beneficiaries, through economic values or valuable assets that contribute to the trust, and as an obligation when assuring compliance with regulatory laws. In comparison, Native people believe that natural resources and cultural resources are one and the same. "Everything on earth has a purpose" is a traditional teaching of many Tribes. This is also a traditional teaching of Tribes in British Columbia, Canada. As stated in Nancy Turner's 2008 book *"The Earth's Blanket: Traditional Teachings for Sustainable Living* (Culture, Place, and Nature)" [28]:

> *"Everything is One.*
> *The Creator made all things one.*
> *All things are related and interconnected.*
> *All things are sacred.*
> *All things are therefore to be respected."*

—Dr. Richard Atleo, hereditary Chief of Ahousaht

"Essentially, these philosophies from the Raramuri and Nuu-Chah-Nulth confirm and reinforce the idea that humans do have a kinship

> with all the other elements of their world. Ceremonies, customs and
> stories recognize and validate this connection as do the ways in which
> people relate to their lands and resources." [28].

Tribal members must be able to gather, hunt, plant, and harvest while simultaneously engaging in spiritual activities in order to pass down their knowledge about how natural resources are respected and sustainably harvested. The actual practice of a subsistence activity is as crucial as the end product derived from the activity. It is very important for Tribes to pass down their traditional practices and teachings to their younger generations and this is becoming increasingly difficult with lack of access to natural areas. Today many Tribes are studying their own traditional practices when developing natural resource management plans to improve Tribal environmental management.

3.2.2 What Is a Cultural Resource?*

Cultural resources are the remains and sites associated with human activities, are non-renewable and include the following: prehistoric and ethno-historic Native American archaeological sites; historic archaeological sites; traditional cultural properties, which include such resources as traditional resource gathering areas; sacred sites; historic landscapes and buildings; elements or areas of the natural landscape which have traditional cultural significance; and museum collections. However, Native people have a much broader definition of cultural resources.

Rodney Cawston making a basket.
Source: Rodney Cawston.

---

* By Rodney Cawston.

Tribes have a living history. Many of the fishing sites, camping sites, religious sites, and gathering places are still being used today. Many Plateau Native people would leave implements such as mortars and pestles (because they were relatively heavy to carry) at the same place to use in successive years. Many Native families would fish at the same locations; build scaffolds for dip net fishing at the same locations. Tribes had seasonal rounds for gathering, hunting and fishing. The Makah Tribe made a harpoon with a shaft of red fir to which are attached two prongs of iron wood by means of cherry bark lashing. Each prong is split at the end, into which the bone point is fastened with nettle twine (Waterman, 1920). The Makah Tribe has handed down knowledge of the best time to gather red fir, iron wood, cherry bark and the nettles to make this fishing implement. Everything that was needed was provided through the wealth of the natural resources around them.

The Okanogan people on the Colville reservation have a word "qʷiyula?x" which means all the riches or goodness of the land. Many Tribal elders consider wealth as healthy ecosystems. Today, many Tribal elders are trying to encourage their world views to their younger generations, including respecting natural resources. The elk, deer, salmon, edible roots, berries, and water, are shown respect in many different ways, including through ceremonial dinners such as the first-foods feasts, naming ceremonies, funerals and other important events. During the first food feasts, selected men and women will prepare themselves with spiritual cleansing before they go out to meet the new foods and gather them for a very special dinner of the people. Many traditional songs and teachings about preparing for these feasts and the importance of "respect" in many different contexts are shared. Tribe's respect those animals and plants that give up their lives, so their lives may continue; they respect water that cleanses the lives and souls of everything living. Tribes respect their Tribal elders. Tribal elders and those who cannot go out and gather traditional foods for themselves are respected and others will share their traditional foods with them. When a young man catches his first salmon or kills his first deer or elk, the teachings of many Tribes is to give the meat to an elder who in turn will pray for him. A girl's first roots or berries are also given to a respected elder.

Respect is a very important world view of Native American people. Animals, plants, all living things and non-living things, the natural ecosystem of our world and even mountains, rocks and climate have moral standing and worth and are owed recognition and respect. Elders in our longhouse used to tell us that we shouldn't kill anything on Sunday; this day is regarded as a sacred day. On one of these Sunday's I heard this Elder calling my name from the kitchen. She said there is a mouse trapped in the sink and he is all wet and barely alive. I asked her if she wanted me to kill the mouse and get rid of it. She said, *"No, bring him outside and leave him, let the Creator take care of him, it is Sunday."* They would tell us that everything has a right to perpetuate itself and to be left alone for one day of the week. Their teachings were such that things matter morally and have a claim to respect in their own right, independently of their usefulness to humans. Everything has value. The Elders would also tell us that even though we kill animals and plants for our own food, we also respect them by not using them for trivial purposes or destroying them and then not using them respectfully. When we did something contrary to our ways, such as killing a deer for its horns, our Elders would tell us that we were acting like "*White man*". They felt White men don't respect our natural world and have abhorrent attitudes of domination and exploitation toward all nonpersons and toward our environment. The traditional ways of Native people could benefit a wide variety of human practices, ranging from natural resources management, agriculture, urban development, energy use, technological research, and may have a profound alteration by a recognition of respect to nonpersons and our natural environment.

3.3 Keeping Deep Culture in Two Worlds: Interview of Dr. Mike Tulee*

Kristiina: Mike, we want you to introduce yourself, give us a feeling for who you are, where you're coming from, so that the reader can understand what your background is and why they should listen to you.

Dr. Mike Tulee (left), Alexa Schreier (top right), Kristiina Vogt (bottom right).
Source: Mike Tulee, Alexa Schreier, Kristiina Vogt.

---

* By Alexa Schreier and Dr. Kristiina Vogt.

Mike: *I'll just start by saying that I come from very humble roots. I've lived on four different Indian reservations, all my young life. I grew up in a family of ten. My father is deceased, he was a BIA person for twenty-five years plus.*

Kristiina: Why did you live on four different reservations?

Mike: *Being that my father was a BIA person, he was transferred, almost like government workers and similar to the military. You're in it for so long and then there may be a demand for your services at a different reservation where there's a federal office. Sometimes it works out that you're going to be transferred where a more powerful need is and that was the case with my father. I actually lived on the Colville reservation, the Spokane Indian reservation, the Yakama reservation, and the Tulalip reservation as a youth.*

Kristiina: But you were born in Yakama?

Mike: *I was actually born in Coulee Dam, which is on the Colville reservation. We lived there for probably about at least eleven or twelve years. We lived on the Spokane reservation for about fifteen years and then we lived on the Yakama reservation for about five years. Then I went into the military and bounced around for a little bit trying to find my calling and was a vehicle mechanic. It was a lot of fun working on vehicles, but somehow that wasn't quite me. Even the commander said, "Somehow, I don't think you belong here, I think you should go into teaching or something?" I thought about it and I think he was right because not only was I tinkering around, but I was always trying to figure out how to get things done. Other people would kind of chime in and all of a sudden, the next thing is that we were discussing what the best way was to fix this particular part. Once we gained those little pieces of knowledge, it was easy for me to explain it to others. It kind of resonated with the commander who thought I was probably better suited for the education field. So, when I got out of the military, I knew I wanted to go to school.*

Kristiina: How long were you in the military?

Mike: *I was there for four years. I made sergeant, as an E4, which means I was still low grade enough to be bossed around quite frequently. But it had its ups and downs but was certainly still something that taught me a great deal about discipline. Getting out of the military, I immediately enrolled into College. I still didn't quite know what I* *wanted to do, even though teaching was in the back of my mind. I ended up taking a lot of geography courses and history courses which really was a strong attraction for me. Then one day I decided that since geography was kind of like a social service field, I decided that geography was my discipline of choice. I finished getting that degree and then decided if I wanted to teach, I should probably go into the teacher's certification program. So I immediately enrolled into that program and it took me about a year and a half to finish that degree.*

I went into teaching in the Seattle school district and worked in the Seattle school district for about 12 or 13 years. During that time, I also decided to pursue my master's education and finished that in about 1999. Then I worked in the education field for 10 years. I worked in about three different professions, all in the education field. But I eventually decided to go for my doctorate, I was convinced by this professor on a cold rainy day that being a PhD was a way to go.

Kristiina: What if it had been a hot day? Then you would not have been attracted by my encouragement to pursue the degree?

Mike: *Not that I think about it, it might have been a hot day. But it was an outdoorsy day. I enrolled into the University of Washington in January of 2009 and so it fell into place. I had a thirst for knowledge and finished my doctorate in 2015. Ever since, I kind of went back to teach in the university system up until 2017 when I was hired as the Executive Director of an organization known as the United Indians of*

Two Dr. Mike Marchand (left), Dr. Mike Tulee (right) at an ATNI meeting — both good Huskies from U.W.!
Source: Mike Marchand.

all Tribes Foundation. I now oversee about 56 workers and we conduct 7 different social services programs as well as deliver education programming for the Native communities in the greater Seattle area.

Kristiina: Going back to what you mentioned at the beginning of this interview, and I've heard it from several different Tribal members, they live on more than one reservation. It's not unusual for Tribal members to in fact live in other places. How do you think that impacted your culture? Do you remember the different reservations where you lived? Was it difficult to go to a new reservation? Was it a fluid transition? Was it a normal thing, like people were pretty used to it? How did you react and feel about it?

Mike: *One thing I find that's similar amongst them is that it's a very different world. It's a very isolated system where you're really cordoned off from the mainstream society. It means that when you go into a city, you're going to have a culture shock because you're not conditioned to converse with people who look different than you. So even though the reservations were relatively close to Spokane, going to Spokane was always a new adventure for me. Simply because we were always conditioned to be very careful, don't be very trusting, always stay close to mom and dad, always count your pennies to make sure you aren't getting cheated. It was this kind of world of distrusting this other dominant culture.*

So I never really became familiar with everyday norms and customs of Caucasian race except for watching Channel 8, ABC, CBS, NBC and PBS back in the 1960s and 1970s. Back in the days, you didn't really get very pragmatic information because you were watching Archie Bunker, Lucille Ball, Happy Days and shows like that. They had more of an entertainment focus. They didn't really educate you on what was out there and the outer world of mainstream society. So, I didn't actually gain a strong footing, practical knowledge, of what everyday life is like outside of the reservation.

Being in an isolated system, everybody is brought up in the same way. We're in close proximity to each other and our customs are very similar. What I mean is that you have Plateau and Salish cultures, which were Colville, Spokane, and Yakama. There wasn't very much variation, the language base was different, but the religions were actually very similar. They all have long houses, they have Catholicism, and Protestant denominations there which I was exposed to. So, in that sense I was able to gain a religious thought process that was similar to others. They had cultural norms that were very similar. All three of them did what's called, war dancing.

Kristiina: Were you doing that even as a kid growing up?

Mike: Yes, as a little kid, my mother was very powerfully ingrained into native culture, language, as well as dancing, singing, and drumming. In fact, one by one, she would steer us each in that direction. Oddly enough, none of the boys ended up being dancers except one of us. There were ten kids, seven boys, and only one of them ended up being a dancer as an adult. All three girls were dancers and they seemed to enjoy it, being a part of that culture.

Kristiina: What is the importance of dancing?

Mike: For some people, for some Tribes, it's a really strong identity model. What I mean by that is that back in the old days, there were no such things as POW WOWS. Each particular tribe had ceremonies and part of those ceremonies was dance and singing and drumming.

There were times when you were fostered, conditioned, and taught to have a certain purpose in your village. One could be that you're going to be a medicine man, another is that you're going to be a leader of your people, or that you're going to be a warrior. Another one is that you're going to be a weather man or a person who specializes in hunting. All of those specialties you had to have, you had to be given power and power was given through special songs and dance and medicine men praying over you. So, when you had gone through these rituals you were taught these certain avenues, these certain lanes of knowledge from elders who specialize in these particular disciplines.

In other words, if you were taught to be a hunter, the village hunter would work with you. He would teach you all the ways how best to bring down that deer or how to hunt that elusive elk that can run 45 miles per hour. How to protect yourself from bears who are more powerful than a normal man, or how to watch out for a cougar who may be stalking you as you're doing the hunting. Or there's another avenue, say the warrior, where your number one purpose is to protect the women and children. So you're taught hand to hand combat as well as the ways to know if there's danger afoot. You're constantly taught by the elders through these different avenues.

Kristiina: What age would that normally start at?

Mike: *You're normally taught at a young age the generic items by the elders. Usually the parents and young adults would take care of the village. As a youngster you're taught by the elders the primary items of how to gather berries, how to mind the homes or teepees or sweat lodges or the long houses, or how-to bring fish to the village. As you got older, which usually happened around adolescent time or as a young man, the elders let the village know you're ready for the next step. That's when you go through these rituals of what you're going to be. Ironically enough, those things have all kind of gone to the wayside as we've become assimilated to a great extent into the mainstream culture today.*

Kristiina: You said you like to tinker, were there any jobs or professions in the tribe where you could do those kinds of things? Was this something where you would have to go outside the tribe to do?

Mike: *Well, it was strange that the BIA had jobs, but they were very limited. Back in the 1960s, 70s, and 80s, the unemployment rate was probably about 55% on the reservation. BIA jobs were very few and far between. It was a very downtrodden system in that the U.S. government rarely or never provided ample financial resources to Tribal people in order to train them to better themselves. That was actually supposed to be the purpose of the BIA, to prepare us for mainstream living. But, to a larger extent, we were severely underfunded starting as early as 1849 when they first started putting money aside for Tribes. The money was very few and far between, there was a lot of corruption. Sometimes the money would be just barely enough to feed families, opposed to be trained to better ourselves.*

So, even though a person like me, early on in my life, was interested in mechanics, the thought of me being a mechanic would be very difficult to pursue. There would probably be only one or two mechanics for our whole reservation. Back then, we probably had 7,000 Tribal members and we probably had about 100–150 government vehicles, but there wasn't enough money to hire me even as an apprentice to go into that field.

My dad looked at me one day and said "Son, it's great you're in forestry, you're doing forest development, you're cutting down trees, but what do you really want to do with your life?" He actually really made me think about what do I want to do? I was only making $4.15 an hour in the year 1979, so I didn't think I wanted to be doing that kind of *work for the rest of my life. I just barely made above minimum wage. So, one day I walked into the military station and I decided there's got to be a bigger world out there for me. They asked me, "What's the first thing you want to do?" I said, "I think I want to be a mechanic." That was a done deal. But I didn't realize you had to jump through a lot of hoops to even pick up that first wrench. I did a lot of jumping jacks, calisthenics, push-ups and being yelled out and being mentally abused, but it was something that every soldier or Navy person has to go through.*

Eventually it conditioned me and my thought process, and it conditioned the way I wanted to learn. It just really made me appreciate going into the military — not that I would join again. But at that point in my life it was something that had a good payoff for me. It set me in the direction of being a curious person who wanted to learn a trade that I believed could help me. Even though I loved being a mechanic at the time, I didn't think I wanted to be a mechanic for the rest of my life. As I indicated earlier, all these small incidents of arguments and agreements and discussions led me to want to teach than to perform as a mechanic.

Alexa: Going back to when you mentioned moving around to all the different reservations, a few of the people we've talked to have mentioned that growing up on a reservation provided them with a strong place-based perspective and knowledge of that landscape. Do you think you lost that at all from moving around to different reservations? Or do you feel like you were still connected to those places?

Mike: *It's really strange, I actually still have dreams about when I was a little kid of the exact places where I grew up. Even though I haven't lived in some of these places for decades on end, I can still remember these trees we used to play on when we were kids. So, I still have a strong attachment to these places where I lived. In fact, when I go visit Nespelem, the first thing I do is take a little walk around what is called the campus, which is where we lived. Then I go to Wellpinit. I actually have visited the campus there, which is also where we lived. I retraced all these particular places that I have very fond memories of, and it brings back a lot of great memories actually. I feel like I still have a pretty strong attachment towards some of these trees that are still standing. I imagine some of these trees are well over a hundred years old. I can still, to this day, remember climbing these trees and even the one I fell out of. I remember I hit a home run at this baseball field in*

1972. Those are the kinds of things that I can immediately recall that have strong potent memories for myself.

Alexa: When you joined the military, how do you think that impacted the culture you had developed growing up on the reservation since you were going into a very westernized system? I have heard that Tribal people are fierce fighters.

Mike: *It seemingly should've impacted me. But one thing that made me survive, or it made me more grounded, was that it really made me miss my homeland. Being in the military, was a totally different structured area where there were aggressive actions based on the mainstream society's philosophy of discipline. It made me revisit all these great memories and thoughts I had about my upbringing. It seemed to really keep me attached to my background. Even though I was in a totally different cultural environment, it didn't impact me because of my attachment or affinity or my strong memories of my upbringing on the reservations. I didn't feel like I missed a step.*

Kristiina: Did anybody else from your family join the military?

Mike: *My father was actually in the military, he was in the Korean War back in 1950–1953. He was a first lieutenant. My sister, my younger sister, was also in the army back in about the mid-1980s. I was in the military in the early 1980s and got out in the mid-1980s.*

Kristiina: Tribal people serving in the military doesn't seem unusual. This is interesting since the military is a very structured institution that wants to take away any culture you have. They want to make you into what they want and someone who responds to a command immediately. We've also been at Tribal ceremonies and there's always an emphasis on having the Tribal people who have been in the military to stand up to be recognized. So, this is not an unusual thing for Tribal people to do and they are not forced to do it. I think Tribal people have also been fierce fighters, so I think there may be some link there. What are your thoughts on that last comment?

Mike: *Tribal people, even before the western culture took over, were conditioned to do these specialties and one of them was the warrior. But as the assimilation process progressed over the years and when they started putting us on Indian reservations, those kinds of disciplines were pretty much disappearing because we were confined to a small area. We were provided these living arrangement, we were provided food, and clothing. They took away a lot of the structure of the Tribes. One thing many Tribes hung onto was the warrior mentality. So as* *early as World War I, various Native Americans volunteered to fight in the U.S. Army when the Austrians, the Germans, the British, and the Russians were fighting.*

Native Americans weren't even citizens in the early 1900s, 1917–1919, which is when we entered World War I. We weren't even U.S. citizens but the mere fact that we were on reservations for decades on end, somehow, they felt that there was a drive or an innate need or drive to be useful in the warrior sense. So many of them easily volunteered to go into the armed forces. In fact, data shows that 70% of all Native men volunteered to fight in World War I and World War II, compared to the mainstream society which was about 10%. So, there was a very wide gap in the amount of determination and desire of Native men opposed to mainstream society to fight.

Having that warrior mentality and going forward to help fight happened during World War I. But during World War II, more Tribes actually thought "Wait, we see this Adolf Hitler, whose number one mantra was that the Arian Nations shall rule the world. What will happen if we lose the war, we'll all be exterminated, our women and children. We all have to pitch in, we all need to fight, and we all need to enlist to make sure our reservations, our Indian villages, are not taken over by this Arian Nation that may actually conquer the world." So, in 1941, December 8th, a lot of Native Americans freely enlisted into the armed forces with the determination to protect their Native villages.

Kristiina: Were there no problems with that? I don't know anything about it, but don't you need to be a citizen to be in the military?

Mike: *Now you do, well even now you don't need to be quite a citizen.*

Kristiina: That's right, because there are some that are coming from Mexico and Central America and they are can be on the fast track to become a citizen after serving in the U.S. military.

Mike: *It is kind of seen as a path to citizenry. We actually weren't recognized as American citizens until 1924. Back in 1917 we served in the military because of this innate need to be warrior like and feel like we're useful even though we were a conquered people by the dominant society. Later, we didn't want to be exterminated by Adolf Hitler and his Nazi regime.*

Kristiina: Mike, when you were living on the reservation, what was your perspective of culture? What is it that you felt like when living on the reservations? One of the discussions we've had with a couple of people is that the reservations are in fact important because that's really where the deep culture is maintained (*see* Section 3.2). They mentioned that it is harder to maintain one's culture as soon as you get outside the reservation and you're living in a city. You've got so many pressures trying to force you to conform to the current society. To you, what is culture and what do you want your kids to have if you think about it?

> **Mike:** *First, number one, to me, culture is everything that captures your life and every single thing you do within your immediate living arrangements that defines you. It's your learning, the way you act, and particular traditions that your mother and father do every single morning. It's the particular worshipping you do, what reverence you have, certain landmarks and how you dress. It's how you treat other people every single day. These are the customs that define you, the languages that you learn, and how do you approach life every single day. That's what culture is to me.*
>
> *Number two, for my children, my desire is for them to freely choose how strongly they want to be enmeshed into our traditional culture. If you go back to my reservation, there's still language that's spoken. I can speak the language, but not fluently, and I can understand broken phrases. I can catch words, but I'm not a strong speaker. When I go back to the reservations, I can still hear people my age and younger and they can actually converse with each other which I think is really wonderful. The language is becoming stronger, they're beginning to reteach it. As far back as 1607 when the English first settled in Jamestown, one of their first orders of business was to break us of our language. So, that resonated every single year up until about 1964 when* the Civil Rights Act *was signed. Before that, their goal was to assimilate Native people. They thought the only way Native people are going to survive is if we break them down and then build them up in the way that we think they can survive.*
>
> *Well my thought is for my children is that I want them to be able to learn as much language as they want. I want them to be proud of their upbringing. I want them to be proud of what our religions hold, of the particular roots we have, how our elders speak highly of the whole natural environment and how our immediate natural environment sustains us and what we need to do to keep it as such. So, those are the kinds of things I hope my children will appreciate. Hopefully they will be able*

to keep that phenomena going that the land comes first no matter how long we live.

> *The land will always be there, and we want to make sure it stays as such for our children. That way, even if you go to virtually any reservation today, there's very little development because the number one desire is to protect natural resources. We don't want things to get destroyed. We don't want our way of life to dissipate in totality. We believe that if we lose the land as it is, we're going to lose our culture, lose our language, we'd lose our calling in life. What is it that our life is all about? Well if we don't have all this sustainability of our resources, cultural resources, that's essentially what's going to happen.*

Kristiina: Do you think that going into the military was in fact an easier transition for you because it provided a shared community? A 2016 book written by Sebastian Junger entitled *"TRIBES"* [29] suggests that the military creates a sense of community where you take care of one another, you share. In fact, the books author was shocked to find out that the hardest thing for soldiers returning from a war zone was to lose their sense of community when they returned to the U.S. and had to interact with people who don't understand or care about what happened in the war zone. So, do you think that going into the military is an easier transition because there is a sense of community among a group of people? Part of the military, even though you're fighting wars, you learn a lot from one another and know you can depend on your fellow soldiers during rough times. This creates a sense of community. What if you had left the Yakama reservation and you had gone to live in Seattle instead of joining the Armed Forces, would that transition have been much harder for you? By going into the military that transitioned you into more of a community, with some attributes of a Native community, was it an easier transition for you then ending up at Antioch where you would have entered a more structured academic institution? Would that have been a very difficult transition for you?

> **Mike:** *Yes. I think the military strongly prepared me for mainstream kind of interactions. It was pretty much trial by fire, you are thrown into a mix of fifty young men, like me, and every single moment of your life for four years, you're taught to work with each other. You're taught that in a field of battle, you really have to work together or there's a possibility your life is going to end. So, in other words you have to watch your friends back and he watches your back and you watch other peoples' backs. More of us will survive if we can all accept that notion. Well that's essentially what happens in the military and is why they break you down and build you up to create this sense of*

community where we're all going to protect each other. So, when you go out of the military, as you suggested after the Vietnam War, there was a loss of sense of comradery when they came home. Many committed suicides or went into severe alcoholism problems. Some of them you still see here in the city who are in their upper sixties and are still homeless. All because their hearts never left Vietnam. They feel lost, there's something missing from their lives.

If you go into the military, you'll see that young Native men and women that go into Iraq or the Persian Gulf War, come back and there's still a degree of hurt, PTSD, a lot of trauma. But the one thing that can be assured of is that when they come back to the reservation, they know that there's a lot of medicine men and people who will really look after them to help them find their way. So, a lot of our veterans actually are very well cared for. We don't have the kinds of medicine they have in mainstream society. Our way of medicine is done differently, a lot of praying, a lot of herbs, a lot of mental therapeutic sessions with these elders who are able to help talk you through all these traumas. The reservations have medicinal approaches that are not Western science but are extremely powerful and very helpful.

Kristiina: So, it sounds like what you've just said is that if you had not had the military and had just jumped into the world of the white culture that would've been pretty devastating transition for you. Even though you had the tribe that you could fall back on and you know they're always there.

Mike: *In fact, I'll even give you another example, if a person like myself didn't go into the military and I just went into a big city like Seattle, it would've been a very daunting challenge to survive for four years and get my degree at U.W. Regardless of how smart I was, it would have been such a culture shock. I would not be used to 2.1 million people who live in this immediate area of Seattle, whereas if you go to my reservation, you're talking about 22,000 people. In my reservation you know virtually 2/3 of the people, whereas in Seattle, you know virtually nobody. Seattle is a fast-paced life, concrete jungle, where nobody is going to care about Mike Tulee because they don't even know Mike Tulee. So, you get this notion, "Wow, I don't belong here, I don't have any friends, I know I'm supposed to study, I know I have a purpose here, I want to succeed, I want a better life for myself, but I feel like I'm lost here. I don't feel like I really have a whole lot of connections with the city of Seattle. I know! I'll make up an excuse that I'm going to go home to say "hi" to my cousins and I'll be back by Sunday and*

I'll start studying then." *Well, one Sunday turns into two weekends which turns into another. The next thing you know, you've gone back about 4 or 5 weekends during the first quarter of your freshman year. The next thing you know, before the first quarter is even over, you're dropping out because you have this comfort zone on the reservation, and it was a cold world in the city of Seattle. You find out the city of Seattle isn't for you.*

Whereas, when you go into the military, it's trial by fire, you will adjust. You will understand people of different cultures. You will work together. You will be in different lands. You all will work together. You all will have a comradery. It's all structured. So by the time you get out and get thrust into a city like Seattle, it was not much of a challenge at all. In fact, I recall that when I came in the University system, I was very determined to succeed. I just didn't let anything hold me back. Studying was the least of my worries at that point.

Kristiina: I've always had the feeling that Tribes are extremely adaptive and they like other cultures, they aren't scared off by cultural differences. Growing up in Finland, there was nobody that looked different, everybody was blonde, blue-eyed just like me. I was the norm. You didn't see people of different cultures and most of those people don't want to leave Finland, they wanted to stay there. Then there's a few people like myself, my dad and a few others who leave.

I've been thinking about what makes the Native people, especially in the Americas, different. I think one attribute is their ability to be adaptive and they're interested in a lot of different things. They're not just interested in a narrow set of things. Many people I meet, they have a focused interest and that's what they care about. Most of the Native people I've met have very eclectic interests, and sometimes it's things they're doing and not even thinking about it. I think you're that way. But you also have what I call in Finnish, the SISU. I watched you when you were in the PhD program. You're persistent, you have the guts, you're not going to let little things sideline you, but that doesn't mean you don't have a lot of other things that you are also interested in. Because of that, you're going to succeed, you don't let it get in the way. So somewhere along the way, something about the culture does allow Native people to be successful. They have the guts, and the interest, so they're able to think broadly, they don't think narrowly and egocentrically. So, they're aware when things are going to come up that they need to be careful about that could become a problem.

I think that this is something that other people need to learn. They need to open up and be interested in people from other cultures. If they are not interested in other cultures, they are BORING and don't have the SISU to address complex problems. I'm not sure that you've thought about this, but it is interesting to think about what adaptability is. This is one of the big take home messages I've gotten from a lot of the interviews and discussions that we've had for this book.

Mike: *Well I guess first and foremost I'll say that in terms of eclectic interests, in the old days they did have to multi-task and have a multi-thought process. I'll just give you an example. My grandfather, his name was Samson Tulee. He fought a case all the way through the Supreme Court in 1939 that he won in 1942. It was "Tulee vs. Washington", which enabled Native fisherman to fish off reservations without a having to buy a permit in the usual customary places. He was multi-talented as he could speak four languages fluently. He could speak Chinese, Tagalog, English, and our Yakama native language. Somehow, he felt like for him to be successful he would have to learn all those languages. It was traits like that which helped him that he passed to my father.*

My father was one of the first Tribal people to get a degree and he was a very highly intelligent man. In fact, he was forced to go to first grade when he was only four years old. He graduated from high school when he was 16 because they gauged his IQ as relatively high. He had many talents. He liked to tinker around with his hands, he loved to read, he was very mathematically inclined, and he loved to argue. He probably should've been a lawyer, I think. He and my brother, who is a lawyer, I've heard really argue issues. He was more conservative, and my brother is a far left liberal. During the Vietnam War, I can remember them distinctly arguing the merits of the Vietnam War, or the lack thereof.

So, it was just really interesting to see how our family was constantly thinking on what it is ahead of them and what adjustments we should make, how determined we should be to succeed, am I going to be the valedictorian, or I want to be the highest scorer of this basketball game, or I want to hit this next home run for the baseball team. There were just numerous things that made us have these different tools at our disposal that helped our family succeed. In fact, six out of ten of my siblings have a bachelor's degree or above and it's all because of the way we were conditioned by our parents. Even my mother, who only has an eighth-grade education, really pushed us to succeed, even more so than our father. She herself speaks three languages and she only has an eighth-grade education.

Kristiina: There's a Professor at Harvard — Howard Gardner — who has written a lot on multiple intelligences [30]. One of the intelligences he identified is emotional intelligence and how you have to have this intelligence to succeed, e.g., it isn't enough just to be smart. One thing that's going through my mind, and we haven't talked to anyone else about it, but Tribal people have the intellectual capacity and emotional intelligence. You can memorize things, you can memorize a lot of facts, and you can spout it back, maybe you can make all kinds of connections, but is there a balance to this intelligence? I'm wondering if emotional intelligence is not the right word for the adaptiveness that Tribes have. This is a term that scientists use but maybe there is a better term that Tribal people use. I'm wondering if what the Tribes have is not only the intelligence but then they have the emotional intelligence. We met a lot of people at Yale who didn't have the emotional intelligence and they couldn't succeed. They ended up leaving. They ended up flunking out. So, they had to be sort of balanced. Maybe Tribal people are better at getting this balance in life so that you've got both sides of your brain taken care off. It may be part of growing up is that there is this better balance. Dr. Nancy Maryboy likes to call it balance, that there's a balance in how you look at things and do things. What do you think about that?

> **Mike:** *Well I think a lot of Native Americans have emotional intelligence as a learning tool. I think a lot of them are kinesthetic learners. Those are two very strong ones.*

Kristiina: What do you mean by that?

> **Mike:** *I was drawn to the mechanical aspect of working with my hands, that's kinesthetic learning. Unfortunately for many of our Native people, in my sense emotional learning for me was a negative, but it turned into a very powerful positive because I used this anger of events or incidents that were a hindrance for me. What I mean by that is that I can recall some of my teachers really not pushing us forward very strongly to pursue college as the ultimate goal to achieve. Many of them just went through the motions of* "Open up your book, read page from 11–21, then answer the questions at the end of the chapter," *and that was the extent of our learning. There was no emotional drive there, it was very boring.*
>
> *I come to find later that I just loved history, but my history teacher was a very terrible experience for me. In fact, there were many times I was falling asleep and I just couldn't seem to get excited about going to this class, learning about Julius Caesar, or learning about the Greeks,*

or learning about WWII. Somehow, I just thought, how is this all going to pertain to me? Why am I learning this? How is this really relative to what I should be learning? I can recall that I barely even passed these classes because I wasn't emotionally charged to learn it. In fact, again, I was more of a negative that we were only pushed to achieve so much. When I got to the adult stages, I felt like, "Wow, I've been cheated."

It just made me mad thinking I wasn't provided these platforms or schemes of knowledge in a real positive fashion that made me really want to learn. So, I used my anger as a tool to drive me to learn. Sometimes if I didn't understand, I'd read something three or four times to make sure I got the point across. In fact, I can recall that on one of my tests the teacher kind of got upset with me, because I was writing stuff in my answers — this was my history test for College — that she disputed. I said "Oh no, this happened for a fact, and let's look it up. I'll get the book for you." The teacher said "Wow, you know more about history on this one than I do." But it was only because I was so angry that I didn't learn it in the first place, I wanted to make sure we got it right.

In that sense, it was kind of a flip-flop for me, using this negative emotional tool for me to learn and I would dare say that more Native Americans fit into that role simply because there are a lot of negatives that happen on Indian reservations. It's not Polly Annie. There's a lot of sadness on Indian reservations. We have a high unemployment rates, we have high alcoholism rates, and opioids are now being introduced on the reservations. We have higher incarceration rates. We have the highest suicide rates in the United States. Just a lot of negative elements there. So, the social, emotional energy has to be targeted in a way where you turn it into a positive rather than use it as a negative to be self-defeating.

Kristiina: The thing that's really critical and important is how you have a certain way you've learned indigenous and Western forms of knowledge. You came to work on a PhD and not a lot of Tribal people have PhDs. So this puts you in a unique category since less than a percent of Tribal people have PhDs. It's really low. How difficult has it been for you to balance this different way of learning that you learned as a kid and growing up and then all of a sudden matriculate into the linear learning process found at universities. The teaching at universities is so counter to what Tribal members learn when growing up. Tribal way of thinking is circular, you don't have an end point, and it can be a very different kind of end point. How would you change graduate university programs to teach Tribal ways of forming knowledge? How would you change the program you were involved with at Antioch? How would you redo those programs?

> **Mike:** *The one thing is that they're very well-intentioned. They want goodness to happen for Native education systems and institutions. However, they use their money to advocate the dominant society tools — establishing hypotheses and experimentations that go from experiment A to experiment B, do a comparison model, and use scientific models. This approach is something that Native people aren't really conditioned to do. A lot of Native Americans, and I'll say probably virtually all Native Americans, probably traditionally learned by observation over the millennium. They learned by observing all these trends that occur generation after generation. These are passed down through these certain events that happened and so when these occur again, they're able to systematically utilize these learned knowledges to their advantage. Whereas, western society thinks that if you just have these hypotheticals you can actually put them in a jar and somehow do this little contained kind of experiment and you're supposed to be able to prove your hypotheses.*
>
> *That's not the way Native people think. There are all kinds of elements we believe that are constantly at work in our natural world and even our unnatural world. There are metaphysical things at play here, you know you have a spiritual sense that helps drive your learning process. What I mean by that is that some Native Americans, and I'll even use myself as an example, I'm thinking, "Well I've learned this but I'm really having a tough time understanding this," and I let it bother me so much I'll fall asleep and I'll actually have a dream about something that's somewhat similar but it's not quite the same. I'm thinking, "I wonder if this is the right solution" and I'll apply that process into my reading and even though it doesn't match up I'm thinking "I wonder if there is another way of looking at this knowledge that's established by this empirical scientific experiment." I'm thinking differently and thinking there might be another way of looking at this and that maybe gives me a right answer. This is also based on this alternative way of looking at the same particular issue, but I have a different take on it.*

Kristiina: You were teaching at the American Indian Studies (AIS) program at U.W. for a while. Who were you teaching, they weren't all Tribal people, right?

> **Mike:** *Right.*

Kristiina: So, how did you have to adjust your teaching? Or did you? Did they understand when you would bring up things in a very different way or if there are stories? Did they put up with those? I've seen in classes where students just want the facts, this is where I'm going, it's very linear, this is what I've got and I'm going to end up over there. So, how did you adjust anything you taught, or did you just teach the way you thought it should've been taught and then not worry about what the non-native students were doing?

> **Mike:** *Well, my initial approach was just to make it interesting. There was no one central cultural way that I could state knowledge based on my background. But I would say that my approach was just to make it as interesting as possible to the non-native students, especially when I'm talking about something that's specific to Native culture. I'll just bring up an example. When I was teaching an education history course, I would talk about these kids who would be handcuffed. The students couldn't fathom that, they thought, "What? There's no such thing as handcuffing, there's no such things as jails for little kids, there was no such thing as whippings or sexual abuse or sexual molestation. That's not allowed in schools, there's no way that could've happened. These are schools, why would teachers do that." So, I had to bring in actual pictures of handcuffs, I brought in actual pictures of these boarding schools that had jails in them. I actually brought in these films of hours of testimony and I said you've got to watch this but watch it at your own peril. There were actually students that came up to me after with tears in their eyes, they said it was just so powerful. They just couldn't believe that had happened, but after viewing this it was terrible what happened to Native Americans.*

Kristiina: So, they started realizing there was a whole different story out there and why maybe Native youth might need a slightly different kind of education. Of course, they used the extreme punishments in the boarding schools. Mike Marchand was talking about that, I think it was his father-in-law when he was a child who was put into a church basement with no clothes on because he had acted up in some way according to the priests. It was the middle of winter and he was just put down there in the basement. Their attitude was tough luck since you had not obeyed the priests.

> **Mike:** *Yeah, a lot of them were put in a dark room with no windows and they would turn all the lights off. Can you imagine doing that to a little six, seven-year-old kid? There were just so many ways of trauma, especially from 1879 all the way to 1965, that's pretty much 75% of all native children had gone through this process, so that's why they say,*

"Why can't Native Americans have higher graduation rates? Why can't they have better jobs? Why does it seem like they struggle all the time? Why does it seem like they have a seemingly high alcoholism rate? Why do they have a high suicide rate?" *They don't realize that there was a very sorted history that our whole peoples were put through for hundreds of years, especially from 1879 to 1965. So, again, I tried to approach it from not just the facts here, these are what the statistics show, but I want to show qualitative information that made them think,* "Wow, this is way too far-fetched, but if it's true I'm just really blown away. I can't believe this happened to you people and now I can begin to understand why things are the way they are for Native Americans."

Kristiina: There's a core strength here expressed by Tribal people. I see something similar with Finland even though they did not experience anything like what Tribal people were exposed to. But the SISU character shows how Finland was controlled for over 500 years by Sweden and then Sweden had a war with Russia and Russia won, so Finland was given to Russia. So, over 100 years, Finland was controlled by Russia. You start looking at this situation and ask why did peoples' culture survived and persist when the big melting pot requires you to blend in. So, at some point, there's still this strength of character that I think is part of the culture and maybe it's that emotional intelligence. I think that in itself is really interesting because you don't just blend into the melting pot [13], you don't just become part of the conquering nation. This has been going on for how many hundreds of years for Tribal people? I think the fact that you

Dr. Mike Tulee at a United Indians of all tribes Foundation gala in 2017.
Source: Mike Tulee.

still have culture and a different way of forming knowledge really demonstrates to me the importance of culture.

> **Mike:** *Natives are in large part determined to hang onto their culture. They want to make sure it persists. Some accept an enculturation or acculturation, but many of them still feel like there was so much abuses and liberties taken that they are very protective of what little we have left. So, a lot of Tribes you'll find, they will try to institute and protect all these language bases we have, the religions. In fact, if you go to any reservations, even myself, I can't just walk onto a reservation and say, "Hey, where's the longhouse at? I want to go to your longhouse and learn your ways." They're going to look at me, even though I'm a Native American, and say, "That's not the way it works, buddy." We've kind of got to get to know you first. Let alone a Caucasian person would have a very, if at all, would be hard-pressed to be welcomed into that longhouse, especially because of the powerful religion that they practice even today. So, things like that, they're still determined to make sure that stays intact and they're really conditioned and fostered to be kind of isolated in that sense.*

Alexa: The last thing we're really touching on, is if you could talk a little bit more about the challenges you face now being a part of both your Native community but living in Seattle and being fully immersed in the Western world.

> **Mike:** *Pretty much, my wife and I are very well acculturated. We can bounce in and out of Seattle and going back to our reservations quite easily. The only thing that I would say that's a challenge for me is that it would not be very easy for me to go home and get a high-scale job. Even though I have a PhD, I've been away for almost 40 years, they're going to say, "What? You can't just come here and think that you can teach us something? What can you actually show us? You don't even have a good command of our natural environment. You haven't been a strong practice of our religion. Even though you speak a little bit of our language, we have a lot of fluent speakers here who know not only our language, but our delicate custom ways, our traditions, and you can't just come here, show up here and think you can be a Tribal leader." So, a challenge there for someone like me, although my wife's actually okay with it, she's a Makah Indian, which is in Neah Bay. She's content with moving to my reservation and us just settling in over the sunset, which means we're just going to be good citizens, just trying to help out our youth and try not to be a leader. I have a feeling that may not happen, because sometimes when you have*

> a college education, there's going to be people saying we need people like you to be in those positions. But again, there would be a strong pushback for me, even though that's not my desire or design, I'm quite sure that will be thrust upon me after I've lived there for a certain amount of time.

Kristiina: That is interesting though because it gets back into the whole idea that the reservation is sort of like a place that the culture and practices are maintained. So, if you're outside, then you change. I've found out that I still have something of Finland, but I'm also different. Every place I've gone to, when I was at Yale, I was a little different. At Yale I was going around the world to different places. It's almost like what John McCoy said in our first book [13], he had to learn how to be an Indian again. So, here he was, he was going to be going in to take charge of the Quil Ceda Village, getting that built and managing it. He needed to learn how to re-talk, he said. He used too many of the western words. I think academically we tend to do that, we have words that don't mean the same thing to others, but we use them with authority. It's a way of keeping everybody else out because its people don't understand your lingo, then they don't deserve to know it. I think in some cases, it is an interesting challenge because how do you balance both cultures? You will change just by experiencing another culture, so you never go back to what you were when you were growing up. You are different.

> **Mike:** Yeah, I can state that even with my own family. All my family still lives on the reservation, even though they went to college, and they don't view me the same. I feel like we're not really that close and I don't know if that's a bad thing. I used to think we were very wholesome; our family was very close-knit. Somehow over the years when I go home, even though they're still cordial, I sense there's not quite a strong emotional attachment as much as I'd like, as much as I'm used to. I don't think its jealousy, but it's just that I'm different. It's like, "Oh this guy, he's not really one of us anymore." I'm actually kind of a different person now, that's my take. I mean, I'm sure they really love me, but it's just not quite the same.

Kristiina: We've noticed that people that work in the academics and have been married for a while frequently end up splitting up as each person changes to becoming a different person. Back then it was usually the husband who had the job and was meeting interesting people all the time. The wife worked at home and immersed herself in other activities. The men would go home, and the wife wasn't as interested in what he was doing because he was talking scientific jargon. Instead she talked about problems she had to deal during the day, "*The* *crocus lost their flowers today or the dog and I had a good run today."* But the guys are looking at these big global issues or addressing the world's problems. She's not really into that because her sphere doesn't revolve around that but is locally contextualized. At Yale, these marriages typically didn't last. So, you have to adapt and grow together, and it doesn't always happen. You have to have that balance in life for both of you.

> **Mike:** *Well, fortunately for myself, my wife is very well read. She and I argue about national politics quite a bit, even though she's far left, and I'm left of center. We still have disagreements.*

Kristiina: What I still think is interesting, and I don't know if all you guys are unique, but I just feel like all the Tribal people I've met are really smart and they're interested in a lot of different things. I just keep thinking, *"Am I just meeting an unusual subset of Tribal people and everyone else is different?"* But it's been pretty amazing for me in terms of interacting with Tribal people and learning about what you do and your culture. I think the fact that they're so willing to really accept different people and cultures.

In Finland, it was very formal in the olden days. You couldn't call somebody by their first name until you'd known them for five years. More recently my father, when he was still alive, was really upset because he'd go visit Finland and they'd call him "Isä", and "isä" means dad and he'd say, *"I'm not your dad, don't call me isä."* It's like you're not supposed to do that, you've lost those rules about what you're supposed to follow.

So, the culture almost seems to give Native Americans an ability to adapt, because they can kind of pull back when they want, but then they can be involved in a lot of other kinds of activities and they are. Yakama Nation, and many other Tribes, have been working at mitigating the impacts of land-uses and dams on salmon. Tribes do not shy away from confronting environmental issues that impact salmon. Tribes are the only reason we have environmental laws regulating activities impacting salmon.

3.4 Culture Defined by Nation-Level Melting Pots[*]

In America, a nation of immigrants, there are generally two primary theories about assimilation—the melting pot and the salad bowl. The melting pot theory assumes that all immigrants come to America to find a better life and must give

---

[*] By Melody Starya Mobley.

Melody Mobley when working for the U.S. Forest Service.
Source: Melody Mobley.

up their native language, culture and values to fit in and be successful. However, in the salad bowl theory, immigrants celebrate their home-country's language, heritage, and culture, their native religion, and make America better by adding their uniqueness to the whole. The melting pot theory originated in the 1700s, the salad bowl theory in the 1960s.

Since the 2016 presidential election, proponents of the melting pot theory have become more and more outspoken and prominent. If immigration is allowed at all, they feel true Americans must completely lose their original identity and assimilate into the ethnic and racial majority by abandoning traits from their home country. Religious freedom is greatly denigrated for the Christian majority. Neo-Nazis, Ku Klux Klan, and other terrorist groups feel free to speak out and up for a national level melting pot without penalty. Police can target certain races for discriminatory profiling. Voting rights for people of color are threatened by gerrymandering and other means. The President seeks to limit immigration from Muslim-majority countries.

But people who support diversity of race, ethnicity, and thought have not given up without a fight. Public protests and marches occur with regularity. Celebrities speak out in support of unique identities. A few billionaires place advertisements in public places and speak out against the current Administration's apparent support of the other side. Democrats are winning elections in once all-Republican states and are fiercely gearing up for 2018 midterm elections.

Who is right? Both theories have good and bad points. A nation cannot succeed without some unity of thought. *The Declaration of Independence* and *the Constitution* are cornerstones of American democracy and unite us all. As Americans we all share inalienable human rights that transcend racial or ethnic identity. The United States remains the primary destination of choice for people around the world seeking freedom and opportunity.

Census data show that people of color, specifically Hispanics, will soon be the majority culture in the United States. They typically hold tight to their native languages, heritage and festivals, and predominate Catholic religion. They tend to have larger families than the current majority race so their dominance will only increase over time. We truly are a nation of immigrants and will continue to be. Today, the United States continues to be the most inclusive and tolerant nation in the world.

3.5 Tribal Peoples' Cultural Context: Interview of JD Tovey*

Kristiina: First, the way we're going to start is we want you say something about yourself as a person. To introduce yourself, so we have a feeling and understanding of where you grew up, what influenced your thinking, and the roles that you've had in the past.

> **JD:** *I grew up in southeastern Idaho, the middle of absolute nowhere and I grew up off the reservation. So, my grandmother moved away from the reservation and she was a nurse. She grandmother was Native American, and she was going as a sergeant, as a nurse, in the Air Force. Before she enlisted, she went to visit her friends who were doing a residency in southeastern Idaho. She went down there, met my grand-*

JD Tovey climbing Mt. Everest.
Source: JD Tovey.

---

* By Alexa Schreier, Dr. Phil Famcett and Dr. Kristiina Vogt.

Dr. Phil Fawcett (left) while working at Microsoft, Alexa Schreier (middle), Dr. Kristiina Vogt (right).
Source: Phil Fawcett, Alexa Schreier, Kristiina Vogt.

father, and he talked her out of enrolling. But the only way they could get her out of her discharge papers was to get married. So, they got married within a month of meeting each other, which is strange. But that's where I came from and that's how she ended up in southeastern Idaho.

But actually, it was my mother's side that is important. I'm a quarter Cayuse and Wallowa Band Nez Perce, Wallowa Band is also known as Joseph Band Nez Perce, which is out of the Wallowa Valley. They're the ones that did the flight and tried to escape over to Blackfoot Territory and then up through Yellowstone and then tried to escape up and were stopped in Bear Paw. It's about 20 miles from the Canadian border —it's a location but there's no city there. On my mother's side I'm almost entirely Welsh and my hometown of Malad was settled by Mormon settlers from Wales so we have the highest concentration of Welsh ex-pats in the country. There are still people who swear in Welsh, it's this tiny strange little town.

It was funny though because they're all Mormon so they're super proper and everything. So, it was proper swearing, if it was actually swearing, because you're doing it in Welsh. But it's a weird flavor of Welsh. You grow up where you don't think it's anything but family traditions. But it's very strange in that little town because everyone looks like they're little Welsh people that just came out of a coal mine. All then men are 5'4" to 5'6" tall and they just run around. They're angry all the time. They wear little caps and they just look like they're right out of Wales.

Phil: You're not 5'4".

JD: *No, somewhere I got some Scottish ancestors. I'm not sure where the height came from.*

Phil: Yeah, cause you're a tall guy.

JD: *My mom's 5′ 8″ and my dad was 6′ 1″. In the whole county there were 3,000 people and in sixth grade I was 6 feet tall. As a freshman in high school, I was 6′ 1″ or 6′ 2″. I was the tallest person in the county except for one man that was taller than me. He and I are now the same height, he's like 6′ 6″.*

Kristiina: Did it impact you at all? Did you get special attention or was it a negative/positive reaction?

JD: *Oh yeah. It was weird because even in kindergarten there were all these little kids and then me. I was just kind of this bump out. It was interesting actually because a lot of people thought I was held back a couple years, because I was so out of the norm because of my height. I was off the charts for the normal height scale. For the most part it wasn't too much of a problem, but the flip side of that was that I was never really picked on because I was the biggest kid in the county. I grew up really reserved though. I was that kid that, even in college when I was 22, a professor or teacher would call on me and I was that kid who turned bright red. I wanted to crawl under my desk.*

Kristiina: Really? Well you're not that way today. You're very articulate, you have a lot of insight.

JD: *I have this little process I have to go through every time I do it. First of all, this doesn't matter. As soon as I get over the fact that it if even matters at all, then I'm usually okay. On Monday, I'm working on this legislation — the State of Oregon passed the Transportation package, House Bill 2017. It is this massive thing worth billions of dollars for transit and transportation in the State of Oregon. There's one little thing in there that's kind of screwing us up with the Tribes and so we've been going back and forth on this. I had meetings down there this past Tuesday. I was driving down there when our lobbyist, our legislative affairs and our legal counsel, Oregon DOJs legal counsel and the legal counsel director for ODOT were all flurrying emails back and forth to each other. But I'm driving so I don't see any of it. I'm going down there, I get there at 5 pm and there's a joint committee for transportation going on at 5:30 on Monday night. So, I go walking in there and I just wanted to go there because I knew a bunch of the*

> ODOT people were going to be there that I wanted to talk to. I go walk in there, causal, at like 5:30 and our lobbyist, Phil, comes running up to me says "Oh thank God you're here." He's asking like what did you think of the language and I go, "What language? I was driving." He says "oh" and pulls out this sheet and it was legislative stuff that we thought was okay but apparently it wasn't. I was like, okay and started reading it and said it seems okay, but I'm just off the fly looking at it. He says, "Okay, are you ready?" I'm like, "Ready for what?" He said, "I put you down and you're testifying in 3 minutes." No pressure! So, I got up there and just BSed it for 15 minutes.

Kristiina: So, when you think back to when you were growing up, were there any things that were happening that kind of made you what you are today? Any experiences?

> **JD:** Honestly it is probably more the rougher things. My mother was 17 when she had me and then my mother was a teen when they got married. My mother was divorced when she was 18. She met my stepfather when she was 18 and then got married and had another child before she was 19. It sounds awful but when growing up my stepfather was the salt of the earth, but he was a decent man. But he was not intelligent at all. He is very blue collar. I remember we always clashed. When my brother was 3 or 4, we would go water skiing. If you were not in church, you were water skiing. My step father would do things like throw us out as far as he could throw us and then dive in after us. He would do this with my brother when he was three. He was just rough, and I was never into that.

Phil: So, it was sink or swim?

> **JD:** Yeah, he was really into that. My mother had a hard time with that cause, I would always go to my room. When I was in 5th grade and I was doing homework, he used to make fun of me doing homework. The thing was that I loved doing my homework. He was so irritated with me that I would do hours and hours of homework so he would say "Let me help you do this." He sat down and I realized that I was more intelligent than he was, and I was only in 5th grade. I remember looking at him like you don't get this. It was weird because what it did was then as a kid, I realized that why do I have to answer to someone I can out think or out smart. My mother told me many years later that is when he and I started having problems. This went all through high school. Every time we got into an argument or a fight, she said it

scared him more because he was intimidated since I was just there. It was also frustrating because my half-brother does things in a different way. He is the state's number one florist in Utah. He makes amazing things but he can barely read. He is the same mentality as my step father but it was frustrating for him to go through school. He was held back one year.

Kristiina: He probably had to respond to the fact that he followed you in school so he was probably taking classes after you had already been there. So, the teachers compared him to you.

JD: *Yeah, I was super calm. I barely worked and I would get A's and my brother struggled with C's. So, there was a lot of tension in the family. They were also the nuclear family and I was the black sheep always.*

Kristiina: You know what is interesting about what you are talking about is that when we interviewed Mike Marchand, he also read anything and everything he could get his hands on. It sounds like you are similar to him. I used to hide behind the bookcase and read books I was not supposed to read.

JD: *I just had this weird recall that I consume information. Even over Christmas we were playing board games and there is a board game called "Smart Ass" which there are donkeys and you have to go around the board answering questions. If you get them right, you win. It is supposed to take an hour or an hour and a half to go around the board. I went around it in 1 minute and won. Everyone was like we are not doing this again. They made me go around twice. I had to answer every single question. I had to answer twice as many questions, and I had to answer two questions before I could roll the dice. Then I won again, and my friend said I hate playing games with you. I just know the most stupid stuff which help me win these games.*

Kristiina: So then as you were growing up how much did being a Native person affect you? Where you exposed to this culture?

JD: *Hardly since there is this generation that happened around 1950 to around 1975 where some Indians were of the reservation and were going out and doing their own work. There was not a lot of economic development on the reservation. This was a whole era of the civil rights movement where you had a lot of African American who could pass for white were definitely passing as being white if they could. There were a lot of Indians and if they could pass as non-Indian they were doing* *it. My father's name in high school was half breed and he hated it. He hated it so much that he would get into fights in high school. There was an association if you were an Indian since you were associated with all the ills of the community at that time. There was poverty, alcoholism and other substance abuse. So there was a whole generation that were impacted by these views. It was my grandmother's generation in the 1930s that were encouraged to leave because they realized that the only way we are going to survive are if we got out of the reservation. My grandmother's generations parents were still tied to their parents who were put on the reservation. So, their kids were kind of agrarian and held on to their culture. But then they asked, "What do we do now?"*

My father's generation was like we are not part of it. A lot of the people who left, like my grandmothers' generation and my father's generation, when they left no one thought these people were going to come back. There was a termination period of reservations going on at that time due to Eisenhower. So, people were flooding away from reservations. But then in the mid- to late 1970s, all the people who left were now coming back to the reservations to take care of their elderly parents. Now you have all these people who went out and became doctors, lawyers and accountants and needed jobs while taking care of their parents. All these people with knowledge came back and they created jobs. It wasn't an organized effort thought.

You had some kid who was working as a lawyer and was reading the treaty. This person said, "Does anyone realize the power of this converse law in the treaty and it wasn't selling baskets along the road." We can do this. Starting in the 1980s, change happened. In the mid-80s we got the first casino and lots of economic development happening. So, there was this whole thing happening in the 1970s and 80s because a lot of the kids were coming back. It was a movement called the **Indian Awakening** *but I have never seen anyone do a profile of that time period.*

There was a whole Indian Awakening that happened, and we are dealing with the repercussions of it today. A lot of the people that were my age were born in the 1970s. They knew about the movement and they kind of lived it. But now we have kids (I am talking about high school kids) who were born after 9/11.

Kristiina: So that is when their history started.

> **JD:** *They are entirely post 9/11 so they heard all their grandparents' stories. So now you have people who wear all the best clothes and have decent economic development. They are driving really nice cars, but we still have a lot of the ills. There also were all the social movements, e.g., gay rights. A lot of these cultures were cohesive because of oppression. It's like China Town. The social forces that caused China Town were oppression and racism. But now the question is how do you preserve that as a unique space? But the very social mechanism that created it no longer exists. So how do you preserve something when the social pressures are not as strong?*

Kristiina: So, are reservations important? Do you need a place where people can go to hear stories to retain their culture? If you didn't have reservations, what you are saying that you would start losing.

> **JD:** *It is very strange that I did not grow up on a reservation and only spent a couple of weeks every summer there. I don't have any cousins on my father's side and I have one second cousin who is 9 years old. So, it is a very narrow family. Even when I was there for the two weeks, I was way up in a canyon all by myself with a dog. I was just fishing.*

Kristiina: It probably had a big impact on you?

> **JD:** *I am very comfortable living alone. I spend a lot of time alone. Sometimes I prefer it. It is one of the challenges I face because I am living with my grandmother helping to take care of her. So I am used to my own space and it is hard. I am incredibly introverted so even when I am giving a lecture, I can be up front and talking to people but I don't get energy from it. I am expelling energy the entire time. When it is done, I need my space.*

Kristiina: It is interesting side comment that you have an ability just like Mike Marchand of being in a room and you get attention. People will look at you and you come across as credible. When you were in the IGERT you maintained a balance and you are not just pushing that you are an American Indian and need to be treated differently. Some Indians talk like the White people owe them something because of the past injustices. I thought one of the most striking things was, when you were in the program with all these other non-Tribal PhD students, you did not push that. Balance is one of the things we have talked about on leadership that you have to have this balance where people will trust what you are going to say. You are not trying to make people feel bad because of the past.

In fact, we did have someone come talk to the IGERT class and it was you who spoke up against what she was saying. She was trying to manipulate the White people in class to feel bad about colonialism and its impact on Tribes and you spoke up against this. You come across really well and as someone who doesn't want to blame today's generation for what happened in the past.

JD: *I think some of where this comes from is going back to my childhood and my great grandmother on my mother's side. My mother was young — 17 — and my grandmother was 18 when she had my mother. She was 36 when I was born so I remember my grandmothers 40th birthday. It was a really young and compressed family. My great grandmother was 60 when I was born. They all were working, and I was the first grandchild. I am actually closer in age to my two of my uncles than to my cousins. I am kind of like this weird island generation. Like even now all my cousins are in their early 20s and doing all that early 20s. I am in this weird middle space. I grew up in a very small and very highly religious monoculture town dominated by the Mormon Church. I grew up thinking that nothing ever sounded right so I thought I was nuts. I was thinking "Is this how life is supposed to be?" Anytime we left and went to Logan, Utah and Ogden Utah. We were going deeper into the well. Again, I wondered "Is this what life is supposed to be?"*

My great grandmother, on my mother's side, was never religious. She was the second youngest child of 13 children in a depression era family. But they took her out of the 3rd grade and put her in the 5th grade and she graduated valedictorian. She was a writer for AP and Reuters, and she lived most of her life in a small town. But she was a humanist atheist through and through. She never flaunted it and went to the Presbyterian Church more out of fellowship than anything else. But I remember when I was about 13 or 14 years of age and I had all these questions that were burning. I remember her eyes just twinkled and we had this long conversation. She talked about how you just don't talk about this in this town. That is when I realized that there was something else.

Kristiina: It does sound like she had a big influence on you. Having certain relatives helps.

JD: *It is weird stuff because it was the depression era and she lived in a really tiny town. She was a writer, so she was literate. I remember my brother running around in the rafters and having a good time but even at the age of 5 or 6 I was playing scrabble with her. That was our* game. I played thousands of hours of scrabble from when I was five to 15 years of age. She had an original scrabble board from the 1950s that I still have today.

Alexa: Was there a certain point in your life or experience that sparked you to be more interested in Native culture or history? Did it happen at a certain point or develop over time?

JD: *I don't know actually where it happened because I was always passively interested in it. I knew who I was. But my mother was working so I only got a splattering of language exposure. My father graduated from College in 1988 right at the time when the Savings and Loan (S&L) bust occurred. My grandfather's family are all bankers. So the S&L bust collapsed my family finances. My father was going to school in finance because his uncle had promised him his own bank branch. But this all fell apart. He was unemployed for about a year. When he got the job at the tribe he was crying because he thought it was a failure. There was no place for him to go. He was never closely tied to the reservation. This was in the late 1980s and he was part of the Indian Awakening. I like to think my father was pretty instrumental in the starting of the economic development on my reservation, but I know that where a tremendous number of people who were involved. I think he was the point person.*

Kristiina: The view from the outside is that your family are really good business people, with a head on your shoulders.

JD: *Sort of. We joked that the family has been rich especially on the white side of the family. My great, great grandfather were millionaires in their teens. The stock market crashed in 1907. He owned a dam and all the irrigation canals in this portion of Idaho. He sold the dam and a bunch of other stuff. He ended up gambling and lost it all. I think he committed suicide. It has happened three or four times where the family would lose it all. It happened in the 1980s with the S&L bust. My grandfather spent over a million dollars in legal fees after the collapse. In some ways my family have all the skills. I am the Planning Director. My father was the Executive Director of the Tribe, but he is now working with the Indian Land Tenure Foundation. My uncle is the Director of Economic and Community Development. My other uncle is the General Manager of the casino. So, we are a powerful family, but we are not an expansive family. If we went together and did real estate, I think we could be millionaires. But there is this tepidness from all the stories in the family.*

Kristiina: One of the things that Mike talked about with Cal was what kind of car do you own (*see* Section 5.4.1). You don't go and get yourself a fancy car and live on the reservation. They mentioned about one guy who was quite well off but had a beat-up vehicle that he would drive. He would work on it and fix it up but could have afforded a really nice car. But you are not supposed to be ostentatious. It sounds like you have some similar things in your family. You don't get all dressed up.

> **JD:** *Well I learned that from my boss in Florida. He would not let us be too well dressed up or driving fancy cars since we were in the private sector. A lot of our clients were old farmers, so we did not dress better than our clients.*

Phil: A joke in tech was could you dress as close to a homeless person. We had contests to see who could dress that close.

> **JD:** *Going back to other questions, I was on the reservation, but I was never really close. In my grandmother's house I found a bible written in Nez Perce. I remember going through it with my grandmother when I was 11 or 12. It was the first time I realized my grandmother spoke Nez Perce. It was her first language and she didn't learn English until she was five years old. She was really impressed because I could naturally learn the language, but I was also learning German at the same time.*
>
> *When I was at the University of Idaho in Physics. I will never forget in my first year I was living in a fraternity and there were a whole range of people. Physics is in Engineering so I was an engineering student and would have been during the first couple of years. It was boring and I remember being in class and just not carrying. I was the social chair of my house and all these guys from my lab would be having a party and I would show up. There were nine guys and one person showed up with a 6 pack. I thought what is this? I am not doing this. I was interested in theoretical physics and I am not going to end up in a 600 foot hole in the Mohave Desert and be with these people for six months out of the year.*
>
> *Clearly in my head I want to learn how everything works. That clicked in my head when I finally took calculus. Math did not make sense to me. I was not very good at math since being in second grade. Because I was going so slow, they thought that I was slow in math. But then I clearly remember getting help from Mrs. Gunnell, her husband was my physics teacher in high school. She is the one who clicked in my head that I wasn't an idiot or stupid. This happened in second*

grade. She said the reason I was slow was because I was actually doing the work. I am doing the calculations instead of just memorizing that 2×2 is 4. That helped me because I realized everyone else was just memorizing it and had no concept of what two times two meant. They just knew when you see that you write four.

Phil: That is pretty good insight to have.

JD: So getting back to the Indian culture stuff, after Physics I went into Philosophy since I still wanted to have the arts. Physics was missing all the arts. It was missing anything metaphysical. It was hard numbers and I loved it and it made sense to me but it wasn't everything. I went into philosophy with the expectation that I was going to go into law. But there was a whole other conflict with the people in philosophy. It was a pop cloud when you walked into your Asian philosophy class. I didn't get along with any of the students because I looked like a frat boy with my pack back on and they were a bunch of hippies. So, I took a year off. Then I came back and was visiting a friend on campus. I was walking across campus and needed to use the restroom and ducked into a building. It happened to be the Architecture building and it was the middle of summer, so the door was open. I went in and sat looking at the projects on the wall and this gentleman came and started talking to me about the projects. We ended up having a conversation about how cool some of the projects were. He was the Chair of Landscape Architecture Department. I came back to school the next fall.

In Landscape Architecture I could learn everything. They have physics, sociology, anthropology and how people interact with each other, and biology, ecology, soils, engineering, construction. Everything was in it. It just fit. It just so happened that I had a project that I was interested in looking at on rural spaces. Our senior capstone project on my program was on a project I could get a lot of information on in a short time period. It was on my family's own property. It was the traditional lands for the Shoshone-Bannock Tribes. My father was selling the farm because they owned the allotment of farm lands after my grandfather passed away. My father and uncle had built the casino and wanted to do some economic development here. It was a rural space. I was interested in what was the cultural impact of having this massive economic impact on a rural space. I really got into ruralness, i.e., "What does it mean to be rural?" "How do you do development in rural spaces without ruining the rural character of the place?"

> Then I graduated with the Tech Bust so I surfed for a year. Then I got offered a job in Florida on an exurban area outside of Orlando on a land development project. It was huge housing projects. It was an insane time to be working in Florida. I killed a lot of orange trees in building a lot of subdivisions. Then I ended up getting into planning because of in Landscape Architecture I was doing all this design work but you never learn about the rules, e.g., land owner regulations, the philosophy behind the process. I was hitting all these roadblocks and it was really frustrating because we couldn't design without including planning information when writing all these designs. We couldn't design their projects because of the regulations. So that is why I ended up back here. I loved my job and my co-workers but I was trying to find a guilt free way to come back to Seattle. Because I am a more northwest person. That is when I got into the University of Washington in a Master of Urban Planning program. Because my undergraduate was in rural spaces and my professional life was a lot of exurban, conversion of rural into urban spaces.

Kristiina: But you still have a lot of stories when you give talks in the Forest and Society class. You know a lot of traditional stories, you have a grasp of them and weave them into your talk. So even though you may feel like you grew up elsewhere you do have a lot of knowledge about Tribal history and stories.

> **JD:** *I think I have a perspective on the history. It is what we even fight with our elders. There is a woman from Yakama and she has a quote that she doesn't say often* **"Always respect your elders, and always try to incorporate what they tell you in your everyday processes but never forget that even idiots get old."** **(Yakama Elder)** *There is a perspective that I see a lot of elders where there comes an authority with being an elder. This is sometimes bestowed even when it is not earned. There is this power that may be the only power and will try to stop things because that is their only power they have. It is like little kids that destroy things because that is all they can do, they can destroy things. It is the only power they have. It is like blocks. It takes more effort and thought to build things but it is easier to destroy them after someone has built it. I think we get a lot of that with some of our elders.*

Kristiina: Did you see the video where there was an eight-hour flight from Germany to New Jersey and there was a kid that had a tantrum during the whole time. It was night time, and someone recorded it. So, every hour they would say it is hour 2 and then hour 4 etc. They couldn't seem to get the kid to quiet down. There is clearly out of control. Can you imagine being on the plane as a passenger?

Phil: Impressive that he had that much energy to continuing going along for that long.

> **JD:** *I am thinking of one specific elder. This is an issue that a lot of Tribes have with a lot of the people that worked outside of the reservation who did not want to be a Tribal member or were forcibly removed from an Indian reservation and sent to other places like San Francisco for work. Seattle had a lot of people. This was all in the guise of helping us with all the poverty issues on reservations. They said, we will take your best and the brightest and ship them of to Seattle so they could get work. Then they could remit funds home or whatever. What it ended up doing is expatriating everyone. He moved to San Francisco in the 1960s and now he is an elder. So all these people that were part of that and didn't want to be an Indian because of its poverty. They left the reservation and then when they came back they are now our elders. So someone who was in their 30s is now in their 80s.*
>
> *It is very interesting seeing him and see the romanticizing of Tribes as the lost race. They portray Tribes as being all peace-loving types, smoking Indians, and that there was peace on this continent before all this happened. The idea is that the Anglo-American brought fighting to the Americas. It is all this romanticizing about being the lost race. I see him being heavily influenced by that. So now even he and I get into arguments. I was even called on because I was disrespecting my elder. But I was challenging my elder because he was wrong. The one thing he would say is we have all these Tribes and* "We did not have Tribes back then" *and everyone just knew everyone else. But Tribal people had clans and divisions. They had clans, bands and Tribes. We wared with some people. He says that the war with the Paiute never happened since we would never do that.*

Kristiina: Mike Marchand talked about all the fighting that occurred between Indians and that they killed all the members of a group that tried to take over the Kettle Falls, except one person.

> **JD:** *Yes to send him back as a warning. We knew exactly where our territory and where our resources were. We managed and defended them. There is this mentality among our elders who grew up in the 1960s that romanticize and don't accept this reality. Another thing he did when we went on a forestry tour and I was sitting in the back seat.*

> *He asked,* "What did the forests do before you had all your plans?" *to the forest managers. The forest manager who is a great guy but is not an Indian and with the BIA, agreed with this elder. I said,* "Hey wait."

Phil: What is wrong with this picture?

> **JD:** *I said we managed it and he said* "No". *He said the earth took care of us. I said,* "That is not true and we actively managed everything. The only way this forest was created is because of the fact that we planted it and managed it." "Every aspect of this was managed." *We had plans and we had schedules, but he was absolutely adamant that this didn't happen. It frustrates me that the notion that the Indian was a recipient of this bountiful cornucopia is absolutely wrong and feeds into the western notion that taking the land since you are not using it. You are not improving it. You are not tearing trees down. You are not doing something to the land so it is not your land and you can't have it. It was never your land in the first place. I got into a little bit of trouble because I absolutely challenged him that he was wrong and should stop saying this.*

Kristiina: It does show that there are conflicts within the various Tribal groups and Mike talks about it and all the differences that exist. Everyone wants American Indians to fit a stereotype, so you put them in a box where they are all the same, they act the same. But they are very diverse.

> **JD:** *I don't know how many times people tell us how to be Indian. My father was working down in Coquille and they were putting up a billboard. It is a beautiful area. They put up this billboard and non-Indian people said,* "I can't believe you would put up a billboard since you are Indians. You shouldn't be doing this. It is upsetting the view shed." *My father said,* "Are you telling me how to be an Indian?" *Many people were telling him what he shouldn't be doing. It was in the newspapers saying you should be better than this and you shouldn't be doing this. Don't tell us how to be an Indian. We don't tell you how to be White.*

Kristiina: It goes back to the past when you had the Buffalo Bill shows and how they portrayed Indians. Buffalo Bill even went to Europe and held his shows there with mock battles between Indians and the cowboys, just like what happened in the western U.S. He would have battles and you can guess who always lost. The Indians.

JD: *This occurred even in the Pendleton Round-up when they first held it in 1907. There was always this history between cowboys and Indians on the reservation. It was 1907 so people were literally fighting in the wars. The Indians fought in these wars and they remember these wars. So these new settlers in Pendleton were going to do some exposition and have these big parties. So, they were having this big round-up and they wanted to re-enact a battle on the main floor of the round-up grounds. They came out and asked the tribe to participate but they told them they had to lose. Apparently the whole Tribal Council just sat there and stared at them. They just backed out the door and left. Apparently that was their answer. So, they moved ahead without talking to the Indians.*

So now all these people from all over the northwest came when they heard there was this huge gigantic rodeo thing was being held. Then all the Indians got into their war regalia and came down to the rodeo grounds just to participate. We never did participate in an Indian battle. What really scared the non-Tribal people was that Indians were not even allowed off the reservation at that time and you needed special permission to get work leave. It terrified people because they thought something awful was going to happen. The Indians just camped and nothing happened. Where they camped is still reserved as an Indian Village. During the round-up, the Indian Village is Tribal grounds and the Tribal chief is in charge of that and keeps the peace for that location. Families had their own location for their teepee for generations.

Kristiina: Mike talked about when he was in prep school on the east coast and then everyone went home during Thanksgiving except the Indian kids (*see* Section 2.2). Some woman came from the BIA came and picked them up and said let's go to Boston. He said they were re-enacting this whole Thanksgiving dinner when this pretty ragged group of Indian kids showed up. He was dressed up as a preppy since he came from a prep school. He said he was sitting there and all of a sudden someone called Mike. He turned around and apparently some food went flying by. He ducked but a food fight started. The people were dressed up as pilgrims and their eyes got big. They couldn't figure out what was going on. This was not playing out the way it was supposed to be. Then the Indian kids left and walked to the Mayflower. The guy at the ticket booth saw them coming and got really big eyes like saucers and he fled. He said you guys can have it. Mike said they took over the Mayflower for a while.

JD: *It was weird growing up since I was the only Indian kid. My grandmother was my only connection to the family. But she was a full-*

> time nurse. It wasn't like we sat around and talked. I wasn't learning a lot about my family.

> I was always the example on Thanksgiving Day. I was asked to raise my hand. I even remember my step-family and someone saying something like a "red bastard" or something like that. It didn't register with me but someone mentioned it later. It was a rough family. I didn't know what it meant and I was only five then. It never really registered, and I was never that close to the whole community anyway. As for the reservation, I never had an expectation that I would ever go there to work. It was never in the cards. Even though now I did end up working here.

3.6 Cultural Foods and Food Security

American Indians have observed the cycles of growth or abundance of plants and the animals that were important for their life. They have been doing this for thousands of years and passed down this knowledge through the generations. The ceremonies were keyed in to these cycles of plant and animal growth. This knowledge is what kept plant and animal populations levels high during the pre-colonial period. When Tribes lost their lands, this knowledge was no longer applied on customary lands to determine when to harvest or not harvest cultural foods.

Alaskan Inuit describes six dimensions of food security: (1) availability; (2) inuit culture; (3) decision making-power and management; (4) health and wellness; (5) stability; and (6) accessibility. It describes the many elements of food security and it is not just having something to eat at your table. This approach to food security uses the knowledge held by different disciplines to inform whether food security has been reached. They clearly show how all parts of the system are connected — from nature, wildlife to snowmobiles. This quote from the Alaskan Inuit food security report summarizes this well [23]:

> "We are speaking about the entire Arctic ecosystem and the relationships between all components within. We are talking about how our language teaches us when, where and how to obtain, process, store and consume food; the importance of dancing and potlucks to share foods and how our economic system is toed to this. We are talking about our rights to govern how we obtain, process, store and consume food; about our IK (Indigenous Knowledge — not traditional knowledge which suggests is of the past and stagnant) and how it will aid in illuminating

the changes that are occurring. We are talking about what food security means to us, to our people, to our environment and how we see this environment. We are talking about our culture."

Hunting was an important activity for many North American Tribes. It was both an important practical activity and also a sacred activity. Native Americans considered plants and animals as sacred. Both plants and animals had spirits and were to be respected.

If a plant were gathered and used, it would be thanked. Often there were songs and ceremonies associated with the plant gathering. Camas was an important root for Northwest Tribes and eaten during ceremonies. Also, thanks were given during its collection.

Animals likewise would be thanked. The animal provided food for the tribe. A young boy's first game animal killed was a big event. The clan would have a feast. The boy was honored as an important person in the tribe. Thanks were given to the animal.

Today, many Tribes honor the sacred traditional foods over store bought foods. So, whenever there are special occasions such as weddings or birthdays or any Tribal gathering, special attention is given to having these sacred wild animals and plants on the menu at these dinners. There are songs and ceremonies associated with the natural foods provided by the Creator for Native Americans.

Food security has been challenging since Tribes have to follow rules and regulations set up by state level agencies. These rules are spatially and temporally incompatible with what Native people would practice. The major issue for Tribes is being able to hunt when they need to instead of hunting during a time period that hunting allowed. In Washington State, the hunting season depends on the animal being hunted. Further, there is a separate hunting season for hunters over 65 years of age, disabled people and youth and by specific location (*see* Table 3.1). Since Tribes are located on a smaller piece of land, they cannot collect all their cultural foods on the reservation. When they need to collect from the customary lands, frequently they cannot access and collect their cultural foods from these lands, despite treaties giving them the rights to them.

Table 3.1 Deer hunting using modern firearm [30][1]

	General Season 2017
Black-tailed Deer	Oct. 14–31
White-tailed Deer	Oct. 14–27
Mule Deer	Oct. 14–24
Hunters 65 and Over, Disabled, or Youth General Seasons	
White-tailed Deer	Oct. 14–27, Nov. 11–19 (Hunting in different management units)

---

1 Accessed March 24, 2018 at https://wdfw.wa.gov/hunting/regulations/summary_hunting_dates.html

The following Treaty Hunting Rights FAQ fact sheet published by the Northwest Indian Fisheries Commission describes well the issues associated with Tribal rights to hunting and why Tribes do not follow the same hunting season as non-Tribal hunters [31][1]:

"What is a treaty?"
A treaty is a constitutionally recognized contract between sovereign nations. These legally binding contracts are protected under the U.S. Constitution, which states that they are the "supreme law of the land".

Why did the federal government sign treaties with the tribes in Washington?
In the mid-1850s, the federal government wanted to make the Washington Territory a state. The federal government determined that the tribes were sovereign nations with title to the land. The United States government approached individual tribes, in the same manner that it would approach another sovereign nation, and negotiated treaties to acquire the lands held by the tribes.

What did the treaties say about hunting?
Under the terms of the treaties, the tribes ceded millions of acres of land to the federal government. However, the tribes retained certain rights that would enable them to provide for themselves. Among these reserved rights was the "privilege of hunting on open and unclaimed lands". The Washington State Supreme Court has ruled that there is no legal distinction between a tribal "right" or "privilege" regarding hunting.

What are "open and unclaimed lands"?
That treaty term has not been clearly defined. Federal courts have ruled, however, that certain public lands (such as National Forests) not set aside for uses incompatible with hunting can be considered open and unclaimed.

What does hunting mean to Indian Tribes?
Indian tribes allow their members to hunt to meet their ceremonial and sustenance needs. All tribes prohibit hunting for commercial purposes. Deer and elk meat are elements of feasts that are part of tribal ceremonies and other cultural events such as potlatches, funerals and naming ceremonies. These occur throughout the year. Tribes harvest only a small number of animals for ceremonial purposes. Tribes also depend on hunting to feed themselves. On some reservations unemployment reaches 80 percent. Indians hunt after the fall fishing season to provide food for their families. Deer, elk and other species provide important nutrition.

Why don't tribal members have the same seasons and bag limits as other citizens?

---

1 Accessed on March 1, 2018 at https://nwifc.org/about-us/wildlife/treaty-hunting-rights-faq/

As sovereign governments, the tribes exercise their right to set regulations that may be different from those established by the state Fish and Wildlife Commission. The tribes set seasons based on the needs of their hunters, and take into prime consideration the ability of the resource to support harvest. Most tribal hunters do not hunt only for themselves. The culture of tribes in Western Washington is based on extended family relationships of grandparents, parents, aunts, uncles, cousins and other relatives. A tribal hunter usually shares his game with several families. Many tribes issue a "Designated Hunter" permit to allow a tribal hunter to harvest an animal for an elder or family who cannot provide for themselves.

<u>How are tribal hunters regulated?</u>
Each tribe develops its own hunting ordinances and regulations governing tribal members. Regulations, including seasons and bag limits, may vary from tribe to tribe. All tribal hunters are licensed by their tribe. If a tribal hunter is found to be in violation of tribal regulations he is cited to appear in tribal court. Penalties include fines and loss of hunting privileges. Many tribes have hunting regulations that are virtually identical to those set forth by Washington State Department of Fish and Wildlife, such as minimum caliber requirements and safety regulations.

<u>Are tribal hunters overharvesting the deer and elk resources?</u>
No. Tribal hunters harvest about 2 percent of the harvestable population of deer and elk. According to Washington State Department of Fish and Wildlife statistics for 2012, non-Indian hunters took approximately 29,154 deer; treaty tribal hunters harvested about 495. In that same period, non-Indian hunters took about 7,236 elk; treaty tribal hunters harvested about 365.

Most elk herds in Washington are healthy. Loss of habitat poses a far bigger threat to the health of elk herds in the state than the small number of tribal hunters.

<u>How are the tribes working with Washington State Department of Fish and Wildlife and others to protect and manage wildlife resources?</u>
The tribes want to work with Washington State Department of Fish and Wildlife. The state and tribes have held a series of meetings to share information and discuss management and enforcement needs. More meetings are planned. Tribes participate in a variety of cooperative programs, such as population surveys and habitat enhancement projects that aid wildlife management."

In the past Tribal members would travel long distances to seasonally collect food. Colville youth would travel to Yellowstone to hunt bison and learn hunting skills. MacKendrick [32] described the pre-colonial hunting and collecting of cultural foods along the Oregon Coast as:

"... permeability of indigenous territories on the southern Oregon Coast in the past as people moved seasonally following food. There weren't formal agreements between coastal Tribes in the past, but it was accepted that different Tribes along the coast crossed each other's territories through the seasons following food. Coastal Tribes in the area used to move a lot depending on the season for hunting, fishing and harvesting."

"Individuals described that coastal Tribes enjoyed a bounty of food sources, including berries, game, and lots of fish and seafood: eels, clams, oysters, salmon, flounder, and crab. One individual noted, our tribe were fish eaters. When the tide was out the table was set. They ate clams, cockles, barnacles, mussels."

One individual described, "My Tribal relatives used to do a lot of clamming and fishing. People would clam cohogs, empires, and butter clams in Charleston in the deep mud and razor clams in the sandy beaches in Bandon. Individuals described the social practices surrounding food and relationships."

Another individual described, "When harvesting or hunting natives never took more than they needed for 1 to 2 days. Indians still do that when fishing, hunting foods. They'll ask themselves, who am I catching for? It's fun to catch and give away. It's connected to the Tribe's tradition of potlatches — getting together with others and giving gifts; the more you gave at a potlatch, the greater your esteem. Potlatches helped to develop social bonds and friendships among people and different Tribes."

This has been replaced by the Indian food distribution programs set up by the federal government as they provided surplus foods produced by farmers to Tribes (*see* Box 4). The surplus food was distributed to Tribal communities by the Office of Indian Affairs which was in the War Department. This started soon after the Indian Removal Act was signed into law by then President Andrew Jackson on May 28, 1830 so the

"... President could grant land west of the Mississippi River to Indian Tribes that agreed to give up their homelands. As incentives, the law allowed the Indians financial and material assistance to travel to their new locations and start new lives and guaranteed that the Indians would live on their new property under the protection of the United States Government forever."[1]

---

1 Accessed on June 1, 2018 at https://history.state.gov/milestones/1830-1860/indian-treaties

> **Box 4: History of Surplus Food Given to Tribes**[1]
>
> "In the mid-1800s, rations — food "staples" like beef, bacon, flour, cornmeal, coffee, sugar, salt, and rice — were distributed to tribal communities by the Office of Indian Affairs, which was maintained by the government agency that was then named the War Department. These rations were quite literally tools of war, included in treaties signed between tribal governments and the U.S., that left tribes on land bases too insufficient to feed their people and without sufficient animals to hunt after the intentional destruction of buffalo herds. These rations were often delivered late or in poor quality, due to insufficient government oversight, and sometimes in an intentional attempt to starve communities into submission. Rations were sometimes withheld from uncooperative families who wouldn't give up their traditional religions or send their children to boarding schools designed to strip them of their language and culture. In 1933, the federal government developed the commodity program, in an effort to stabilize farm prices by removing price-depressing surplus food items from the market through government purchase and then distributing them to needy families. *The Agricultural Act* of 1949 made funds available to process and package these foods in order to give them a longer shelf life and to then make them available for distribution through the Bureau of Indian Affairs as well as through public welfare systems. *The Agriculture and Consumer Protection Act* of 1973 gave the USDA the authority to purchase foods for hungry families in response to a decline in surpluses. In 1977, under *the Food Stamp Act*, the Food Distribution Program on Indian Reservations (FDPIR) was authorized to provide access to food for low-income Native American households that did not have easy access to grocery stores to use food stamps (Native American people have the option to use either FDPIR or SNAP). The USDA purchases and ships food to Indian Tribal Organizations (ITOs) in each recipient community; ITOs then distribute the food to families in need. For most of the history of the program, the content of the food boxes that each Native American family received through this program was not flexible. Every family received the same box of canned and preserved foods, which included foodstuffs like canned meats, vegetables, and fruit; powdered milk and eggs; bags of dried beans and pasta; bottles of oil and juice; and a brick of commodity cheese — whether or not they liked these foods or could even eat them."

When Tribes lost most of their customary lands, they lost access to their cultural foods and became dependent on the federal government for food. They

---

[1] Accessed on June 1, 2018 at https://www.teenvogue.com/story/food-boxes-have-already-failed-for-native-communities-why-would-they-work-for-snap

became "prisoners of war" when the Western Europeans occupied their lands and prisoners are typically not well fed. The US Department of Agriculture has provided to food aid to Tribes for decades which was mostly canned food and foods that were non-perishable [33]. As Godoy [33] reported: "*Jernigan says recent studies have found that 60 percent of Native Americans who receive food assistance through the program rely on the government program as their primary source of food.*" These are not very healthy foods that Native people became depend upon.

Dr. Nancy Maryboy talks about many Navajo people eating spam because it is available and cheap to buy. Allen [34] summarized this shift for the Navajo people from healthy diets to one that is causing them numerous health problems:

> "*Their food would have been the envy of any modern dietitian as they foraged for pinyon nuts and wild potato, and nibbled on sumac berries, yucca fruit, prickly pears and beeweed greens. Today, life is very different in the Navajo Nation, the largest American Indian reservation in America, which covers an area of Arizona, New Mexico and Utah that is the size of Scotland and has a Tribal population of 300,000. The modern Navajo, who prefer to be called the Diné, are facing an intensifying health crisis fuelled by a complete transition to a diet based largely on fried potatoes, tortillas, cookies, crisps, sugary drinks and spam. According to the Navajo Area Indian Health Service, around 25,000 people on the reservation have type 2 diabetes and 75,000 are pre-diabetic.*"

Today, knowledge of cultural foods is not just knowing what foods to eat but when to gather or collect a wild plant or when to hunt a deer or collect oysters. Hunter-gatherers also needs to recognize when natural cycles will cause boom-and-bust cycles in the abundance of a food.

Native Alaskan hunter-gatherers continue to maintain their knowledge of food and adjust what animals are hunted or what foods are collected based on their knowledge of their lands and waters. Native Alaskan hunter-gathers have been called the top predator because they are able to keep the ecosystem stable by keeping track of how all the animals in their food web are doing [35]. They shift to another species to eat when they realize that their current species being hunted is endangered and experiencing a down turn in their population size. At this point, they shift to eating another species.

In contrast to the Native Alaskan hunter-gatherers, non-native hunters continue to catch and eat species that are endangered. This is where Native people's knowledge of climate change impacts on the timing of species growth or the lack of growth is critical to maintain stable populations of animals hunted for food. Non-Native Alaskans are not sufficiently in touch with the species they eat to know when the species are experiencing cycles of collapse and becoming endangered. They are dependent on agencies managing these species to tell them when they can hunt. These agencies monitor the species and use models to explore future population levels.

3.6.1 Loss of Food Security: Chemical Contamination

Today, Native and Non-Native people face new challenges due to scarcity of cultural foods (*see* Section 6.1) but also that their cultural foods are microbially and chemically contaminated. These factors contribute to health problems among Native people. As was written in the High Country News [36]:

> *"For anyone consuming large quantities of fish, one of the greatest risks comes from methylmercury, which builds up in the fat of carnivorous fish like salmon. Mercury poisoning can lead to vision loss and other neurological problems, and can cause severe developmental damage to infants exposed in the womb. Most mercury is deposited into water from the air, with coal-fired power plants, smelters and waste incinerators as major sources."*

The landscape of the past is not the landscape of today — according to research published in 1998, *"less than 50 percent of Washington's salmon stocks were considered to be healthy."* (Governor's Salmon 55 Stock is defined by Governor's Salmon Recovery Office, 1999, pp. II.9 – II.10) Despite this, cultural and subsistence gathering, hunting and fishing continue since they are an important part of Native people's routine diet. You only have to look at the consumption rate of salmon compared to the average American citizen to see how important cultural foods are for Indigenous people.

When you live in a smaller land or water area, you are more cognizant of how climate change is impacting species that live on your land or in your waters. If you eat a lot of fish that is part of your critical everyday life and culture, you have to worry about whether the fish you eat will slowly poison you or cause severe developmental damage to infants. You have the right to fish but if your fish are contaminated or toxic, what does this "right" mean? If you eat 175 g of fish a day but the standards are set to 6.5 g/day for Washington State by EPA, the amount you can eat is significantly lower than what is actually consumed [36]. Twenty treaty Tribes of the Northwest Indian Fisheries Commission have been working several decades to convince state politicians to revise Washington's fish consumption rate.

According to a State of Washington 2013 Report, chemical contamination of the Puget Sound has been occurring for over 150 years (Box 5). In the Colville

Reservation, Tribal members do not eat as much local fish because of concerns of the upper Columbia River and Lake Roosevelt areas are chemically contaminated [37]. These areas have been *"affected by contaminants from Teck Cominco lead-zinc smelter operations for over 100 years."* The Swinomish Tribe consume locally harvested seafood at a consumption rate of 260 g/day [36] so one needs to be concerned about the amount of chemicals you are ingesting. For the Swinomish *"finfish/shellfish contaminants that contributed the most to human health risks were PCBs, arsenic, and dioxin/furans"*. In the 1980s, the Environmental Protection Agency (EPA) set a threshold of 6.5 g/day of fish (this equals two fish servings per month) consumption based on the amount of mercury and other toxic chemicals found in fish; in 2000 EPA increased this to 17.5 g/day [37]. By increasing the consumption rate, it only increases the risk of people eating fish getting cancer-causing toxins; this includes all people who eat a lot of fish such American Indians, Asian and Pacific Islanders and sports fishermen.

Another issue that Tribal people face is how many of their cultural foods have now become part of the global commercial industries. This means that what they used to be able to collect in sufficient quantities for ceremonies is becoming scarce. Not only are the cultural foods decreasing in abundance but they are being collected unsustainably which reduces future harvest of these cultural foods (*see* Section 3.6.2).

Box 5: Some Chemical Contaminants of Concern in the Puget Sound

- metals (inorganic contaminants)
- organic contaminants
- lead
- cadmium
- tributyl tins
- copper
- mercury
- arsenic
- polychlorinated biphenyls (PCBs)
- polycyclic aromatic hydrocarbons (PAHs)
- dioxins and furans
- selected pesticides
- phthalate esters
- polybrominated diphenyl ethers (PBDEs) Hormone disrupting chemicals (bisphenol A)
- petroleum and petroleum by-products

(Sources: Puget Sound Action Team, 2007, Table 4-1; West et al., 2011a, 2011b)

3.6.2 Loss of Huckleberries and Tribal Culture: Interview of JD Tovey*

JD Tovey.
Source: JD Tovey.

Kristiina: So cultural foods were one of the questions I wanted to ask you about. You collect huckleberries and you have given me huckleberries as a gift during the NSF IGERT funding at U.W. Is it harder to get huckleberries today because they have become a commercial product collected by many non-Tribal people? So, what do you think about your culture being linked to huckleberries or salmon that are scarcer and less available? Last year (2017) several Tribes decided not to fish for salmon because the runs were so poor. Suquamish stopped their fishing. What do you think when you have all these other people that are coming in and are commercializing your cultural foods?

> **JD:** *It is scarce. In the past this was managed by an elected person from several Tribes who was called the Salmon Chief. This was not hereditary even though the sons might be involved with this. This person's job was to keep the peace but to also set the fish limits. We knew we needed to manage fishing or people would over fish and we knew in 14 years we would not have any fish since that was the cycle — from spawning to returning. There were these rules that were being lived by on managing that natural resource.*

Kristiina: So, you are saying that you can still have these cultural resources.

---

* By Alexa Schreier, Dr. Phil Fawcett and Dr. Kristiina Vogt.

JD: *Let's talk about huckleberries. They hate being taken care off. As soon as you start taking care of them in an agricultural way, they die. If you try to grow them in a pot or in your yard, they don't like it. They need to be cared for by doing the burnings but don't try to take care of them too much. They are closely related to blueberries, so huckleberries cost $100 to $110 dollars a gallon. They are a pain to pick and I usually have my face covered in huckleberries because I am eating so many and don't put much in my bowl. A very traditional stance is not to cultivate them, we should not be messing with them culturally or genetically. But the problem is that their value is so high, and you don't cultivate them.*

We have a lot of groups that are not ethically-based and we have seen them out there with rifles to keep others away. They go out there with rakes and have tarps on the ground. They completely destroy the plants so that fields of huckleberries are gone for five or six years. They are destroying it but there is so much value there. The question is "We collect huckleberries for a reason." *It may or may not be economic. We do trade them. We have a cultural reason for collecting huckleberries. Is there a way we can help preserve huckleberries by taking that value out of them? If the Tribes owned this process then is it ok? We can take huckleberries and cross them with blueberries and get them to grow commercially. Then we could grow huckleberries at a lower elevation so we strip the value from them. So, then huckleberries would go to $10 dollars a gallon instead of a $110/gallon. Now people would not be raking them at the higher elevations and we could still do that because we do it culturally. Since the value is not there, we are not worried about being shot while collecting.*

Kristiina: But you can't keep people out.

JD: *I worry about it. Even camas which someone someday someone is going to find out about its value.*

Phil: Or its properties.

JD: *Someone is going to make some marketing claim on it and before you know it will be the biggest thing. We would not even be able to afford camas then because it will be $300/pound. If that happens we should own that. But we take care of it and don't own it. To claim ownership of it, especially doing a very large economic development on it, is a conflict for our culture. But in some cases, it may be the only* *way we can protect it. When I have brought up the huckleberry thing and you can see people really thinking about it, but it is not what we do. We don't mess with the plants but it might be a good way to save them.*

Kristiina: See that container of açai over there on my table? It is now sold as açai and blueberries covered in chocolate. But there is not enough açai now. It is a palm that grows along the edges of rivers in the tropics and the local people are totally dependent upon it for their food and health. It is tasteless also but it is full of protein.

3.6.3 Skokomish Litigation for Rights to Hunt by Indian Tribes*

The Skokomish Tribe has filed a federal lawsuit involving the rights of Indian Tribes to hunt game on "open and unclaimed lands" with the State of Washington. The lawsuit states that actions by state agencies and officials have denied Tribal members access to their legitimate hunting areas. The lawsuit further claims that state officials have imposed civil and criminal sanctions on Tribal members and promoted a "discriminatory scheme" of hunting regulations that favor non-Indians. To date, a federal lawsuit has not been filed that deals with access for treaty hunting and gathering rights. This litigation could define the extent of treaty rights related to hunting animals and gathering cultural resources by Tribes across Washington State.

Listed as defendants in this federal lawsuit are the Washington State Departments of Natural Resources, Fish and Wildlife and the Attorney General's Office; also listed are the county prosecutors in Mason, Kitsap, Jefferson, Grays Harbor, Clallam and Thurston counties, who are charged with prosecuting hunting violations.

> *The 1855 Treaty of Point No Point preserves the "privilege of hunting and gathering roots and berries on open and unclaimed lands". That treaty — signed by representatives of the Skokomish, S'Klallam and Chimakum people — ceded to the United States Tribal lands around Hood Canal and the Strait of Juan de Fuca.*
>
> *In its lawsuit, the Skokomish Tribe contests state maps that purport to identify those ceded lands, as well as a Washington State Supreme Court ruling that limits "open and unclaimed lands" to those not in private ownership.*

---

* By Rodney Cawston.

> *The lawsuit argues that State Supreme Court cases fail to define the extent of the hunting and gathering rights of the Skokomish Tribe. That lack of definition has caused state officials to "unlawfully interfere with plaintiff Skokomish Tribe's privilege of hunting and gathering on open and unclaimed lands as guaranteed by Article 4 of the Treaty of Point No Point. ...*
>
> *"This unlawful interference ... resulted and continues to result in denial of lawful access to plaintiff Skokomish Indian Tribe's territory and use of resources located thereon."* [38]

The Skokomish Tribe also seeks an injunction to prevent state officials from interfering with Tribal rights to hunt and gather roots and berries.

In the *Washington State Legislature House Bill 1496*, enforcement offices will provide a training tool in order to best protect Tribal members' treaty-protected hunting rights and curb inconsistencies in the application of state law to Tribal members. If a law enforcement officer stops a Tribal member and the Tribal member produces an identification card from a federally recognized tribe, the officer will refer the incident to appropriate Tribal law enforcement officials. This bill passed the House of Representatives and was referred to the Senate. The Senate asked for an amendment but it did not make it out of committee. However the bill is still alive and if approved could strengthen relations between Washington State Natural Resource Agencies and federally recognized Tribal governments.

For Tribal government and state/federal government at any level in the United States today, relationships are challenging and complex. There are many federal, Tribal, state and local laws and regulations, multiple interconnections and interdependence that complicate these lively and vital relationships. Natural resources management for Tribes in Washington State is again at a threshold of resolution either by legislative measures or litigation or both. Tribes and Washington State want to use resources effectively and efficiently, they want to provide comprehensive services and a safe environment for both Tribal members and Washington citizens, protect natural environments, and sustain healthy ecosystems. Tribes have learned to become very adaptive to changes all throughout their histories. Tribes and state and federal governments have also worked together to conserve natural and cultural resources. Almost all Tribes in the State of Washington are engaged in very large and small conservation projects and making recommendations on the management of shared resources.

3.7 Holistic Nature Knowledge not Decoupled from Nature and Religion

As written by Kawagley [39], holistic knowledge is not something that can be experimentally tested in a laboratory setting. This probably explains why Western science is reticent to adopt holistic knowledge frameworks. They can't figure out how to form knowledge using a framework that differs so drastically from the scientific method (*see* Section 4.1). Western science frameworks follow a linear process to form knowledge that tends to result in one consistent and testable solution. It is a compartmentalized science where knowledge and theories are used to develop solutions to problems. It tends to assume that knowledge is fixed and doesn't change in contrast to Native ways of knowing where there is recognition that constant change is the norm [38].

The Western science framework is predominantly built on information derived from the natural sciences and implemented using a disciplinary focus. It is decontextualized from the natural ecosystem and the people who live in its environment in which a problem plays out. Using a Western science framework approach, it seldom includes information from the social sciences. Therefore, the knowledge framework is not holistic but fragmented. It also does not include "Deep Culture" in the decision process even though these are the behavioral drivers of why decisions are made (*see* Section 3.2). Also, scientists are the main providers of expert knowledge and frequently are ignored by the decision-makers who don't really understand the complexity inherent in scientific knowledge. In contrast, Indigenous knowledge frameworks connect people and their "Deep Culture" as an integral part of their knowledge-forming process of nature. This contrasts the Western science framework that is still struggling with how to connect people, i.e., packaged as the social sciences, to make credible decisions related to nature and the environment.

An underlying difference between Indigenous knowledge and the Western knowledge-forming process is whether the spiritual or religious aspects — fundamental to Native people's culture and art — should be part of a science assessment. Western world scientists have not been comfortable including spiritual or religious aspects as a credible part of their scientific inquiry. A 2009 Pew Forum poll of AAAS scientists [40] found that half of the scientists

> *"view science and religion as distinct rather than in conflict, with each attempting to answer different kinds of questions using different methods."*

Since the spiritual aspects are important elements of the Indigenous knowledge-forming processes, Western world trained scientists need to recognize that spiritual aspects are not defined in the same way by the Western science world. The Western world has a very formal definition of spiritual aspects of knowledge. If this is not possible, it will be very difficult for the Western knowledge framework to include Indigenous ways of knowing in their knowledge framework.

Western scientists need to become comfortable with Einstein's definition of religion as *"evaluations of human thought and action"*, e.g., are people behaving ethically. He saw no conflict in saying that religion and science are needed to improve human life. If Einstein's view was more commonly accepted by Western scientists, it would be easier to interconnect these two forms of knowledge and recognize that both are needed to achieve environmental and social justice when making decisions related to nature. Einstein wrote:

> *"Science without religion is lame and religion without science is blind"*
> He further wrote: *"If we want to improve the world we cannot do it with scientific knowledge but with ideals. Confucius, Buddha, Jesus and Gandhi have done more for humanity than science has done."* [41]

This Einstein quote reflects some of the perceptions found in the Medicine Wheel of Native Americans.

When the Western knowledge framework considers integrating spiritual or cultural aspects into their framework, art replaces spiritual and culture. Part of this is probably because culture is harder for Western world trained scientists to measure and put into a scientific equation. Recent calls for art to interpret science do not address these different ways of knowing. In 2009 Edwards wrote his book *"Artscience"* [24] where artists interpret science using their chosen medium. This approach results in art objects that can be viewed and interacted with by a visitor to museums, but it does not have artists and scientists collaborating on connecting knowledge from different fields. The Artscience Museums have an important entertainment and experiential role for the visitor but is not the tool that will link Indigenous and Western science knowledge frameworks. ARTSCIENCE museums still keep the causality of relationships on separate tracks and do not address scarcity of knowledge. This is the subject of this book.

If we want to mitigate or restore damaged environments, we do not have the luxury of continuing to make decisions using silos or compartmentalization of knowledge by disciplines. We need to stretch our knowledge beyond a disciplinary focus to stimulate our imagination for creative solutions. Taleb's book on *"Learning to Love Volatility"* [42, 43] gave the example of Thomas Edison who was a prolific inventor but most of his innovations were the result of tinkering. We need to begin to include all the components of "Deep Culture" as shown in the iceberg example (*see* Fig. 1.6).

An important part of "Deep Culture" is language. Therefore, it is important to discuss why language is important for Native Americans (*see* Section 3.8).

It is worth thinking about what is embedded in a language for the people who speak and communicate in their Native language. It also introduces an important concept that the loss of language for Indigenous people is more than not being able to communicate in their native tongue. Language supports the idea of the Medicine Wheel since Indigenous languages are not just words on a piece of paper or today on a computer screen, but it stores the complexity of knowledge on nature that goes back many generations. Language is part of the culture and links to nature. These tools also need to be language based since that is core to Indigenous knowledge and the transmittal of knowledge. Our moral judgments depend on what language we are speaking and if a foreign language is used people make less ethical decisions since people do not have an emotional connection to the problem.[1] Therefore, language loss means a loss of knowledge and how a group of people remain resilient in the dynamic natural environments. The book entitled Landmarks documented well how the loss of words means that people are less able to read the ecology of the land and understand the changes that are occurring in the Scottish moors [44].

3.8 Languages and Indigenous People*

The extinction of any language is the loss of not only the language but the essence of the culture that the language expresses. Lloyd Pinkham writes eloquently of significance of the experientially grounded and cosmically connected language of his people, the River People of the Northwest Columbia River Basin [45]:

> *"From a Native American research perspective, when we examine the foundations of language, the River language path traces back to the lived experience of the native psyche relationship of the conscious connection to the environmental experience. This includes the ancient generational cosmic relationship, seasonal interaction, ceremonial involvement of sun, moon, stars and Earth Mother. The fundamental nature of the River language relates to a living feeling connectedness to the ln'chi wana (the Great River, the Columbia River), based upon the ancient two-legged's participation in the seasonal shifts and cosmic relationships. The foundation of the River language of who we are extends from ancient generational relationships and personality of the land. Our language both represents the ethnocentric foundations of who we are as a people and at the same time separates us from other people, cultures,*

---

[1] Accessed on July 14, 2017 at http://www.nytimes.com/2014/06/22/opinion/sunday/moral-judgments-depend-on-what-language-we-are-speaking.html?_r=0

* By Dr. Nancy Maryboy and Dr. David Begay

Dr. Nancy Maryboy and Dr. David Begay.
Source: Nancy Maryboy.

> and other Tribal ways of knowing. Our language connects us with the river and the spiritual consciousness pathways of the far reaches of the stars, 'the land of the light.' The words and western phrase 'far reaches' is used to emphasize the ability to transcend and reach. In the River People's language, regardless of the distance or time, one can reach and still be consciously connected. This ancient knowledge is still within the reach of the post-modern native philosophy, cosmology, and the group socialization of the Wyam Pum and Paloos punx cush E wa (that's the way he is)."

From our experience, the gap between English and Navajo language is substantially wide. Many anthropologists and linguists have written of the extreme inaccessibility of the Navajo language and mind to a non-native. The Navajo tongue is so radically different from ours, says John B. Carroll et al. [46], citing authors Clyde Kuckhohn and Dorthea Leighton, that an understanding of Navajo linguistic structure is virtually a prerequisite to understanding the Navajo mind. They cite the tremendous translation difficulties existing between Navajo and English, and imply that the two languages almost literally operate in different worlds [46]. Along with the need to understand Navajo linguistic structure and the world view of the Diné, it is easier to grasp the holistic qualities of the language and culture if you actually live within the culture—an admonition easy to write and far more difficult to do.

Other linguists see language as a kind of code. Gary Witherspoon describes language as an encoding process [47]. Language is, among other things, a symbolic code by which messages are transmitted and understood, by which information is encoded and classified, and through which events are announced and interpreted. Furthermore, Witherspoon elaborates,

> "Like language, culture is a symbolic code through which messages are transmitted and interpreted. But, more than a code, culture is a set of conceptions and orientations to the world, embodied in symbols and symbolic forms. Through the adoptions of and adherence to particular concepts of and orientations to reality, human beings actually create the worlds within which they live, think, speak, and act."

Native American languages, such as Navajo, have the ability to express complex and holistic ideas, but they go much further than that. In their very essence, they are themselves complex, polysynthetic and at the same time holistic, containing elements of process, relationship and wholeness in virtually every utterance.

This remark is strikingly similar to the traditional Navajo consciousness in which the Navajo world is experienced as a process of constant interactional motion. Although, Navajos recognize the fixed order of the natural elements, they also know at the same time the world is going through interaction, flux, processes and regeneration. From our experience, we know that Navajos are able to discuss extremely complex subjects without resorting to abstract intellectual artificialities. In accordance with the Diné cosmology and spirituality, Navajos utilize certain cultural conventions in order to properly discuss complex concepts. Cultural conventions may include explanations of natural forces and processes, on multiple levels of meaning, such as physical, spiritual, emotional, intuitive, or psychological, among others, focused and articulated through one complex image of being.

Trying to write from a Native American holistic consciousness into a linear format using English terminology is a very challenging task. There are severe transcultural and translinguistic limitations in the attempted transposition of a holistic concept. The moment we place words on paper, in English, a process of dissection, fragmentation and alienation begins. This in turn leads to significant loss of the original intrinsic meaning: A separation from the holistic consciousness which leads to de-integration and, even more seriously, to marginalization and trivialization.

Navajo language needs to be explored more in terms of relationships, rather than through static nouns and permanency. Like almost all other Native American languages, it is a language of verb, process and motion. In the Navajo language, nouns are often constructed from verbs, in direct opposition as to how verbs are often constructed from nouns in English. For example, one can put *igii* or *ii* on the end of a Navajo verb and it will become a noun. Contrast this to the English present participle where one can put *-ing* on the end of a verb and it becomes a noun. Navajo ceremonial names are generally translated as nouns, but a more accurate translation by traditional people reveals them to be words of action and interaction expressing continuous movement and transformation.

A further example of the importance of motion in Navajo is suggested by the verb "to go". "*In the Navajo language,*" suggests Martha Jackson, Professor of Navajo Language at Diné College, "*there are probably more than 200,000 ways to say go.*" (Martha Jackson, personal interview, May 4, 1998). On the other hand, there are very few ways to say *be* and then only under certain circumstances.

The Navajo emphasis on process and movement is exemplified by certain terms, for example, *nanit'a*: The cognitive process and the articulation of the human consciousness. The significance of the term *nanit'a* comes from the motion that expresses the dynamic interrelationships and processes that regenerate and transform life through extraordinary complex further interrelationships with a vast cosmic order. This process is expressed through regenerative and interactive *alk'inagish*, natural forces, such as clouds, rain, and climatic changes, in conjunction with time and space alignments. Ultimately and spiritually, nothing is permanent in the Navajo consciousness, yet everything seems permanent. The movement as expressed through *nanit'a* operates by interactive harmonic cosmic laws.

Movement can be caused by many factors and discussed within a cultural context, from the macro to the micro movement of natural phenomena. For example, at the micro level, Navajos refer to the mist and dew that facilitate the growth of the most minuscule vegetation as *to alta naschiin to biyaazh naninse' yilth althk'inagish*. The growth implied here is a reciprocal movement in terms of the relationship of the mist to dew, and vice versa; the movement is in the growth. *To alta naschiin* and *to biyaazh* in this case refer to the dew and mist which condense and vaporize on the vegetation as a lifegiving force. The *naninse'* (vegetation) in turn reciprocates by creating and releasing its own pure mist and dew into the atmosphere. Science might explain this as a purification of the air and creation of oxygen by plants. Depending on one's world view, the Navajo or the scientific focus may become one's primary explanation, both being equal to the task of explaining natural phenomena.

It may be difficult for a non-native to understand the strong emphasis placed on language. From a monolingual English point of view, it seems that different languages are just different labels for the "same things" in the world. From the Native side, on the other hand, language is of extreme importance, carrying a world view which is rich and different from the Cartesian world view. What do we mean by "carrying the world view"? A culture's world view cannot be expressed in its richness and complexity without the words and language needed to express and support it. When you lose the language, you lose the intellectual and spiritual complexity of a world view. Among Indigenous people, language is much more than a way of expression: It is also an internal manifestation of one's essence, life and consciousness.

Speakers of dominant languages such as English have had little experience facing the possible loss of their language, not having been threatened with the potential extinction of their language for nearly a thousand years since the Norman Conquest in England. Yet even today there is a bitter political struggle surrounding the concept of "English Only," primarily in relation to Hispanic and Asian languages. There is a huge difference between an English person learning French in school and that of a native person learning English. In the first case, one is remaining within the Indo-European language family and one retains the Western world view. In the second case one is making a transition to an entirely different language stock and an entirely different non-Native cosmovision. For example, the Native is naturally talking as a participant in his universe, whereas the English speaker is talking as a non-participant, an observer who is separated from nature.

The vital importance of retaining one's language is not confined to Native Americans, of course. Indigenous people all over the world are struggling with similar concerns. Harold Gaski, Associate Professor in Sami Literature, at the University of Tromso, Norway, has written about the importance of the Sami language to the Sami people [48]. What is it, then, that makes the Sami language unique? It is, of course, *"the language of the heart"* to those for whom it is a mother tongue, but it is also one of the most developed languages in the world when it comes to describing arctic nature and conditions of life in the north [48].

Sami, similarly to other arctic native languages, has a highly specialized and detailed vocabulary dealing with landscape and family. *"The terminology for animals and landscape, snow and conditions for traveling on snow, family and kinship, is so precise that no scientific system can be more so* [48]." Yet this language, from which the Norwegian and English languages could learn so much, is in danger of being lost. The living situations from which the language developed no longer exist to the same degree. Sami are living increasingly in a Western-type society which separates them from nature, so there is less need for specialized terms regarding reindeer herding, weather, traveling on snow and ice. In addition, the acquisition of the Norwegian language, the language of power in Norway, is facilitated at the expense of losing the thought patterns of traditional Sami. Gaski [48] explains:

> *"When we have learned the language of power, we may begin to forget the thought patterns that form the foundation of our own language. Then our 'differentness' can develop into purely a rhetorical veneer, turning us into a kind of political actor without a cultural base. We may ourselves begin to regard experiential knowledge as inferior to scientific knowledge. Science, in the formal sense, has status because it is 'rational', while the Sami precise observation and terms are regarded as mere empirical and typological knowledge — and consequently of lesser value."*

Yet even the dominant societies occasionally recognize the value of empirical and typological knowledge. The United States Marines employed hundreds of Navajos during World War I and World War II (*see* Box 6, Box 7).

Box 6: Code Talkers from the Choctaw Nation and the Congressional Gold Medal[1]

they received for their service during World War I.

Box 7: Members of the 3rd and 4th Division Navajo Code Talker Platoons of World War II[2]

They dressed in their units uniform, pose for a group photo during a commemoration of the landing on Iwo Jima, 02/21/1987.

---

1 Unites States Mint - usmint.gov, Public Domain, https://commons.wikimedia.org/w/index.php?curid=43130196
2 National Archives Identifier: 6428371.
https://www.archives.gov/research/native-americans/military/scouts.html

Navajo Code Talkers used their language, in the form of a secret unbreakable code, throughout the Pacific. The Navajo Code Talkers played an important role in winning the war in the Pacific against the Japanese. The paradox is that for several hundred years the federal government worked to eliminate native languages and cultures. Native people are keenly aware of this paradox. A Navajo serviceman fighting in Vietnam described his experience:

"As a Navajo, I look at war as if fighting alongside the whiteman, who tried desperately to annihilate the American Indians, fighting for what was once ours and now to protect what is no longer ours [49].*"*

Native languages have evolved from a non-western consciousness and as such are not dependent on constructs such as static nouns and objectivity, non-relationship oriented being and non-participatory methodology. Therefore, holistic languages such as Navajo can naturally and more fluently discuss process, relationship and human participation at the subatomic level.

Even though the new sciences of wholeness (such as Quantum Physics, Chaos Theory, Complexity Theory and Systems Theory) are holistic and process oriented, they still exist to a large extent within the construct of a Cartesian frame of reference and objective methodology. Native American thinking can be objective while at the same time it can also involve a subjective consciousness in regards to human and spiritual relationships. Native Americans often express their thoughts through non-restrictive metaphysical terminology, which does not easily coexist with the more restrictive terminology and principles of the scientific method.

Most Native American Studies courses are content-based, by this we mean subject-oriented. They focus on learning <u>about</u> something, not necessarily learning the essence of <u>being</u> of something. Traditional Natives are holistic, with spirituality as the vital dynamic. The traditional language expresses itself in terms of verb-based flux and process.

Native students who do not challenge the culturally biased system in this respect are able to meet the university requirements but when they return home with a degree, they may often find that the methodology and trainings that they received are irrelevant and meaningless when applied to community culture and concerns. This unfortunate situation is not new. It has been a major and disturbing factor in Indian education since the days of compulsory education in government sponsored residential schools in both Canada and the United States. These lasted well over 100 years.

First and foremost, the Native person is a holistic thinker. Native languages are intrinsically holistic, being verb-based and focused on processes and relationships. Finally to be relevant, native-based research must be applicable and integrated into the life of the community.

When traditional Navajos talk of story and ceremonial song, their thinking encompasses the complexity and totality of intellectual knowledge. They are making reference to a cosmic organization of knowledge while at the same time acknowledging ancient and natural sources. On the other hand, the concepts of story and song have seldom been associated with deep and complex thinking in the western tradition, although this seems to be changing. As a result, say Native people, western thinking is not equipped with the proper cultural context with which to process and deconstruct the information contained in native story and song.

Linguistic contextualization is vitally important, due in part to the considerable difference between Navajo, a holistic Native American language and English, a reductionistic language that has evolved from many roots, including Sanskrit, Greek, Latin, German, Norse and Old English. It should be immediately obvious that the two languages, Navajo and English, come from widely different origins and have developed independently of one another.

These similarities in language and culture concepts could lead one to assume the Navajo and Apache world views are one and the same. However, this can be misleading since each culture is distinct and built on its own relationship to local geography, sacred landmarks and environment. The world view of a people is shaped by many forces, among which one of the most significant is arguably language. *"One of the principle ways in which we experience and interact with our culture is language."* says anthropological linguist Dan Moonhawk Alford. *"World view includes not only our own habits, but those of our culture as well, in dealing with the world — the world as background against which we operate, the world as culturally modeled habits of doing and being, perceiving and sensing, thinking and speaking."*

It is extremely difficult to think about language in terms of world view, if one thinks solely in terms of relationship of one to another. However, if one discusses language and world view in terms of a holism, further complexities are revealed. Several of these complex associations intersect with a native interpretation of consciousness.

3.9 What Is Your Real Name? Dr. Mike's Wolverine Encounter*

As a young boy I was walking along a creek in the mountains on the Colville Indian Reservation. I was fishing. The creek mostly had small trout. It was in the middle of the forest. The fish in these small streams are very wary and wild

---

* By Dr. Mike Marchand.

Dr. Mike Marchand.
Source: Mike Marchand.

and you must be quiet and sneak up to them or else they will scatter and hide. So, I had a caught some fish that morning. It was a nice hot bright summer day. I rounded the stream and heard a big crashing sound of water. There was a tree snag sticking up out of the stream bed and a wolverine had run to the top of it. The wolverine was eye to eye with me and was about five feet away. He glared at me and was not afraid. I was stunned. I had never even seen a wolverine before, except on TV. They are pretty rare in this part of the country. He looked at me for what seemed like an eternity, but really it was likely only a few minutes. He started talking to me. He said this was his home. He said that I was welcome to move in here if I wanted. He said if I did choose to be there then he would take care of me and protect me. Then as suddenly as he appeared, he jumped back into the creek and disappeared. I sat there for some time stunned and just pondering what had just happened.

    Eventually I got older and became an adult, got a job and was married and had three children. I needed a home. It took some years to do so, but eventually I did move my home to this original scene where I had seen the wolverine as a young boy. It has turned out to be a very nice home site. In the mountains. In the forest. Wildlife passes by pretty often. So I see animals like deer and elk and moose. There was a porcupine living in a hollow log in my front yard. Many birds such as shrikes and eagles and herons and ducks and geese sing their songs and make their homes there too. A paradise for anyone who likes the outdoors. My personal life had been very lucky. I went from a young boy to be chairman of my tribe, something that I consider to be a high honor. My mother believed in education, so I eventually obtained a PhD in Forestry from the University of Washington.

At a family dinner one time about 30 years after I had seen the wolverine, actually more like a clan dinner with a lot of extended family and friends, I was talking to an old aunt. We were not talking about anything specific, but I eventually told my aunt about my youth encounter with the wolverine years before. My aunt listened with great interest. She was very traditional. Spent her whole life studying and working with plants and medicines and had a lot of knowledge about traditional things. She said that was a special encounter, that the wolverine was an animal spirit helper. Animals can have special powers. In our language this power is called "sumechth". She said that should be my Indian name. So that is how I got named "**Qualth-a-meen**". That means wolverine.

Every animal is very unique, wolverines probably even especially so. They are not the biggest animal, but they are very strong and have no fear and they will fight the biggest animals and take what they want. A very strong animal. A good one to have an alliance with. I don't know if this is always true, but it seems to fit well with my life's work and with my personality. I am often fighting for my tribe. Often the odds are against us, we have big powerful enemies. We don't always win, but we win more then we lose.

In our culture, everything has a spirit. Whether it be an animal or plant or rock or Earth or sky. Everything is interconnected. Everything is sacred. Elders are to be respected. My aunt had a lot of traditional knowledge handed down to her from her elders from one generation to the next. Native American people have a tremendous amount of knowledge about nature. Perhaps it has been learned from thousands of years of trial and error, through observation and reason, or sometimes maybe by intuition. I don't really know if the wolverine spoke to me or not. But the thoughts got into my head. Maybe it was my imagination. I have no way of knowing. But my thoughts were validated by my aunt. She knew they had meaning. She knew what to do with those thoughts. She knew the proper ceremonies and practices to deal with the matter to improve my life.

3.10 Sports and Games Invented by American Indians

Before the arrival of European colonialists to the Americas, gambling and playing games was an integral part of Native American life. American Indians were not just focusing on survival but had time to pursue other activities like playing games or creating art. There is an erroneous idea that semi-nomadic people spend all their times hunting or gathering food and just barely surviving at a subsistence level. This has been disproved in several indigenous communities who may spend half of their time on games or the arts.

Games were played for recreation, as part of their economic system and also as part of their religion[1]. This paragraph summarizes well how wide-spread playing games, that included gambling, was for American Indians before the arrival of the European colonialists [50]:

> "Stories of Indian gambling are found in Tribal oral histories and Native American literature, and archeological evidence supports its occurrence. For example, the Navajo played stick dice and arrow and moccasin games as well as gambled, informally, on foot and horse races. The Dakota played a hand game called "hitting the bones" (hampa ape' achunjpi) and other forms of gambling that were used for healing and decision-making purposes, among others. The Crow Indians' favorite gambling game was "hiding", a guessing game; the Crow also had dice games, usually played by women, and gambled on various athletic contests like shinny, or "ball-striking". Native Americans gambling took many forms in premodern times, just as it does today; it could be ceremonial, social, or charitable. Native Americans also incorporated non-Native American games like poker and monte into their rituals. The overarching theme of Native American gambling before the 1980s, however, was that it served a social purpose rather than a financial purpose—it was cultural rather than practical."

Several examples of games played by Native Americans are summarized next:

(A) The Stick Game:
The Plateau Indians were known and liked playing games, gambling and any type of competition. They had invented a game called "The Stick Game" (Fig. 3.2). Almost all Tribal members played this game so it was not restricted to certain members of the tribe. This YouTube shows the Stick game was played by the Muckleshoot in 2013.[2] The following YouTube shows how this game was played as a singing contest.[3]

This section summarizes how to play the stick game[4]:

Game pieces: 11 sticks used to keep score, and 4 round pieces of bone (or small stones) used to play the game. Two pieces of bone are plain and two are marked somehow to be dark — either wrapped with string or dyed or marked somehow.

The Play: Two sides. Each side has the same number of players, usually 4–8 people on each side. Each side is given 2 pieces of bone —

---
1 Accessed on March 1, 2018 at http://nativeamericannetroots.net/diary/1405
2 https://www.youtube.com/watch?v=MRT83g0wXO8
3 https://youtu.be/ng8cRCwr8UY
4 http://nativeamericans.mrdonn.org/plateau/handgame.html

Fig. 3.2 (a) PIMA Indian women from southern Arizona playing stick and ball game using a small ball made out of buffalo hair covered with buckskin[1]; (b) Indians gambling on a Puget Sound beach, ca. 1884. Photographer: CE King. Negative Number: SHS 3488[2]. "One of the many games, called the bone game, or slahal, was often played between villages. A team member held up his hands, each of which contained a small bone, and the people on the opposite team tried to guess if a marked bone was in the right or left hand. Slahal stakes were often high, and the games might last for hours. In this photo, taken sometime before Washington became a state in 1889, Puget Sound Indians play the bone game near their beached canoes. The two teams sit opposite one another in the sand. Some team members hold sticks which are used for scoring and for keeping up a rhythmic beat." (c) Stick game being played on the Colville Reservation ∼ 1900.

one plain, one dark. Each side folds their hands into fists. Hidden inside a fist is one of the bones.

Object: To guess which bone piece is in which hand on the opposing team.

Game Play: By turns, one person from one team guesses what bones are where. If they are right, that team gets a stick. If they are wrong,

---

1 Accessed on May 20, 2018 at http://hockeygods.com/images/10271-Pima_Indian_Women_Stick_and_Ball_Game__Taka_or_Shinny. Colville photo from the Spokane city Historic Preservation Office collection.

2 https://digitalcollections.lib.washington.edu/digital/collection/loc/id/64/rec/562

the opposing team gets a stick. Then it's the opposing team's turn to guess. Men play against men. Women play against women.

The Prize: The winning team gets all the gifts brought by both sides. Each player has to put in one gift to play. Gifts might be a knife, a mat, a basket, a fishing spear — something of value.

(B) Shinny:
This was a game played the women of the Great Plains before the arrival of the European settlers. It was called the shinny. This game used two straight or curved sticks. The point of the game was to get the ball past each other's goalposts.

(C) Lacrosse:
In the southeast, they played lacrosse, also referred to as "The little brother of war" because of its dangerousness [51]. Lacrosse had Native American origins and was originally played as the stick game[1].

> *"With a history that spans centuries, lacrosse is the oldest sport in North America. Rooted in Native American religion, lacrosse was often played to resolve conflicts, heal the sick, and develop strong, virile men. To Native Americans, lacrosse is still referred to as "The Creator's Game"* (Fig. 3.3).
>
> *Ironically, lacrosse also served as a preparation for war. Legend tells of as many as 1,000 players per side, from the same or different Tribes, who took turns engaging in a violent contest. Contestants played on a field from one to 15 miles in length, and games sometimes lasted for days. Some Tribes used a single pole, tree or rock for a goal, while other Tribes had two goalposts through which the ball had to pass. Balls were made out of wood, deerskin, baked clay or stone.*
>
> *The evolution of the Native American game into modern lacrosse began in 1636 when Jean de Brebeuf, a Jesuit missionary, documented a Huron contest in what is now southeast Ontario, Canada. At that time, some type of lacrosse was played by at least 48 Native American Tribes scattered throughout what is now southern Canada and all parts of the United States. French pioneers began playing the game avidly in the 1800s. Canadian dentist W. George Beers standardized the game in 1867 with the adoption of set field dimensions, limits to the number of players per team and other basic rules."*

---

1 Accessed on February 4, 2018 at http://www.poyntonlacrosse.co.uk/lacrosse-basics.htm

Fig. 3.3 Painting of Sioux playing ball by Charles Deas 1843. Courtesy: Wikimedia[1].

Gambling did not arise when American Indians started building casinos on reservations. The first casino was built in 1979 in Florida and this had increased to 150 by the year 2000. The US government tried to stop any form of gambling by American Indians. This also meant stopping American Indians from playing games like foot races, canoe races, lacrosse, wrestling and archery — all activities which they also betted on.

The idea of setting up casinos for gambling was not really an unusual activity for the Tribes to consider. However, it faced serious challenges from the non-Tribal communities. This summarizes the challenges American Indians have had in including country or state level gambling on reservations[2]:

> "In 1987, the United States Supreme Court in *California versus Cabazon Band of Mission Indians* held that state gaming laws have no force in Indian country if the type of gaming does not violate the state's public policy. In this decision the court acknowledged the retained right of tribes as sovereign nations to engage in gaming without state interference as long as gaming of some form was legal in that state. If the Tribe's gaming is to be regulated from the outside, it must be regulated by Congress, not California. Justice Byron White, writing for the majority, finds:
>
> > "Tribal sovereignty is dependent on, and subordinate to, only the Federal Government, not the States."

---

1 Accessed on May 20, 2018 at http://www.stmuhistorymedia.org/the-creators-game-native-american-culture-and-lacrosse/
2 Accessed on February 4, 2018 at http://nativeamericannetroots.net/diary/1405

Twenty-one states supported California against the Tribes in this case. Federal courts and the Supreme Court have long recognized two basic things: (1) under the constitution of the United States, Indian Tribes are sovereign nations which are not under the jurisdiction of the state in which they are found, and (2) the states are often the worst enemies of the Tribes which are found within their borders."

Chapter IV
Western Science ≠ Indigenous Forms of Knowledge

> *"It is this subtle dimension of understanding that marks the southwestern Indian peoples from other religions and separates Tribal peoples from the world's religions. Somewhere in the planetary history religious expression changed from participation in the sound, color and rhythm of nature to the abstractions of man outside this context pleading for temporary respite and hoping in the next life to return to the Garden."*
>
> *"Nature not really owned by a group, even though they may think they own it."*
>
> ∼Vine Deloria. *Red Earth, White Lies: Native Americans and the Myth of Scientific Fact* (1997)∼

Many differences exist between the Western science and the Indigenous knowledge-forming processes as well as the frameworks that each uses to explore nature and environmental problems. One approach to explore these differences is to group the knowledge into local (Indigenous) versus what can be called the "airplane science" (Western science) view of the environment. The airplane (or satellite) view of knowledge is difficult to generalize when addressing a local problem because the information is generated at a lower resolution and lacks local contextualization. In general, decontextualized knowledge mostly fails to include the specific spatial and temporal factors that are needed to derive solutions for local problems. The airplane scale science lens can be localized by linking it to the local variations that exist in nature [52] but this approach is less able to also include culture and other social factors that drive our natural resource use decisions.

Historically, Western science has not respected local (or Indigenous) knowledge because it is frequently anecdotal information and difficult to model. In contrast, Western science generally needs to be quantitative and testable to be credible in the eyes of the disciplinary leaders and therefore acceptable to use to solve some problem. In fact in Western science, the conclusions produced by practitioners utilizing modeling or quantitative methods are considered by most scientists as reliable enough to use to solve problems.

Respect for Indigenous forms of knowledge began to emerge after it become clear that the Western science practices have not resolved many complex environmental problems. A former Dean from Yale University — John Gordon (*see* Section 5.3.1–5.3.2) — wrote about the need for new leadership models — or Essential Leader — to address environmental problems that persist for decades. This supports the need to learn from practices of people living close to nature. People living close to the land or water using holistic approaches are Essential Leaders. They are increasingly recognized as essential co-managers of land and water if the goal is to stop exploitative deforestation practices or to revitalize oyster populations in the Chesapeake Bay [6] [53]. There are many global examples of increasing efficacy of land management and conservation when the local knowledge is included.

Even though we are developing the rationale for linking the Western knowledge framework to the holistic approaches found in the Indigenous knowledge-forming processes, we are not saying that future land management models will end up looking like the Indigenous knowledge framework. If the Western framework to form knowledge only had to transition to the Indigenous framework, this would be a simple change that could be easily incorporated. But this incorporation or transitioning is generally very difficult because the holistic approach to environmental problem solving is only possible when Western scientists learn to form knowledge that does not follow a linear process. Today the Western science knowledge is more commonly designed to assess landscapes and the impacts of climate change at larger scales. This landscape knowledge needs to be part of the framework that is linked to the local knowledge (and sensitized local parameters) held or understood by Indigenous people. Once this occurs, it allows for the contextualization of resource and environmental problems at both scales.

It is important for the reader to recognize that Western science knowledge frameworks cannot become holistic by adding tools and methods from the Western trained social scientists. Such a combination does not produce an Indigenous knowledge framework (*see* Section 4.1). Indigenous knowledge is not just placed-based knowledge but it also interconnects information from several disciplines that are typically not connected by Western trained scientists. The Alaska Inuit Food Security report summarized Indigenous knowledge as [23]:

> *"Indigenous Knowledge (IK) is a systematic way of thinking applied to phenomena across biological, physical, cultural and spiritual systems. It includes insights based on evidence acquired through direct and long-term experiences and extensive and multigenerational observations, lessons and skills. It has developed over millennia and is still developing in a living process, including knowledge acquired today and in the future, and it is passed on from generation to generation."*

How Indigenous people form knowledge will be discussed in greater detail later (*see* Section 4.4–4.7).

To make our case that Indigenous knowledge-forming processes have a lot to contribute to holistic problem-solving approaches, we need to understand what makes them unique and what are their strengths and weaknesses. This is possible by understanding how each knowledge framework responds to **scarcity of land** or **scarcity of knowledge** since they are not equally effective in addressing each issue. Further since each framework varies in how information from the cultural, social and natural sciences are connected during the knowledge-forming process, it is important to discuss how each obtains facts or data and who are the experts delivering this knowledge. Subsequent to these discussions, it will become clearer why we suggest technology will need to link both forms of knowledge (*see* Chapter VIII). By knowing the strengths and weaknesses of both knowledge-forming processes, it is possible to form a more robust framework capable of simultaneously addressing climate change problems that play out at the larger landscape scale but which are also driven by local scale processes.

One of the common weaknesses of the Western science knowledge-forming process (generally economically derived) is that data are generated from a few sites but then extrapolated across the landscapes using models [52]. This makes this science knowledge decontextualized from the land. Western science builds on theory that is based on earlier research findings but this information is frequently derived from just a few core research sites, so it is fragmented or disjunct knowledge. So Western science focuses on theory despite the theory being formed from this fragmented information. It is common for graduate students to be asked *"What is your theory?"* The other question a graduate student has to address is *"What new insights did you develop from doing your research?"* If you cannot articulate a theory to support your research, the student's research is not considered to be credible and accepted by the scientific community. This gives one the perception that there is a body of work that supports the hypotheses being tested by a graduate student that will result in new knowledge being formed. Being able to articulate a theory suggests that the foundations are strong and have been repeatedly proven by researchers in different parts of the world.

Thus, the Western science knowledge approach is less able to inform which solution(s) to implement at the local scale since it is not designed to address problems occurring at this scale. Another issue is that the Western science knowledge approach is mostly based on information from the natural sciences. To make the Western science framework holistic is not just a process of merging "society" with the natural sciences using Western social science practices. The Western social sciences tend to focus on the political, social, and economic indicators of a group of people but are decontextualized from a holistic view of people's behavior or their "deep" culture. The Western world view has to move beyond the lens where the Western world ideas are superior to those of traditional societies. We need to understand how both forms of "knowing" differ from one another before any attempts are made to identify their potential interconnectivities or how to make Western science holistic.

This book is not saying that Native people and Western science ways of forming knowledge have to be integrated. In fact, we would not encourage such an approach. It would face the same challenges as experienced by the US National Science Foundation's Coupled Socio-Ecological systems. Researchers have very different definitions of what coupling social and natural sciences really means. Since we live in such highly altered and fragmented lands today, it is important to include knowledge from both ways of forming knowledge — from the local context and culture as well as from the landscape where climate change impacts are being recorded. Further, we are not talking about an amalgamation of knowledge by combining both forms of knowledge. This would make the assessment process linear. Western science cannot take, or select, pieces of what Indigenous knowledge offers, or take what a scientist likes because it is their field of specialty. This seems to be the approach that people are pursuing today because it appears to be logical and doesn't require the scientist to understand another form of knowledge. To really move beyond this approach will take a revolution in how youth and students learn science knowledge-forming processes (*see* Chapter VII and VIII).

A Native people's science approach recognizes that there is potentially a "spider web" of knowledge (*see* Section 1.2). Native people are the ultimate interconnectors in nature and recognize that every plant or animal is dependent (either directly or indirectly) on each other to survive. This means that Native people are a "tribe" and not an individual who perceives that they, as an individual, will "save the world" through the knowledge they produce.

4.1 Knowledge-forming Processes: Western Science ≠ Indigenous Ways of Knowing

It helps to discuss the practices and frameworks used by Western science approaches to form and test knowledge because it begins to highlight the challenges they face to form holistic knowledge as practiced by Indigenous people. The Western science framework is a methodical approach to research and to organize our understanding of natural processes. An issue with the Western science framework is that it is very disciplinary focused. It is not holistic and struggles with how to link information from both the natural and the social sciences. This makes any assessment decontextualized from society and the other natural science disciplines.

In the Western science knowledge-forming process, the problem formulation process is fundamentally different between the social and natural sciences. This is one of the reasons it has been difficult to couple both forms of knowledge. Ecologically-trained scientists studying nature formulate testable hypotheses that determine the structure or design of each research project. In contrast, social scientists prefer to ask larger-scale questions that are testable using different statistical tools to focus their research. This makes it difficult to integrate information between both types of knowledge. Commonly they are temporally and spatially disconnected from one another.

"Western-world" trained natural scientists use frameworks similar to one shown in Fig. 4.1 to test and form knowledge. It is a very linear process for developing hypotheses and testing for significant relationships. It appears to be a circular process but in fact it is very linear. Natural scientists utilize a scientific method that is commonly accepted as being a rigorous and credible approach to develop knowledge. Historically, the tools were quantitative and designed to test causal relationships in pre-selected field sites. In the natural sciences, the scientific method and quantitative tools provide experimental protocols where a phenomenon occurring in nature is explained using controlled and replicated experiments. By developing hypotheses, natural scientists define their research objectives as testing relationships among cause and effect indicators that can be statistically tested. When experiments are not replicable, the scientists conduct another experiment, or they reject their hypothesis.

Except for economics, there are strong biases against the social sciences among many natural scientists. These biases historically were based on their use of qualitative research tools that are subjective and not driven by testable hypotheses. The qualitative tools used by social scientists provide a different type of information and is the causal narrative missing in natural science research. Social scientists prefer asking research questions instead of testing hypotheses. Usually this reflects the need to explore social relationships that play out from the indi-

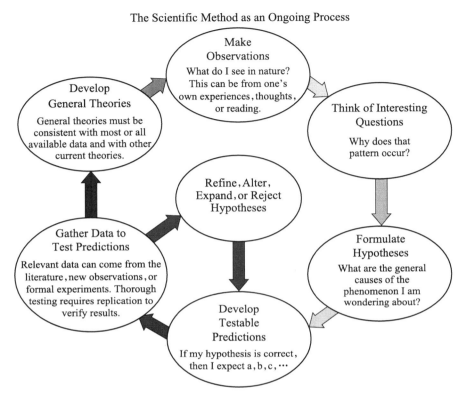

Fig. 4.1 Diagram of the scientific method[1].

vidual to the family and/or institutional levels. Posing questions reflects the fact that there may not be sufficient information to write a narrow hypothesis. The social sciences explore the cause-effect relationships that are harder to decipher. The social scientists are generally not allowed to conduct controlled experiments with human subjects nor is it possible to manipulate all the factors in an experiment. Therefore, they frequently investigate a particular topic from several different angles and triangulate results to tell a story.

Further, social scientists have to conduct research in a way in which the method does not influence the results. For example, a common research tool used in social sciences to gather data incorporates observations or conducts interviews in the community being studied. How these tools are implemented can influence the data obtained (although natural science methods may also incorporate tools that could affect the results). Social scientists also supplement their research by using historical information, e.g., diaries, written documents or reports. The challenge with these historical documents is their fragmentary

---

1 Accessed on January 16, 2017 at
https://commons.wikimedia.org/wiki/File:The_Scientific_Method_as_an_Ongoing_Process.svg

nature and that the writers of the past may not necessarily have described what a scientist may need for analysis and even may have just been wanting to tell another story. Social scientists therefore are decontextualized from the story they are trying to tell and lack the intergenerational transfer of knowledge that is typically found in Native Peoples' stories. Social scientists, however, do explore larger-scale patterns and relationships that cross several spatial and temporal scales that are less frequently studied by natural scientists.

About 30 to 40 years ago, Western trained natural scientists (mostly ecologists and conservation biologists) recognized the need to develop tools and frameworks to include both the social and natural sciences in their research [22]. This shift was driven by the need to use the ecosystem approach to address environmental and resource problems. However, they have not been able to overcome the disciplinary boundaries and preferred research methods as already mentioned. Science knowledge-forming processes need to become universally acceptable to information from different disciplines.

Developing a coupled social and natural science paradigm continues to be an important area of research in the scientific community. It is not uncommon to hear arguments and debates on how to mechanistically link both the social and natural sciences (*see* Section 4.2 for a longer discussion). As long as we don't develop research tools that can couple knowledge from the social and natural sciences, holistic approaches to form knowledge will be out of reach for the Western-trained scientific community. It has been tricky to connect these different knowledge forms because of their dissimilar methodologies and the inherent biases that scientists have for their disciplinary foci.

Today disciplinary-trained scientists are supported by citizen scientists who help expand the temporal and spatial scale of the data base they need for their research. Citizen scientists provide the local context, but the scientists are the ones who design the experimental approach and determine what data to collect. Citizen scientists help scientists to better contextualize their research findings but the lens being used is disciplinary-based. Citizen scientists tend to be very knowledgeable on topics of their own interests so mostly help researchers to collect and deliver specific types of data, e.g., when flowers or trees start blooming, when insects like cicadas emerge, annual bird surveys, etc. They do not concentrate on the theory of a discipline but mostly on data such as the changing patterns of distribution of plants or animals and the timing of their activities. They usually do not have the expensive tools or equipment used by scientists but instead use basic equipment, such as binoculars and write-in-the-rain books to record information, or phones to send information to scientific institutions who collate that information. Therefore, scientific knowledge is held by a group of disciplinary-trained scientists who have years of training on complex issues. Citizen scientists do not have years of education where they have learned to understand how to connect massive data points.

These discussions begin to support the idea that the Western knowledge framework cannot be made holistic by simply linking knowledge derived from the social and the natural sciences. Further, the social sciences do not accurately describe Indigenous knowledge and the beliefs held by a group of people that has been passed down inter-generationally. Indigenous knowledge is not an individual-based knowledge but is a community-or Tribe-based which form or accumulate the knowledge over longer time scales. In contrast, the social sciences document local values and beliefs as an objective outsider who works diligently to not become biased or biasing towards the informants responding to their surveys or interviews. They are not part of the Tribe. So, the quality of their interviews reflects only the facts in which a community is willing to trust with the social scientists.

If both the social and natural sciences follow the scientific method for forming knowledge, they still would not be able to address the elements found in the Medicine Wheel. These negative biases between the social and natural sciences have persisted for more than 30 years and will be difficult to erase by introducing a new framework. These biases are engrained into our scientists during their graduate-training process, that is, when scientists learn what "real" science is. It is not uncommon to hear disparaging words from researchers from each disciplinary group. These debates are further fueled by the designations of each field as being either a "hard" versus "soft" science. These debates were commonly heard by the Vogt's when they were on the faculty at Yale during the 1990s; these debates were at times ferocious and seldom resulted in any consensus. The same debates were occurring before then and are still occurring today in the halls of academic institutions. Our current new scientists and teachers were trained using these disciplinary-bounded approaches to forming knowledge, so their thinking is also bounded and potentially will handicap future scientists they train. Even if an individual researcher wants to conduct holistic research, the universities do not mostly support interdisciplinary research. Since faculty are compelled to be promoted and get tenure in the university system, there are tremendous barriers to teaching and researching holistic knowledge at universities.

4.2 How Knowledge Frameworks Address Scarcity of Land or Lack of Knowledge

What is credible knowledge and who translates the science used to address environmental problems appear to be a simple question that should have an easy answer. But it doesn't. Another way to understand the differences between Indigenous and Western science knowledge frameworks is to explore how each deal with scarcity of land resources and scarcity of knowledge. Dr. John Gordon (the former Dean of the Yale School of Forests and Environmental Studies) wrote that the Western world science process is better able to make choices when dealing with a scarcity of land resources but not a scarcity of knowledge [13]. In contrast, Native American practices deal equally well with scarcity of land resources and scarcity of knowledge.

Another way of thinking about this is to consider two situations. The first situation addresses the allocation of water between different resource users where the total amount of available water is known and finite. Eventually the amount of water will be inadequate to provide for all resource users. In this situation, water allocation will be driven by the need to provide water to those who have the greatest benefit to the community. The second situation shows a clear scarcity of knowledge that may conclude as a smelly and unsanitary situation. In this example, a prisoner is digging out of a jail but the tunnel he digs leads into the septic system of an outdoor toilet just outside the gates of the prison. The prisoner has no clue how his escape will end. The first situation is easier to resolve since abundant data exists on the amount of water available for distribution. The second situation where there is a lack of knowledge which could lead to a disastrous conclusion. The prisoner had the tools to escape from the prison but lacks the other facts that might have made him dig in another direction.

Native Americans deal with scarcity of knowledge because Indigenous knowledge does not separate information into disciplinary tracks that decontextualizes the problem. Also, Indigenous knowledge does not separate society, or spiritual aspects, from holistic analysis of nature when making decisions. Indigenous knowledge is harder to practice since the local context is important and it is informed by many generations of the tribe who retain memory and knowledge from the past. Each tribe has a different local context and therefore the experts are grounded in the knowledge pertinent to their lands and time. Dr. Mike Marchand said just because you are a "Salmon Tribe" does not mean that you manage salmon in exactly the same way as every other Tribe. The context differs between Tribes. This localization of knowledge also means that each Tribe is able to deal with the scarcity of land resources on their own lands. Also they are less able to deal with impacts that seep through borders and impact how they should manage their lands. For example, each Tribe is also greatly impacted by climate change which has no borders (spatially or temporally), and also they could live on lands highly altered by past land-uses. This is where Western trained scientists are ideally suited to contribute knowledge.

Some questions persist on whether Indigenous knowledge really has anything to contribute to Western ways of knowing. One still hears undergraduate university students posing questions like *"Why do I need to learn about Indigenous knowledge today"*. They have read or heard how the Native American culture was lost after European colonizers arrived in North America more than 500 years ago and the colonizers began the process to "civilize" the people living in the conquered lands [13]. Our response to this is Native American culture is still strong and vibrant. It has much to teach us if we are going to sustainably relate to nature. Native Americans may have lost the surface culture but not their deep culture, e.g., values and beliefs (see Section 3.2). If we believed that Native American deep was really dead, why would we waste our time writing this book?

There is a need to identify information or topical areas that are common to both knowledge-forming frameworks. This is exactly what was done by the Alaska Rural Systemic Initiative in 2005. The Alaska Rural Systemic Initiative summarized what the common ground or convergence of knowledge is between the traditional Native knowledge and Western science frameworks in the Alaska context (see Table 4.1). The results highlight that convergence of knowledge was possible when addressing narrowly defined problems supported by specific disciplinary-held knowledge. Research topics did not emerge that were able to address scarcity of knowledge except within several disciplinary fields. This highlights what has already been written: That blending these two different forms of knowledge should not be the goal to develop holistic knowledge.

We need tools to address the complex environmental and social problems that we face today. In the Western sciences, neither the natural nor social sciences address scarcity of knowledge. We know that only a fraction of the worlds land or water have been studied. So a scarcity of knowledge is REAL and our current science knowledge is inadequate to confidently address an environmental

Table 4.1 Information areas or topics that are found in Traditional Native Knowledge and Western Science Knowledge frameworks (Alaska Rural Systemic Initiative [54]).

Weather forecasting	Terminology/concepts/place names
Animal behavior	Counting systems/measurement/estimation
Navigation skills/star knowledge	Clothing design/insulation
Observation skills	Tools/technology
Pattern recognition	Building design/materials/construction
Seasonal changes/cycles	Transportation systems
Edible plants/diet/nutrition	Genealogy
Food preservation/preparation	Waste disposal
Rules of survival/safety	Fire/heating/cooking
Medicinal plants/medical knowledge	Hunting/fishing/trapping

problem. The normal response to an erupting environmental problem is to use information collected in distant locations to predict how to resolve the problem. This ignores the fact that even at one location, the scale of analysis of an ecosystem produces different results and identifies different trigger points causally linked to a problem [55]. For example, different combinations of local soil or climatic factors explain whether productive capacity is resilient to an increase or decrease in precipitation [52]. Thus, a generic model doesn't adequately include these variations and the researcher will conclude that a wrong variable is driving processes. To counter these problems, many statistical methods and data aggregation approaches have been produced to address the lack of knowledge problem [52, 56, and 57]. Even these tools are still inadequate and do not address the scarcity of knowledge and making holistic decisions. To reach that goal will require collaborating with Indigenous people to learn their process to form knowledge.

In the Western science knowledge framework, the scientific method doesn't help the natural sciences to address the scarcity of knowledge. It becomes clear that it is even more difficult for the social sciences to address a scarcity of knowledge because they can't confidently establish controlled experiments to test different causal factors. Not being able to attribute social causality to a problem makes it difficult to identify trigger points that can be manipulated to effectively resolve a problem. Social triggers are seldom discrete events. This explains why it has been difficult for social scientists to connect its knowledge to the natural science knowledge framework which is based on discrete cause-and-effect relationships.

4.3 The Challenge of Culture for Western Scientists

Starting in the mid-1990s, scientists trained in the Western science knowledge framework formally acknowledged the low efficacy of conservation projects by not including the people who lived on the land. Western trained natural scientists recognized that designating conservation areas for protection would fail if they did not factor in local people who are dependent on the same forests for their survival. This new paradigm shift became codified under the name Integrated Conservation and Development projects (ICDPs) by the international conservationists and their funders [58].

Despite the considerable attention paid to ICDPs, the continuing struggles and scientific debates on how to link information from the social and natural sciences paralyzed the implementation process for ICDPs. The continuing debates between social and natural scientists have not helped. This is further complicated by the need to add local culture and spiritual aspects into the decision process. Most natural scientists do not include spiritual aspects in their research because science typically does not include religion in their framework. However Einstein's writings on religion and science are interesting to consider since he sees no problem in including religion as part of science (see Box 8).

Box 8: Einstein Quotes on Religion and Science

"At first, then, instead of asking what religion is I should prefer to ask what characterizes the aspirations of a person who gives me the impression of being religious: A person who is religiously enlightened appears to me to be one who has, to the best of his ability, liberated himself from the fetters of his selfish desires and is preoccupied with thoughts, feelings, and aspirations to which he clings because of their superpersonal value.

If one conceives of religion and science according to these definitions then a conflict between them appears impossible. For science can only ascertain what is, but not what should be, and outside of its domain value judgments of all kinds remain necessary. Religion, on the other hand, deals only with evaluations of human thought and action: It cannot justifiably speak of facts and relationships between facts. According to this interpretation the well-known conflicts between religion and science in the past must all be ascribed to a misapprehension of the situation which has been described."[1]

These ingrained views held by practitioners of the Western scientific method will be difficult to overcome because of the belief that rigorously testing causal relationships is not possible using qualitative methods and spiritual aspects has no place in the scientific method. This view is held despite the qualitative methods being shown to be effective at detecting broad patterns or trends in people's behavior (see Section 4.4.1). These biases are important for our story since it helps to explain why natural scientists struggle with including culture in their design process to form knowledge related to an environmental problem. It also begins to explain why a different framework is needed to connect science and culture.

The challenges of qualitative approaches to evaluate people's behavior is highlighted from the inaccuracies reported by political pollsters. Barone [59] wrote on political polls,

---

[1] Pew Forum 2009 [40], http://www.pewforum.org/2009/11/05/an-overview-of-religion-and-science-in-the-united-states/

"Particular poll numbers are like daubs of pigment on an Impressionist's canvas, which by themselves don't convey a sense of reality but which, taken together with many others, can give the aesthetically sensitive viewer a more vivid sense of the underlying reality than the most accurate photograph. Technological change may be making polls less scientifically reliable, but reading polls has never been entirely a science; it has also been an art and seems to be getting more so."

The poor ability of pollsters, using social research methods and statistics, to assess which candidate people will vote for demonstrates the challenge in predicting people's behavior or identifying causality of relationships playing out in the human landscape. The social sciences category includes a broad range of disciplines with different approaches to studying human behavior under very different contexts, e.g., anthropology, sociology, political ecology, economics. In contrast, quantitative tools are less suitable tools to research people's emotions, intentions, beliefs and/or culture. Therefore, tools to link disparate disciplines requires scientists to think beyond the disciplinary constraints imposed by their classic training and to stimulate their imaginations to develop new tools. As Albert Einstein's is quoted as saying:

"Imagination is more important than knowledge. For knowledge is limited to all we now know and understand, while imagination embraces the entire world, and all there ever will be to know and understand."

We need imagination to integrate both knowledge streams to be located on a common platform.

Economics is embedded in the social sciences and does use quantitative approaches to allocate scarce natural resources. Economics is a branch of the social sciences that is respected by most scientists. This explains why the Natural Capital approach has been so readily adopted by Western scientists and policy makers. It derives the economic benefits from exploiting ecosystem services and then compares this to the costs for mitigating or restoring the exploited natural capital (Box 9). However, this tool still does not factor in culture or art since that knowledge cannot be financially described. It does address the scarcity of resources when it becomes scarce in one location and uses economic tools to substitute or consume resources from other locations. Therefore, the impacts of scarcity at a local level is never addressed nor are the environmental impacts of resource collection.

In contrast, Indigenous knowledge-forming processes successfully integrate Tribal economies with their culture and traditions while maintaining and exercising their sovereignty. This means that there are many insights for how Native people make decisions that can be understood by Western societies if they are

> **Box 9: Why We Need to Think about Natural Capital**
>
> "With financial capital, when we spend too much we run up debt, which if left unchecked can eventually result in bankruptcy. With Natural Capital, when we draw down too much stock from our natural environment we also run up a debt which needs to be paid back, for example by replanting clear-cut forests, or allowing aquifers to replenish themselves after we have abstracted water. If we keep drawing down stocks of Natural Capital without allowing or encouraging nature to recover, we run the risk of local, regional or even global ecosystem collapse.
>
> Poorly managed Natural Capital therefore becomes not only an ecological liability, but a social and economic liability too. Working against nature by overexploiting Natural Capital can be catastrophic not just in terms of biodiversity loss, but also catastrophic for humans as ecosystem productivity and resilience decline over time and some regions become more prone to extreme events such as floods and droughts. Ultimately, this makes it more difficult for human communities to sustain themselves, particularly in already stressed ecosystems, potentially leading to starvation, conflict over resource scarcity and displacement of populations."[1]

interested in environmental justice when allocating scarce resources. Miller [60] said it well when he described the characteristics of Tribal economies: *"Modern day Indian people and nations were not and are not opposed to economic activities, private property rights, and entrepreneurship"* and entrepreneurship is used to preserve and perpetuate *"American Indian tribal cultures and traditions in the face of new economic activities"*. Tribal business ventures and practices can guide how non-Tribal business leaders should allocate scarce resources using market values while continuing to support local cultures and practicing environmental and social justice.

Indigenous people do not follow either of the approaches used by the natural or the social sciences. Instead they have their own unique approaches to communicate knowledge such as telling stories to explain phenomenon occurring in nature and their society (*see* Chapter VIII). It is worth exploring in more detail how Native people form traditional knowledge. The next section (see Section 4.4) will explore what is traditional knowledge in greater detail and how knowledge is formed by Indigenous people. The juxtaposition of the Western and Traditional knowledge will be further explored in Section 4.5.

---

[1] Accessed on January 15, 2017 at http://naturalcapitalforum.com/about/

4.4 Traditional Knowledge: Native Ways of Knowing*

Dr. Nancy Maryboy and Dr. David Begay.
Source: Nancy Maryboy.

A few definitions are important to include so the reader recognizes how they are being used.

Traditional: We take this to mean the primarily oral process of handing down or transmitting native knowledge, ways of living and ways of knowing through successive generations.

Knowledge: In the way of nouns, we take "knowledge" to mean that which is known, or the condition of arriving at an awareness and understanding. This could refer to aboriginal knowledge, the fact or condition of knowing gained through life experience by either direct or indirect means.

Inherent *in* this native *definition* is acquisition of skills and life applications. In native communities one learns to apply one's knowledge. One doesn't just learn something solely to know it; it *is* then only theoretical. This often runs contrary to the expressed desires of many well-meaning non-Native students who come to a traditional community wanting to learn traditional knowledge, with the justification that they just want to know *it* for themselves, for the sake of knowledge. Traditional people are generally not interested in providing time to educate these people. Conversely, in many communities, when Native students go off the reservation for an education, they are expected to bring knowledge home that can be directly and indirectly applied to the well-being of the community.

Native: When we use "native" we are referring to an autochthonous characteristic of ancestral relationship with the earth, and/or evolving or emerging from

---

* By Dr. Nancy Maryboy and Dr. David Begay.

the earth, as well as a characteristic of inborn awareness. By this we mean born into communal aboriginal knowledge, more in the sense of inclusion rather than exclusion. Characteristics of indigenous-ness, meaning that which is traditional and has not been introduced from outside, are also a part of the significance of the term "native." At the same time, within a recent historical and a contemporary context, it is only realistic to include the adoption and adaptation of outside influences selected to fit unique cultural characteristics and needs.

Ways: By the term "ways" we mean pathways, more like the numerous paths of the sun's rays, radiating outwards; numerous means of knowing, multiple and holistic, not just one way. Perhaps the term that Navajos use, *bee*, by means of, is a better term to use. By means of such and such, knowing occurs. Although we are defining ways and knowing separately for the purpose of clarification, nevertheless, holistically speaking, they coexist together as "ways of knowing".

These additional ways of knowing include the spiritual: Dreams, visions, vision quests, sweat lodge, feelings, intuition, stories, moral ethics, and others. Interconnected with the spiritual, one can include cognitive ways of knowing, such as psychological, parapsychological, mental insight, emotional intelligence, transmitted wisdom, values and principles of elders, aesthetics, and logic.

Language is also critically important as a means of learning, specifically through the native language but also now through the English language, as the reality is that many of the younger generation develop with English as their first language (*see* Section 3.8).

Knowing is consciousness. Native knowing is Native consciousness as a holistic, integral way of living. Native ways of knowing are Native ways of being human as manifested through living. This would include the concept of reciprocity, within a context of dynamic balance and harmony. This would also include the social concept of kinship and communal relationship, within a pragmatic ecological consciousness.

Knowing is verby — an ongoing process within Native languages — not a static, fixed nounlike concept. Deep knowing is grounded in the Native language. Because the Native language is expressing a native consciousness which is holistic, both direct and indirect means are utilized. Means and knowing are intrinsically connected in the native language. In Navajo one could say, "*Dine k'ehgo bee hoi bahózin*": by means of the Diné consciousness and ways of knowing, knowing is (or one could translate it as knowing <u>occurs</u>, to be more in accordance with English thinking). Actually the closest translation would be knowing <u>is-ing</u>. not a static state-of-being but rather an ongoing state-of-being or an ongoing flux-of-being.

Each generation experiences different development of life and may utilize somewhat different ways of knowing. Each generation encounters different experiences and responds in unique ways; these developments are widely varied according to location and heritage. Yup'ik life in Alaska followed the same patterns for thousands of years, according to traditional philosopher Angayuqaq Oscar Kawagley 2006 [39], but the last four or five generations have experienced profound change due to the onslaught of western institutions and values. Every generation of Navajos, on the other hand, has experienced change during the past 400 or 500 years.

Today, many middle-aged Navajo have parents who are fluent monolingual Navajo speakers. They themselves are largely bilingual, primarily grounded in traditional Diné consciousness. Their children, on the other hand, are more monolingual in English than truly bilingual. Their primary consciousness is through English; however, many of them can and do intuitively connect with the Native ways of thinking through their associations with their bilingual parents and, in some cases, their Navajo-speaking grandparents. The grandchildren of these middle-aged bilinguals will undoubtedly adapt and assimilate more fully into western society. Each generation seems to experience an acceleration of assimilation and acculturation due to technological innovations, such as the mobility afforded by the automobile and airplane, as well as the mass-media communication infrastructure of television, computer, and the global Internet network.

4.4.1 How Indigenous People Form Knowledge*

Holistic Diné (Navajo) epistemology is significantly different from the Cartesian-based system of categorizing knowledge. Our work, of necessity, deals with multiple disciplines, multiple but at the same time highly interconnected. Navajo cosmology is an awareness that the world exists within a universe of relationships and processes. Navajo traditional knowledge is holistic, integral and all-inclusive. Knowledge holders do not divide their knowledge into segmented areas. They have to know how the real world works, and they learn though careful observation, intuition, spiritual means and oral narratives.

Traditional Holistic Knowledge may not fall within the confines of strict university disciplines. They include but are not limited to the following areas: Indigenous Philosophy, Oriental Philosophy, Global Indigenous Cosmology, American Indian Cosmology, Dine Cosmology and Metaphysics from the Dine perspective, Dine subjects from the non-native perspective, Linguistics, Indigenous Education, Organization Research and Strategic Planning, Native Healing Arts, Ecology, American Indian Policy, Anthropology, History, Psychology, Theology, Pre-Greek Thought, Greek Philosophy, Biology, Geophysics, Astronomy, Western Medicine, Mathematics, Relativity Theory, Quantum Physics, and Postmodern Thought. We wish to emphasize that native ways of knowing are truly

---

* By Dr. Nancy Maryboy and Dr. David Begay.

holistic and integral. Our integrity demands that we include all of them, even though we may not go into all of them in the same degree of detail. The deeper that we go into the Indigenous language, the more a holistic approach is demanded. When one tries to categorize Native knowledge through a western lens, it becomes fragmented and literal. The multiple interconnected levels of knowledge become trivialized, marginalized and lose their rich interplay of significance. They become what earlier anthropologists have loosely termed "folk tradition" existing in the arbitrary realm of "myth".

If we were to take our research to many of the respected universities in the country, a first step might be to place us in one of the existing disciplines which, at face value, might correlate with our work. These disciplines could include departments of Religion, Psychology, Philosophy, Ecology, Astronomy, Anthropology or Native American Studies. In order to further our work within each of these disciplines, we would be forced to conform to departmental requirements and institutionally controlled criteria. However, these requirements and criteria are based on Western methodology, classic scientific reductionism and existing Cartesian pedagogy. We could most probably work within this system, and in fact a majority of Native Americans do, but the system itself is not compatible with, nor totally useful to, the holistic Native mind.

Many Native Americans find it difficult to work within the structure of the university system in terms of their own professional and holistic needs. They generally have very little choice in selecting a program within the philosophy and content of course offerings or department requirements. There is little or no meaningful connection between their education and their community needs. For example, suppose a Native student enters a university and is routed to a department that seems to serve her needs, perhaps philosophy. Most likely she will learn a lot about Western philosophy. However, somewhere in the next two or three years she may begin to question where her own Tribal philosophy fits into the overall scheme of Western philosophy. Depending on the flexibility and innovation of the department and university, she may get support for her request to further research Native philosophies. Or she may not. By this time her training may be almost complete and there is very little support to research Native American philosophy.

Viola Cordova, Jicarilla Apache, provides us with an example from the University of New Mexico [61]. She proposed a dissertation for the Department of Philosophy, comparing the concept of monism of Bernard Spinoza, a 17th century Dutch philosopher, with the possibility of exploring the Navajo world as monistic compared to the Navajo concept of a cosmic oneness of all things. She found that her dissertation proposal was considered a heresy within the department, since almost none of the professors considered Navajo thinking to be on a par with Western philosophy; in fact they felt insulted that she would even consider the proposal and condescendingly told her that Native American thinking could not be categorized under the generic term philosophy. She in turn felt equally insulted that the professors would not consider Native thinking as equal to Western philosophy. *"My 'audacity',"* explains Cordova, *"in using a concept explained by a highly respected (but little read - 'too complex') Western thinker as a comparison with a so-called 'primitive' people was almost too much to bear for many."* One professor exclaimed (vociferously, I might add) "You have mistaken Spinoza for a *primitive!*" My reply, which did little to mollify him, was that *"No, I had not."* And, furthermore, I had not even made the *error* of 'mistaking' the Navajo for a primitive [61].

Rupert Ross, a lawyer who has worked for many years with Canadian Natives in the Northeast areas of Canada, writes of his growing awareness that Native people have ways of thinking and living that are in fact very complex and distinct from the Euro-Canadian culture he was familiar with. He wrote [62],

> *"Our two cultures,"* he explains, *"are, in my view, separated by an immense gulf, one which the Euro Canadian culture has never recognized, much less tried to explore and accommodate. ... I must have carried an assumption with me into the north, the assumption that Indians were probably just 'primitive versions' of us, a people who needed only to 'catch up' to escape the poverty and despair which afflicts far too many of their communities."* ... *"That assumption is both false and dangerous,"* continues Ross. *"We would never carry it into China or any other obviously foreign place. Instead, we would approach the people of those lands with an expectation of profound difference and a sincere determination to learn and accommodate. To date, that has not been the predominant approach of Canadians to Native people on this continent."*

The Indigenous experience in the United States is essentially the same. The international border between Canada and the United States makes no difference in regards to the historical treatment of Tribal peoples. *"What has not, to date, been mutual,"* adds Ross, *"is accommodation between the two cultures (Native and Euro-Canadian). That has been a one-way street, with all the concessions coming from Native people."* [62]

4.4.2 Indigenous Ecological and Spiritual Consciousness[*]

Most Native Americans experience their consciousness in relation to their land and environment, what we would call an ecological consciousness. For instance,

---

[*] By Dr. Nancy Maryboy and Dr. David Begay.

Navajo identify themselves in terms of the Four Sacred Mountains. Their psychological being and Tribal identity connect to specific geophysical places. Their very consciousness is deeply rooted into this specific Southwestern locale. Their relationship with the land provides the language, the conscious and subconscious mind of the people. Even if a people have experienced a forced removal from what they knew as their homeland, their stories and consciousness still relate to their given place of origin. Many Cherokees, for example, still identify themselves through story, clan and place names of the Southeast, even though they may live far from their origins in Tennessee or North Carolina. The Cherokee Trail of Tears and the Navajo Long Walk were not just about devastation due to a physical removal from the land: The removal was a ripping away from one's very essence, similar to ripping a baby away from the breast of a mother, or ripping a tree out of the ground.

It is extremely difficult to discuss ecological consciousness in isolation. It can be more properly discussed in terms of spiritual consciousness or spiritual awareness of the interconnections of all life. Today if one were to ask a Navajo how to say "land," he would probably say "*shikeyah*". That could be loosely translated as "my land". But if one were to look more closely at "*shikeyah*", one would find a more spiritual meaning of relationship connoting an infant-mother relationship of oneself to the earth and cosmos. The consciousness of the traditional Native American is profoundly spiritual, perhaps best expressed through the Navajo phrase, *bee yajoltiih*, literally meaning "with-speaking". However, in a broader spiritual and cultural context or consciousness it could mean a process of relationship with which you speak by means of spiritual consciousness, this relationship which provides your speech process. The concept articulated in "*bee*" (by means of or with) is the connector that relates you to the entire cosmos through your spiritual consciousness.

In addition to ecological and spiritual consciousness, there exists a kind of intergenerational conscience that provides ontological and epistemological guidance. Native Americans have profound respect for their ancestors, in large part because the future rests on the present and the present rests on the past, and even vice versa. Many Tribes think in terms of seven generations, into the past and into the future: Seven generations ago when these traditional beliefs were lived in their entirety, the prayers that were said and the plans that were made: That's who we are. Navajos use the term *iina* (life) *bitsi silei* (that which precedes), which recognizes the importance of ancestral foundation in providing one's unique essence and being. The responsibility and obligations that many Native Americans feel go back to their ancestors; and, at the same time, what is being determined today will certainly affect generations to come. This is being written with great emphasis, as it varies considerably from non-Native thinking.

Native people today have great intuitive and spiritual respect for the tremendous strength and wisdom of their ancestors. Implicit in this respect is recogni-

Chapter IV Western Science ≠ Indigenous Forms of Knowledge — 155

tion of the difficulty of survival, against all odds. The pride Native people feel in the survival of their ancestors is intense. Cherokee people were forcibly marched toward the West in the infamous Trail of Tears. Over one-fourth of the people died in this march. Navajos experienced a similar tragedy with fewer miles but the same brutality and fatal consequences.

4.4.3 Ecological Calendars in Nature Literacy

The Western science-based management approach to ecological calendars is very fixed to a specific period of time. This contrasts Tribal people who look for cues from the land and not fixed dates to determine when to collect or hunt for cultural foods. Tribes have hunted for centuries and did not have a fixed time period when this would occur. Despite this, they were able to maintain viable and balanced populations of wildlife because of their holistic management practices [63]. The key words here are that wildlife management is a holistic approach to managing animals and not designed to maximize recreational hunting.

In Washington State, the legislature mandates that the focus of hunting should be to maximize public recreational hunting opportunities for all of Washington's citizens, whether they are young or old, able or disabled. The Washington State legislative mandates who can hunt and when hunting will occur:

> "As mandated by the Washington State Legislature (RCW 77.04.012), "The Department shall preserve, protect, perpetuate, and manage the wildlife"; "The Department shall conserve the wildlife in a manner that does not impair the resource"; and "The commission shall attempt to maximize the public recreational hunting opportunities of all citizens, including juvenile, disabled, and senior citizens." It is this mandate that sets the overall policy and direction for managing hunted wildlife. Hunters and hunting will continue to play a significant role in the conservation and management of Washington's wildlife." (WDFW 2014) [64]

Western management agencies have published dates during which hunting is allowed and these are usually at the same time each year, e.g., October 14–31 to hunt black-tailed deer (*see* Table 3.1). The hunting season game plan is based on feedback from the citizens of Washington on what they would like to see over a 6-year period. In the State of Washington,

> "Game Management Plan (GMP) guides the Washington State Department of Fish and Wildlife's management of hunted wildlife for six

year timeframes. The focus is on the scientific management of game populations, harvest management, and other significant factors affecting game populations." [31]

They also have special hunting seasons for some animals. There exist hunting seasons for deer, elk, cougar, black bear, goat, moose, bighorn sheep as well as bobcat, coyote, fox, grouse, crow, raccoon, rabbit, hare and wild turkey [31]. These dates were selected based on past surveys of game population sizes, how many were harvested, and other impacts on the population. The hunting season may be shortened if the population sizes shot by hunters are too low to support the hunting season. These assessments are conducted using a compartmentalized and fractured science knowledge base and responds to current knowledge input to determine when to end a hunting season.

Hunting dates are not based on knowledge that is place-based and it doesn't use a historical memory of how the animal populations might have varied over decadal time scales across a larger landscape. Animal population models in management focus on a population of animals, like deer, and not how the predator populations are doing. Since the top predators can cover large tracts of land, it would require these models to link to large scale movement of animals. Instead they focus on the natural history of the animal. There are discussions of habitat, priority habitats and species, fish/shellfish research and management, hatcheries, living with wildlife, threatened and endangered species and good places for wildlife viewing. These are all relevant and important knowledge that address what the State of Washington has mandated as part of their game management. Hunters have to buy a license and there are concerns on safety issues so the type of firearm or bow hunting is also specified.

The balance between conservation and hunting becomes politically driven when there is a powerful stakeholder group. One only has to look at wolf management to see how conservation is not practiced when conflicts occur. WDFW [31] produced a map that shows the location of most of the wolf packs in Washington as of December 2016 (Fig. 4.2).

The map shows that most of the wolf packs were found in Eastern Washington. Wolves are listed as endangered by the State of Washington. Interesting many of the wolf packs in Eastern Washington State are found around the Confederated Tribes of the Colville; at least 6 of the 20 packs are found on the edge or living in the Indian reservation. There are continuous conflicts between livestock and wolves. So even though wolves are endangered, the Sherman and Smackout wolf packs were lethally removed in 2017 because of these conflicts. The pack had killed or preyed on cattle. According to WDFW [31] last "year's survey documented 115 wolves, 20 packs, and 10 breeding pairs." in Washington.

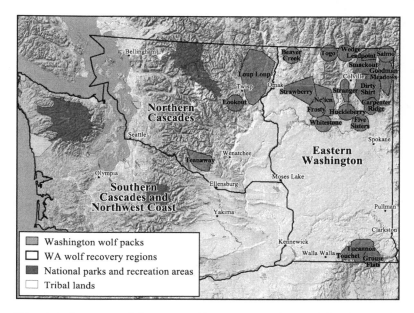

Fig. 4.2 Location of the gray wolf packs in the State of Washington[1].

4.5 Juxtaposition of Western and Traditional Knowledge*

Just as we feel one cannot generalize about the diversity of Native people, we know that Native people can be just as guilty of over-generalization of the Western society. It is a fact that to native eyes, to quote Oscar Kawagley [39], "*Western society often appears as a monolithic entity, despite the fact that it is made up of many diverse institutions and divergent points of view.*" Through our lifelong experience and research we have become aware of the great diversity among Native peoples as well as among Western societies. It is not our intention to create or perpetuate stereotypes in this section. We are only providing a juxtaposition in order to delineate distinctions between Western and traditional Native societies.

One of the least understood aspects of cross-cultural understanding, according to Carlos Cordero, Mayan educator and former President of D.Q. University, currently president of the Alaska Native Institute of Higher Learning, is that an underlying characteristic of native culture "*was a human consciousness that is very different from that of the European*". Cordero [65] goes on to say:

---

1 WDFW 2018; https://wdfw.wa.gov/conservation/gray_wolf/
* By Dr. Nancy Maryboy and Dr. David Begay.

> "The consciousness of the people of North, Central, and South America was based on a definition of being human that was not anthropocentric or hierarchical and that demanded a dynamic relationship vis-a-vis other humans and the environment. By environment I mean not only the planet, but the universe as well. This definition of being human called for that understanding of knowledge as both human and nonhuman."

The fundamental knowledge base of Indigenous people, on the other hand, is holistic, intuitive and spiritual, interrelated with all those areas which we articulated earlier in the section on ways of knowing traditional knowledge. Native thinking is so holistic that few if any native languages have separate words for science, art or religion.

In many ways the differences between Western classic science and traditional Native ways of knowing are incompatible. Western civilization has disassociated from its roots, becoming the product of a schismogenesis, having deviated from a holistic world view to a mechanistic Cartesian-Newtonian mind. Few Americans today have any meaningful conception of the holistic roots of their tradition. They experience little, if any, confusion or resultant inferiority in following the Cartesian precepts of their education. Native people with a high degree of Western education, with limited traditional influences, may have few problems if they use primarily the Cartesian mind, albeit they may intuitively feel some sense of confusion and inferiority as they reflect on the differences between the teachings of their Western education and their Native heritage.

Western society has evolved from a perspective that is completely anthropocentric, and almost entirely reductionistic, or divisionistic. In other words, there exists not only a tendency to separate into disciplines and specialties, but it goes the additional step of viewing this process as the only appropriate and legitimate way of coming to knowledge. The tendency toward separation from the ancient holistic roots of the western mind, and the mental fragmentation that is characteristic of the modern western mind, can been termed dissociative genesis, schismogenesis, indicating dissociation from one's origins with a consequent mental bifurcation.

It is consciousness itself that may be transforming, however, we feel, it is the language that expresses consciousness. For 500 years of western colonization, holistic Native languages have been eroded and marginalized. Western noun-based languages, are simply not equipped to discuss holism, quanta, participatory relationships and processes of flux.

The Native American nations have maintained unbroken practices of earth-based spirituality for more than 20,000 years. Since contact with the European invaders, beginning 500 years ago, Native peoples' spiritual practices have often been targeted for eradication so that their cultural fabric might unravel and the

Indians might then become properly atomized team players in the modern world [66].

In the Navajo language holistic concepts are intrinsically embedded in Navajo phrases. When translated into English they may appear as seemingly redundant. In order to retain an all encompassing holistic meaning, we have added the embedded meanings through the context of Navajo consciousness as we translate these native phrases into English. This phrasing may seem sometimes somewhat awkward in relation to classic English grammar and common usage. However, we intentionally chose to write in this somewhat repetitive vein in English in order to capture the traditional holistic consciousness wherever relevant. The fact that traditional people speak within this holistic context is evidence of their realization of deep spiritual consciousness.

Traditionally, Indigenous societies had models which were an integral part of their cultural and spiritual way of life. These were seldom codified or written in the manner of Western societies. Diné traditionalists may assert that Sandpainting, *ikaah*, is a highly sophisticated form of holistic writing which has been transmitted through the centuries. Whether classified as writing or oral narrative, *ikaah* nevertheless served as a foundation for a way of living.

Patterns of coming and going are also implied in the Navajo language to a large extent, far more than in the English language. Whereas the English language is centered around concepts of being, largely static and nomina-centric, the Navajo language is centered around concepts of movement, such as going and returning, largely, to coin a phrase, verba-centric. In the Navajo these verbs illustrate the predominance of cyclical thinking.

Navajo language places emphasis on the continuing imperfective tense, meaning incomplete or finished. For example, *naasha*, to walk about, (a word implying an ongoing life process of walking and movement, called a come-and-go verb) expresses simultaneously the process of becoming as well as a state of being. The Navajo language has no present tense precisely comparable to English. The verb *deya*, I start to go, is really the perfective mode in Navajo, however it translates into the present mode in English. "The thought of going is firmly established and, barring complications, will be carried out. It is similar to I'm going tomorrow. It is in the present tense but concerns future action."

In Navajo an awareness is provided between things that are not completely finished and those that are finished, and to add further preciseness and complexity, there is an awareness of the state of the thinking process. This may seem like a subtle difference but it is important. In English one might say, I am walking or I walk. In Navajo one stresses the "I am walking" or "*naasha*" aspect, the "ongoingness" of the process. According to Navajo thinking, a present tense might freeze the movement implied in the walking, thus falsifying a state of reality. The Navajo has a built-in expression of ongoing process, a recognition of the present as part of a larger process [68].

All movement verbs, such as come-and-go verbs, are very precise in the Navajo language, far more precise than in the English language. The verb go-and-return is a good example to illustrate this concept. *Niseya* implies to make a round trip, to go-and-return, a roundabout turn, in a perfective mode. Here the cyclical action is completed. In English one would say "He went to town yesterday", but one might not indicate whether or not he returned. In the Navajo verb *niseya*, the return is built into the verb.

4.6 Who Are Trusted for Their Science Knowledge?

Knowledge is communicated by different and credible experts that translate the science used to identify and develop a solution for a problem. Indigenous knowledge is translated by elders who have intergenerational and multi-disciplinary knowledge that includes culture/art [23]. Indigenous knowledge is also communicated in many different ways. Prober et al. [69] wrote about the difficulty for sharing Aboriginal peoples' knowledge with environmental managers since

> *"Knowledge is derived from experience and shared from person to person, encoded in language and artworks (oral, body, and diagrammatic), and not traditionally recorded in written form."*

In contrast, Western science is translated to the public by disciplinarily researchers which frequently means that you do not look for connectivity among disciplines or include culture/art to assess complex environmental problems. Typically, the Western world scientist is trained in their disciplinary strength when they attend college. Most of the future scientists are not immersed in the sciences as a young kid so they mostly learn science knowledge when they matriculate into a university. This means that scientists are less exposed to linking multiple topics but specialize in the knowledge area that they select as the focus area of study within the university.

In the Western science realm, there are always a few outliers where a young kid becomes enraptured with ants or dinosaur bones and later become scientific leaders whose ideas resonate with many people around the world. These unique individuals are capable of discussing not only their topic of research but able to effectively communicate science to global societies.

EO Wilson is a good example of someone as a youth who was inspired to study ants at an early age. He became a prolific writer of books and papers on topics ranging from taxonomy to humanisms (e.g., ants, island biogeography to biophilia and consilience). He has given talks all over the world and people listen when he speaks. These scientists become important spokespeople trusted by society to communicate society's impacts on nature. They are able to talk to the common person with such enthusiasm for their topic that people do not question whether they are trying to convince them to support something that would only benefit the seller. These individuals are enraptured by science from an early age. They tend to become the scientific leaders that we all know about. Their names roll easily from our tongues and are known by the public. This is not the norm.

Mostly the Western world approach to environmental problem-solving worries about getting the science right!! When you have this approach, you are dependent upon experts who learn their knowledge in graduate school and become the credible experts. For Indigenous people, it is not just getting the science right!! This means that you do not just engage with the experts but those who have learned intergenerational knowledge that informs on identifying a problem. In the Indigenous communities, youth with leadership potential were identified and were trained by elders for future leadership roles. This happened to Dr. Mike who was a kid when elders approached him to begin training him for future leadership roles (*see* Section 2.2).

4.6.1 How Citizens of the Western World Get Their FACTS

Western world scientists are seldom effective communicators of their research since they are supposed to be dispassionate deliverers of knowledge and not influence how the risks and benefits are applied by the decision-maker. When decision-makers do not ethically utilize science or cherry pick the science needed to support their decision, scientists do not engage in discrediting these decision-makers. When scientists cross the border to discredit the misuse of science, their ability to communicate credible science to the general public becomes more problematic. Scientists are very aware that for every science fact, there are exceptions to the rule so it is easy to discredit science knowledge. This is, of course, not a holistic approach to communicating science knowledge that is essential for addressing critical environmental and resource problems we face today.

Further, their scarcity of knowledge makes these exceptions to the rule difficult to communicate and to counter a knowledge that differs from that communicated by a scientist. Also, scarcity of knowledge is difficult for many scientists to communicate since they are not trained to put their research in a larger context. The norm for a Western scientist is to only talk about those variables that 95% of the time are statistically shown to explain changes in how a problem is expressed. Also, most scientists do not have the luxury of being trained in an institution where environmental problems are studied holistically. It is cheaper for organizations to fund research that is narrowly designed to address a specific problem. It is too expensive to conduct holistic research even though the problems scientists are asked to address are complex and need to be researched using an ecosystem focus. It also does not help that when a problem arises, its driving factors may lie outside the researcher's expertise, but they still need to conduct detective work to identify the interconnected causal factors that produced the problem.

Further how culture is connected to science also varies depending on whose lens is being used to translate this knowledge. If this is being done by a tourist, it will reflect the aesthetics or visual characterization of culture and not the cultural norms that determine how a group of people behave.

Stager [4] wrote how Western science knowledge missed the insect Armageddon that "amateur naturalists" reported in Germany. Western science has lost its observers or "amateur naturalists" who are field naturalists making observations on changes occurring in nature over longer time periods. The Vogts remember a Yale Professor — Dr. Tom Siccama — who was one of the best naturalists in the world and knew every nook and corner of the entire New England landscape. He was keenly interested in changes occurring in the field and recording his observations. He kept long-term records of everything he observed from changes in tides, flowering of trees to even how long he waited behind a red light when driving his car. His observations provided a local foundation for the research that occurred at the Hubbard Brook Experimental Forest in New Hampshire.

This type of science observation is frowned on in today's technology driven science research. Stager [4] wrote,

> "... A global declines of field naturalists who study these phenomena. Most scientists today live in cities and have little direct experience with wild plants and animals, and most biology textbooks now focus more on molecules, cells and internal anatomy than on the diversity and habits of species. It has even become fashionable among some educators to belittle the teaching of natural history and scientific facts that can be "regurgitated" on tests in favor of theoretical concepts." This attitude may work for armchair physics or mathematics, but it isn't enough for understanding complex organisms and ecosystems in the real world. Computer models and equations are of little use without details from the field to test them against.... In 1996, an editorial in Conservation Biology warned that "Naturalists are dying off." and asked: "Will the next generation of conservation biologists be nothing but a bunch of computer nerds with no firsthand knowledge of natural history?"

Few people have Tom's and the amateur naturalists' keen sense of the changes occurring on the land [4]. We need more people to become keen observers of the

Chapter IV Western Science ≠ Indigenous Forms of Knowledge — 163

land similar to what Native people and many naturalists already practice. Tom Siccama did this with hundreds of Yale FES students who were fundamentally changed by interacting with and learning from him.

4.6.2 How Native Americans Get Western Science FACTS

Western science practices are dynamic on ten year cycles and probably stimulated by new people becoming in charge of research programs. The paradigms and theory change dramatically on a decadal time scale and then need to be adopted by research managers or their management will not be scientifically credible in the research community. Therefore, major changes have to occur in the scientific facts that support management practices every 10 years. The old science is thrown out the window and new science principles need to be used to design and revise the core management practices. These decadal dynamic science paradigm shifts is partly due to research being decontextualized for the local lands and research needs to be conducted to support the new ideas.

Dr. Mike Marchand summarized his experience with how government scientific forest management practices would change and where the old science was no longer valid. He describes,

> "As a child in the 1960s, I would go to Tribal membership meetings held in the local community building on the Reservation. I remember a BIA Forester explaining forest management. He said they were utilizing the latest and best available science to manage our large forest. He said they were utilizing a selective cut policy. In that era Smokey the Bear was also a popular icon too. Put out forest fires is top priority. The community elders listened to the BIA, but some were skeptical of anything the BIA would say.
>
> Then, later in time, the mid 1980s, the BIA came back to the same communities and the BIA Forest experts were explaining their new and most modern forest management practices. The new Forester matter-of-factly explained that the former forestry ideas were wrong. Turns out the selective cut policy caused some problems with some of the forest. The understory was then very shady and this favored certain trees such as Douglas fir. So over time, some of the large Ponderosa Pine forest were replaced by Douglas fir forests, since shady conditions were not ideal for the baby pine trees.
>
> The fir forest was susceptible to some problems such as bug infestations, they were more prone to forest fires, and there was a devaluation

> *since the Ponderosa Pine was worth more than the firs. Also, years of Smokey the Bear style fire control had resulted in the accumulation of fire fuels in the forest. So, the efforts were well intentioned but the results were disastrous.*
>
> *Then in more recent years, the climate has changed also. Warmer weather with less snow pack. All these factors have resulted in massive forest fires the likes of which have seldom been seen before."*

Further, these new research paradigms are still being implemented locally even though the science is not place-based. This was quite apparent in the assessment conducted by the U.W. Climate Impacts Group for the Stillaguamish Tribe. The vulnerability assessment conducted for the Stillaquamish Tribe of Indians reflects this lack of a place-based knowledge informing vulnerability of multiple species to climate change [70]. The focus of this assessment was to explore the climate vulnerability of priority species living on Tribal lands so the Stillaguamish Tribe of Indians could begin to develop their climate adaptation planning efforts and to determine what research to conduct. They describe their assessment approach:

> *"... Assessed as many of these species and habitats as possible, according to their level of priority, data availability, and the time available for the assessment. For species for which adequate data were available, we completed a quantitative assessment of climate vulnerability using NatureServe's Climate Change Vulnerability Index (CCVI). For habitats and species lacking sufficient data for a CCVI analysis, we completed a qualitative assessment of climate vulnerability. We chose the CCVI for our assessment because it is freely available, relatively transparent and replicable, and widely used. These qualities should help facilitate future updates of the assessment as additional information becomes available, as well as comparison of results to other assessments based on the CCVI. The CCVI also highlights the species sensitivities that contribute to vulnerability, offering critical information to guide future adaptation efforts."*

There is value in conducting an assessment to explore and to prioritize vulnerable species but it is difficult to use it as a planning tool. How to apply a generalized vulnerability model at the place-based scale is also challenging. This type of knowledge is categorical, uses universal characteristics of the natural history of many species, and does not link the species to the scale at which impacts may be occurring. Some species may be vulnerable because of a set of conditions that are not found for the same species at a different location [70]. The seven categories of vulnerability are difficult to compare because it is not a holistic assessment that may include different temporal and spatial scales contributing to the vulnerability category. If the Tribe is able to take this knowledge and place it in the context of their long-term knowledge of the land and water, this information is very useful. It can keep or identify priority areas needing management focus. The Tribe can then locally contextualize this knowledge by putting it in the context of their long-term knowledge of the area.

4.7 Women's Role in Passing Indigenous Knowledge Inter-Generationally: Interview of JD Tovey*

JD Tovey (top left), Dr. Phil Fawcett (top right), Alexa Schreier (bottom left), Dr. Kristiina Vogt (bottom right).
Source: JD Tovey, Phil Fawcett, Alexa Schreier and Kristiina Vogt.

Many Tribal members have mentioned that women are important in maintaining and transferring Indigenous knowledge to youth. So this was a discussion that we had with JD Tovey to get a better perspective on this idea.

---

* By Alexa Schreier, Dr. Phil Fawcett and Dr. Kristiina Vogt.

Kristiina: Can you talk about the role of women in Tribes. When we were in Colville, one of the comments was made that women are the carriers who pass on cultural traditions in the tribe. What are your views on this? Is it important generally or just found in certain groups?

> **JD:** *For us, when you have intergenerational groups, when you go into long houses the women are actively talking. While the men are super quiet and don't talk. There are a lot of secondary languages such as head nods that you see men doing. What is the root of all of this? Women who were traditionally gatherers, they had to talk in order to keep bears away. It was like having a rattle. If you are constantly talking to keep bears or wolves away. It was especially effective if there was a large cacophony of people. There was a purpose behind this. This then builds into why stories were being told. Women recited genealogy. If you have to talk, you have to have something to talk about. There was no world news at that time so there was only so much going on. So, you would talk about the genealogy and your family. You would build those relationships and you would remember those relationships.*
>
> *The flip side was that men ended up having a lot of non-verbal communication because they were hunting. They had to be sneaky and quiet. That is always how I have seen it. It was not a division of purpose and there was a reason why women were talking all the time and men weren't.*

Kristiina: You are saying that you think that it is still happening? Mike suggested that it was still going on.

> **JD:** *Yeah.*

Kristiina: So this is probably why you need reservations? If you are living in Seattle, you are not experiencing this.

> **JD:** *It is not like I think that Western Europe didn't do this. I think all cultures have gone through the same stages. Pre-1492, Tribes didn't go through all these stages and developed differently. They were more semi-nomadic.*

Kristiina: But the plants that grew in North America and taken to Europe is what allowed Europe to industrialize and flourish. There is pretty credible evidence that the agricultural knowledge flourished in the Americas.

> **JD:** *But I think a lot of groups went through these at different times*

and things cycled through in different cultures. Even today our elders in the long house are not gatherers full time. That is not their job. Even today in the general council meetings, the women are talking in the middle of the room while men sit at the edges quietly. Nobody went hunting that day. Nobody spent three days out camping and collecting huckleberries. It is interesting how some of the traits of culture persist for whatever reason.

Kristiina: It is almost like Rodney's wife was the Chair in Colville for quite few years. So you would assume she would have a slightly different approach to leadership then like Mike. Or maybe not, I don't know.

JD: *I would think so.*

Kristiina: It is interesting to think about this because the popular media have suggested that African countries need more female leaders to move forward for the good of everyone. Liberia had a female President who did really well for the country and stabilized it. Just now a man replaced her as the President but it was a smooth transition of power. The suggestion was that females were more interested in social issues and therefore better leaders.

JD: *To be honest, that is a gross generalization of gender roles. It comes down to individual people. My previous boss, if I had to stereotype her, she was much more of man in how she operated. So it is an individual. There are super typical female and male roles but there is a wide overlap.*

Phil: It is also what works, right? To keep the tribe going.

Kristiina: There was recently an interesting article on South Africa where they tell their stories to build consensus on the issues they face. The main purposes of stories is for collaborative decision making. So they tell these stories thinking about their lands but it is mostly developing consensus in the group. They have scarcity of resources and so they are setting up rules for how they are going to manage the resources. It seems like stories have a moral purpose, but like the South Africa example, it is a way of almost forming what today is called a "tribe". You have people who support one another and will do anything for one another. There are many different meanings there. This is being talked about today as soldiers are coming back from Afghanistan and lose their tribe when they return to the U.S. They feel lost even though they were shot at in Afghanistan and you could die. They come back to a society in the U.S. where they are not understand and no one wants to hear about what they experienced. They have no place, and many are suicidal.

Phil: Well the external pressures gels them together. It almost creates this bond and then they are ripped apart. They served their purpose and now need to do this.

> **JD:** *Some of the VA (Veteran Affairs) homes, they have figured it out. Look at Rome where the retired soldier used to live in the barracks even if you had a family but you would not live with them. They would be out in the barracks because that was your tribe.*

Phil: It sounds like that is something that happens in the reservation environment where you can always come back home if you have problems. It also allows you to get back to the core values.

> **JD:** *That is a weird thing to. I grew up in the mountains of southeast Idaho. If you drop me off in the mountains, I would last about four days since I wouldn't know what to eat. But there is something about the Blue Mountains. I don't know what it is. I didn't grow up there and hardly ever went there. But there is something about the Blue Mountains that feels like home. It is strange.*

Phil: You have not figured out an objective way to explain your feelings.

> **JD:** *These Mountains feel like home and I have no idea what it is. It is not just the mountains but as a kid I could feel it coming across the Decanter River when we would go from Tri-Cities to Ontario, Oregon. I remember coming home from College and you cross the river and it was like someone took a four pound hammer to me.*

Kristiina: Why would you call it Blue Mountain? In Australia they have the Glass Mountains and it was the way the mountains looked when the sun reflected on it. It was pretty unique. They are so breath taking in a way.

> **JD:** *They were blue because of the trees. When the light shown on the trees you got the image of blue. It is a desert, so you get a lot of hazy air. You get the haze on top of the green trees. And then mixed with yellow so it looks like a blue color.*

Kristiina: I gathered when you were a semi-hunter and gatherer society or semi-nomadic, it was not the same thing as in Europe where it used to be if your father was a blacksmith, you became a blacksmith. If your father was a street cleaner, you became a street cleaner. Mike and Rodney have both commented on this. With Tribes it is more like what the tribe thinks each kid will be good at. The role of the women in Europe was very structured and I would not have wanted to work there in academics. As a female, your chances of getting in as a professor was almost zero. I get the feeling Tribes don't have this structured approach and people will shift into different roles or jobs?

> **JD:** *I don't know about historically. If you think about a population and there were a million people living in an area and you had a lot of cities. I always talk about Strawberry Island where the Snake River and the Columbia River come together at Tri-Cities. It had a village on it that had a seasonal population of about 7,000 people historically. It was a decent size town on a very dense area. Instead of each of us having 10 acres, we used to live 40 to a house. Anytime you get that number of people, even if you get 250 people together, there is this urban thing that happens where they need to deal with clean water in and soily water out. So very basic urban things are happening. If you get 250 people together you start dividing up work where you do fish processing over here, game processing over there. Kids stay over here. These are the things that start happening.*

Phil: It is a functional dispersion?

> **JD:** *Right. Then you start getting people who only do those things so you get specialization. Now the whole community is relying on me being the fish processor because that is the only way I am going to get a basket and provide enough fish to get the basket. I can't do both since I can't fish and make baskets. I have to do my fish really well so I get my basket. The basket weaver has to be really good at making baskets to get fish since they are not going to eat if they are not making good baskets. All these begin to happen with just 250 people.*
>
> *Even the villages had major villages that were down by the river and a whole stream of villages all the way up the watershed. So people would travel between each one of them but each village had full time year round residents. They are all interconnected. It wasn't like Indians would go all over the place. It would be like it is July so these people would go to a specific village. This village was part of a string of villages but they would go to a very specific village. They may do exchanges with other people for social gatherings, but the string of villages was where they were most active. The people who live full time in a village had to be able to take care of a group of people arriving at a specific time of the year. So 80 men are moving into this village to go hunting. The only way I am going to go through the winter is to make sure their houses are stable, and they give me food. There was a whole exchange and trade that occurred here. We don't do that now.*

4.8 Role of Environmental Economics in Environmental Justice*

Jessica Hernandez.
Source: Jessica Hernandez.

Environmental economics research and analyses aim to improve the effectiveness and equitable distribution of environmental policies within the federal government system and structure pertaining to natural capital (see Box 9). Natural capital is the main theory behind environmental economics that refers to how natural resources support human existence in our capitalistic society [71]. According to the Environmental Protection Agency (EPA), environmental economics refers to the "application of the principles of economics to the study of how environmental and natural resources are developed and managed" [72].

Due to its role in the effectiveness and equitable distribution and management of environmental and natural resources, environmental economics plays a significant role in environmental justice. Recent theoretical, empirical, and policy advancements in environmental justice have determined that a community's socioeconomic background will determine their outcome for environmental pollution and hazards that result in greater environmental and health risks [73]. The poorer the community is, the more environmental pollution and hazards they will experience and face.

For the purpose of this writing, environmental justice will be defined as a grassroots movement led by communities of color to protect and advocate for their well-being, livelihoods, resilience, and Mother Earth. Unlike other marginalized,

---

* By Jessica Hernandez.

low income, or communities of color, Tribes have a unique legal and political status in this country: Tribal sovereignty. In addition, for environmental justice to be manifested among Indigenous communities, Indigenous pillars of environmental justice for Indian Country need to be identified. The Indigenous pillars of environmental justice will help policy-makers shift their focus from distributive, procedural, and recognition justice into a more inclusive form of justice that weighs in cultural resources, values, and principles (Fig. 4.3). This will shift the focus from equity and fairness into an inclusive approach that also values indigenous cultures.

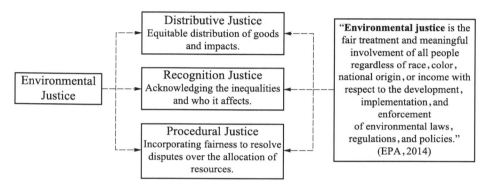

Fig. 4.3 Policy model of environmental justice.

Due to the relationship that exists within environmental justice and economics, it is important to determine the intersection of environmental economics and environmental justice in relation to the Coast Salish Tribes and Nations of the Pacific Northwest. The role of environmental economics in two of the environmental justice cases will be explored in favor or against Coast Salish Tribes or Nations in the Pacific Northwest. The cases that will be analyzed include cases that are documented in the *"Environmental Justice in the Pacific Northwest"* atlas that can be accessed at the following website (www.ejpnw.org). They include the Makah Tribe's cultural revitalization for whale hunting and the Lower Elwha Klallam Tribe's dam removal. Through interviews and case study analysis, the role of environmental economics in these particular environmental justice cases will be explored.

4.8.1 Natural Capital versus Cultural Values

Natural capital plays a major role in the way we govern our natural resources through policies and economics. It is the theory that brought attention to conservation as it utilized human prosperity and wealth as some of the social indicators to advocate for the protection of certain species and natural resources post-colonialism. It emphasizes how natural resources and ecosystem services sustain the human economy [74]. However, natural capital does not incorporate cultural values—the importance of certain species and natural resources for a community or Tribe that are not interconnected to a monetary value, but rather a cultural significance and importance. As we continue to overexploit our Mother Earth, American cultural values will continue to rely on capitalistic ideologies.

As Dr. Robert Goodman describes Americans—reluctant capitalists that desire the benefits of capitalism/money without any costs or sacrifices. However, even natural capital comes at a cost. As we continue to overexploit our natural resources and ecosystem services, we will eventually run out of them, resulting in mass extinctions. Despite natural capital's intention to advocate for conservation, it also advocates for purchasing of lands to extract natural resources or for the purchasing of large boats to establish a large-scale fisheries commerce. It is the same natural capital that we place on culturally significant species like salmon that are leading to the overfishing and overharvesting of this species. Yet little has been done in the Pacific Northwest to protect salmon.

In Pacific Northwest, salmon holds a natural capital role for commercial fisheries. It is a delicacy that many tourists that come to the Puget Sound want to try at local restaurants for a hefty price. However, salmon also holds a cultural value to the Coast Salish Nations and Tribes. While salmon holds a high market value in the fisheries sector, it is a significant natural resource for Tribes in order to maintain their cultural and traditional values alive—despite the years of continued colonization and oppression.

Salmon is also the driving force of the treaty rights the Tribes hold after the Boldt Decision was passed in 1974. The Boldt Decision granted the federally recognized Tribes their treaty rights to 50% of the fish harvest. As we are reminded by many of our Coast Salish elders, *"Without salmon there will no longer be any treaty rights."* While natural capital drives for the conservation of salmon, for the Tribes the conservation of salmon is more significant than their price tags. Since 2016, the salmon runs have diminished affecting the food sovereignty and security of the Coast Salish Nations and Tribes. Urbanization has resulted in high level of habitat loss. As the technology industry is currently booming in the Pacific Northwest, more housing is needed for the years to come. As Fran Wishusen, Habitat Service Director for the Northwest Indian Fisheries Commission stated in an interview discussing the 2016 watershed report, *"Habitats continue to be lost faster than we are restoring them."*

In 2016 both the state and Tribal officials could not reach an agreement pertaining to the opening or closing of salmon fisheries. As a result, sports fishermen held a protest against the Swinomish Tribe. They continued with their salmon fisheries as Tribal and federal officials were searching for an agreement. It is unfortunate that Tribes are now blamed for the low return of salmon when in reality this is a capitalist and colonial problem. The sports fishermen were blaming the Tribes' fish nets for the lower returns of salmon, when in reality that is not the major cause of lower salmon runs. Salmon became overfished once the settlers introduced large scale fisheries and an international export of fish from the Puget Sound to Asia.

Urbanization—a modern day problem—is also the main reason why salmon habitat loss is increasing at a rapid rate. What led to the frustrations of the sport fishermen was that they did not fully understand why Tribal fishermen were still allowed to fish for salmon and they were not. As one of sport fishermen protester stated for the Seattle Times interview, *"What I'm hearing is — if they fish, we fish."*[1]. This demonstrates the lack of understanding of the cultural value salmon has for the Coast Salish Tribes and Nations. What the sport fishermen see is a hobby and a natural capital, Tribal members see as part of their culture and sustainability. Salmon is a crucial component of their food security and is the main source that drives many of their cultural rituals. How these are manifested among the Coast Salish Tribes and nations defers based on their differences and medicine.

Issues that arise from the lack of understanding of the cultural values for certain natural resources by Tribes makes us wonder how natural capital can be more inclusive of cultural values. Native Americans are not overharvesting or overfishing the salmon. Many Tribes are introducing and have already added conservation methods to save their salmon such as salmon hatcheries. They consider themselves co-managers of the salmon, despite the little work that is being done by the federal government to co-manage this species and to address the biggest threat to habitat loss—urbanization. This debate over salmon fisheries and their lower returns leads to important questions, e.g., can natural capital be more inclusive of Indigenous cultural values? Does natural capital incorporate traditional ecological or Indigenous knowledges and sciences? Does natural capital really advocate for conservation as it was intended to do when the theory and concept was established?

4.8.2 Makah Tribe's Cultural Revitalization: Whaling

The Makah Tribe located in Neah Bay, Washington has a long history of legal battles to revitalize their whaling—a key component of their traditional and cultural values. They are the only Tribe located out of the State of Alaska that is entitled to whaling rights after they signed the Treaty of Neah Bay in 1855. The Makah's whaling continued even after the settlers arrived in the

---

[1] Accessed on January 1, 2018 at http://www.seattletimes.com/seattle-news/environment/sport-fishermen-angry-over-Tribal-gillnetting-salmon-catch/

Pacific Northwest. However, in the 1920s, the Makah Tribe took the ultimate decision to stop their whaling to protect the gray whale. Unfortunately, the mass commercial whaling industry that was introduced by settlers was leading the gray whale to extinction. Consequently, the Makah Tribe decided not to resume any whaling until the whale populations were replenished and healthy once again. As *the Los Angeles Times* reported in 1994, the gray whale was removed from the endangered species list[1]. It was 1994 when the Makah Tribe announced it would resume its whaling (Fig. 4.4) [75]. This decision placed the Makah Tribe under a radar, and they became the target of animal rights and conservation groups that utilized hate words and speech against the Makah Tribe to make their point — a point that demonstrates how intolerant these groups are in regards to understanding cultural values.

As noted in our interviews, the Makah Tribe faced opposition and discrimination once they decided to resume their whale hunting. The 1999 Whale Hunt was the last time they were allowed to harvest a whale. Animal rights and conservation groups are intolerant to Indigenous cultural values—they view themselves as the saviors of the majestic creatures, when in reality, it was their ancestors (European settlers) who are responsible for Indigenous peoples' environmental injustices and the loss of whale populations the Puget Sound experiences.

The role environmental economics plays in this environmental justice case is very critical. Whaling became a booming industry in the 1800s, which is why they were over harvesting whales for export and commerce. The Makah have

Fig. 4.4 Photo showing the cutting of a whale in San Simeon Bay.
Source: U.W. Digital Collections. Collins JW, 1888/1889.

---

1 Accessed on October 15, 2017 at http://articles.latimes.com/1994-06-16/news/mn-4742_1_gray-whale

traditions and rituals tied to whaling and there is a big cultural difference between why they whale and why settlers whaled. They do not just go into the waters, but also participate in a ritual that the whalers and families are involved in. However, they were accused of potentially selling the whale meat, because natural capital is what we always associated through the Western lens of environmental economics. It is not their fault that natural capital makes our society associate hunting that is tied to traditional and cultural values to commerce, money, and export. The two images from the University of Washington Digital Collections give us a visual representation of what it means to whale for commercial purposes and for cultural purposes—there is a cultural disconnection in commercial whaling.

4.8.3 Lower Elwha Klallam Tribe's Dam Removal

One of the environmental injustices many Tribes faced was their land being stolen. That included their sacred and creation story sites. This is the same case that occurred to the Lower Elwha Klallam Tribe. When the Elwha Dam was created, their creation site was covered with sediment and the dam structure itself. Despite Congress making the Olympic National Park in 1938, the Tribe was ignored in their requests to remove the dam that was covering their sacred site and threatening the salmon species. In 1978, as it is stated in the Lower Elwha Klallam Tribe's website[1], the dam failed to pass its inspection and unfortunately, neither the State of Washington or the Army Corps of Engineers issued a plan of action to fix it. Despite the advocacy of the Tribe to ensure the safety of the location, they were ignored. As a result, new partnerships were developed that supported the Tribe and ultimately led to the dam removal. It is important to note that as a result of this dam, adult Chinook salmon were no longer migrating through this river. The dam also trapped more sediments in the watershed and this resulted in habitat and biodiversity loss.

In 1992, a new initiative was enacted into law—the Elwha River Ecosystem and Fisheries Restoration Act. This act aimed to restore the ecosystems that had been lost as a result of the building and operation of the dam. This dam had been operating since 1913 and over the years the population of Chinook decreased, eventually being eliminated. The dam removal became a $325 million project that the Lower Elwha Klallam Tribe and the National Parks Services co-led and co-managed. As it was documented in *Indian Country Today*, the removal of the dam allowed for the return of the Chinook salmon and the creation site was eventually uncovered.

---

[1] http://www.elwha.org/decisionfordamremoval.html

4.8.4 Restructuring Environmental Economics to be More Inclusive of Environmental Justice

Environmental economics tends to center itself in the study of how ecosystem services and natural resources are managed and developed under capitalistic ideologies—financial assets and value through cash or funds. However, Dr. Mark Metzler criticizes how materialist analysis can capture only part of the process of capital as it neglects the kind of communicative indication or direction capital truly embodies [76]. This critique denotes why cultural values need and can be included in environmental economics, particularly in natural capital. To dismantle the systems of capitalism that tend to be the root causes of most environmental injustices, cultural and personal values also need to be integrated to move away from the materialistic ideologies.

As we can conclude from both environmental justice cases just discussed, a holistic approach to viewing our Mother Earth needs to acknowledge Indigenous sciences that are not always credited with professional or academic degrees. Academic institutions foster our economists through pedagogies that are disconnected from culture and this is one of the areas that can be improved to train a new wave of economists that do not place a monetary value on all natural resources and ecosystem services. We can shift capitalism from profit and revenues towards nature stocks—which will be investments companies that profit from natural resources and ecosystem services can invest towards conservation. This was the intention of natural capital—to promote conservation and while this idea will not be accepted by major industries, it is a possible solution (Fig. 4.5).

Fig. 4.5 Redesign of the Natural Capital depiction.

4.8.5 Special Acknowledgements

This story was completed and supervised by Cal Mukumoto. I am grateful for his supervision and mentorship. I also want to acknowledge the interviewees who also gifted and honored me with their time and personal memoirs and stories.

Cal Mukumoto.
Source: Cal Mukumoto.

Chapter V
Forestry Lens: Culture-based Planning and Dealing with Climate Change

"Who will find peace with the lands? The future of humankind lies waiting for those who will come to understand their lives and take up their responsibilities to all living things. Who will listen to the trees, the animals and birds, the voices of the places of the land? As the long forgotten peoples of the respective continents rise and begin to reclaim their ancient heritage, they will discover the meaning of the lands of their ancestors. That is when the invaders of the North American continent will finally discover that for this land, God is red."

"Western civilization, unfortunately, does not link knowledge and morality but rather, it connects knowledge and power and makes them equivalent." "Like almost everyone else in America, I grew up believing the myth of the objective scientist. Fortunately I was raised on the edges of two very distinct cultures, western European and American Indian...."

∼Vine Deloria, *"Red Earth, White Lies: Native Americans and the Myth of Scientific Fact"* (1997)∼

5.1 PNW U.S. Tribes and Leadership in Climate Change Planning

The impact of climate change for Tribal people is very clear from collating all of the climate mitigation plans being developed by Native American Tribes. If this was not an important issue for them, they would not be planning on how to address these climatic changes. Other countries worry about climate change but mainly address this by having voluntary goals that each country follow to reduce their greenhouse gas emissions. Countries are impacted by climate change impacts but not at the level felt by most Indigenous communities. They are truly the "canary in the coal mine" and do not have an option to escape the mine when conditions deteriorate sufficiently that people may be unhealthy or die and ecosystems become unhealthy or deliver fewer ecosystem services. They must live with the effects of climate change and do not have an option to move to another location. They are on a fixed land base. They could leave their land and emigrate as many Europeans have done in the past, but it is not the solution that is acceptable. They are planning to mitigate climate change and land-use impacts. Even though they cannot control human activities in resource consumption or land-use changes, they recognize that they aggravate climate impacts.

The Tribes in the US Pacific Northwest are not just waiting for others to develop tools to mitigate climate change impacts. They are actively addressing and planning how to mitigate the continued loss of resources which impacts their sovereignty. The Tulalip Tribes have several concerns related to treaty rights and the treaty boundaries that confine them to a fixed area of land. The Tulalip Tribes write how temperature and precipitation changes are resulting in lower stream flows, shifts in ecotones and the "loss of economic, cultural and religious use of native species ... that would not be replaced by incoming species", and degraded habitats especially for salmon.

The Tulalip Tribes articulated well their concerns with Treaty Boundaries and climate change impacts [77]:

> "For the Tribes, range shifts in native species will threaten their cultural existence. The treaty-protected rights of Tribes to hunt, fish and gather traditional resources are based on reservation locations and usual and accustomed areas on public lands. These locations were chosen to ensure access to culturally significant resources, whose locations were thought to be fixed. If the traditionally significant plants, animals, and aquatic species shift out of these areas, Tribes will no longer have the same legal rights to them. Even if rights to these species could be secured, without proximate access, the use of these species will be virtually impossible.
>
> Traditionally, Tribes relied on moving residence based upon seasonal availability of food and water sources. Modern Tribes are unable to relocate to cope with shifts in the availability of cultural and religious sources. Few Tribes have can afford the purchase of large territories of new land, and federal laws."

The Stillaguamish Tribes developed a climate change vulnerability assessment of their lands that was published by the U.W. Climate Impacts Group [70]. In that report they identified a list of priority species and habitat types that would be impacted by climate change. It also was clear that the species identified by the Tribe as being a high priority for them are vulnerable to climate change. They further recount that inadequate information exists on many of the species to credibly assess their vulnerability to climate change.

In November 2016, 20 of the Treaty Tribes in Western Washington published a report on the impact of climate change on natural resources. In that report they summarized [78]:

> "In the last 150 years our homelands and waters have profoundly changed. Salmon and steelhead runs that are central to our culture and economy are at a fraction of their historical populations. Many lowland old-growth forests have been logged. In some parts of the region, natural shorelines have been replaced by concrete and hundreds of acres of shellfish beds are too polluted for harvest. These changes have contributed to declines in natural resources important to our communities.
>
> Today our environment and the natural resources we depend upon are further threatened by climate change. Virtually all of the activities that our treaties protect are influenced by the effects of climate change."

The Treaty Tribes in Western Washington summarized the impacts that are making it difficult for them to *"exercise their treaty-reserved rights"* that they signed in the 1850s [78]:

> "In exchange for ceding vast tracts of land, the Tribes retained the right to fish, hunt, and gather as we have always done throughout our traditional territories. Major federal court decisions have upheld our treaty rights as the law of the land. The ability to exercise our rights is diminished if species productivity drops too low or if species are no longer available in our gathering, hunting, and fishing grounds."

The following executive summary identifies the broad range of topics that they are worried about [78]:

> "Each species responds to climate change depending on its particular characteristics and the local conditions. Nonetheless, there are overarching impacts that have the potential to challenge our ability to exercise our treaty reserved rights (NWIFC. Executive Summary 2016):
> - **Declining runs of salmon and steelhead** due to changes in streamflow, stream temperature, levels of dissolved oxygen, amount

of sediment in streams, susceptibility to disease, ocean temperatures, ocean chemistry, timing of prey availability, prey type, and competition from warm-water species.
- *Migration of marine fish* away from historical fishing grounds as they seek out cooler ocean temperatures.
- *Replacement of traditional fish runs* with invasive species and new species that have migrated from the south.
- *Declining populations of shellfish* (both mollusks and crustaceans) due to changing ocean chemistry.
- *Closing of shellfish harvest areas* due to harmful algal blooms.
- *Loss of traditional shellfish harvesting areas, forage fish spawning grounds, and important cultural sites* to sea level rise or increased coastal erosion.
- *Declining populations of wildlife and birds* due to habitat changes, loss of food sources, disease, and competition with invasive species.
- *Migration of wild game and birds* out of traditional hunting grounds as they move farther north or to higher elevations.
- *Decreased plant productivity and shifts in species ranges* due to heat stress, drought, invasive species encroachment, or increasing pests.
- *Loss of traditional hunting grounds, plant gathering areas, and sacred sites* due to wildfire, landslides, or invasive species.
- *Changes in timing of key life stages in a variety of species,* such as the migration of salmon, fruiting of berries, or optimal time to harvest cedar bark.
- **Loss of access routes to important cultural sites** due to flooding, bridge damage, permanent road closures, or landslides.
- **Loss of water supplies** for drinking and other needs due to saltwater intrusion from sea level rise, or changes to precipitation, streamflow, and/or groundwater availability.
- **Negative societal outcomes** from poor air quality, heat stress, spread of diseases, loss of nutrition from traditional foods, and loss of opportunities to engage in traditional cultural activities."

Each of these listed changes have significant impacts on the plants and animals that are important for their way of life. They were also promised these rights in the treaties they signed.

Tribes are not waiting for other organizations to produce solutions for these climate impacted resources and environments but collaborating on producing new solutions for problems that historically did not impact them. The other effort important for them [78] was to *"enhances the ability of ecosystems and communities to adapt to changing conditions*:

- *Development of Tribal capacity to assess on- and off reservation climate change impacts and to promote resilience to these impacts at multiple scales.*
- *Developing approaches to natural resources management that consider landscape-scale processes and include innovative solutions.*
- *Working together to restore natural physical processes and ecological function, and to reduce existing stressors, such as water quality impairment, fish passage barriers, noxious invasive weeds, and habitat fragmentation.*
- *Promoting biological diversity, protecting intact ecosystems, and supporting climate refuges—areas where changes are expected to be less severe or to occur more slowly.*
- *Tracking changes to local environmental conditions, including the use of Tribal traditional knowledge of climate patterns and ecosystems as a source for early warning signals.*
- *Promoting cultural resilience through Tribal citizen engagement and education, especially K-12 education.*
- *Sharing knowledge and expertise with Tribes and non-Tribal entities within and outside of the PNW on research, modeling, and tracking environmental trends.*
- *Partnership with federal, state, and local governments to work together on local concerns and solutions."*

Tribes are at the fore-front of addressing local and regional climate problems and their impacts on lands and resources. As what was just summarized, Tribes are developing approaches to respond to climate change impacts on their treaty preserved rights for plants and animals. They are also working with multiple agencies — varying from the federal, down to the state and local governments — to work on solutions to the climate change impacts. The federal government has a trust responsibility to Tribes and as MacKendrick [32] wrote:

> *"Responsibility to uphold Indian treaties, which are legally binding and acknowledge the sovereignty of Tribes, and their rights to landownership and access to natural resources and off-reservation lands. And even though Congress retains plenary power over American Indian Tribes and the ability to abrogate treaties, the federal government also has a duty to act within the interests of Tribes and to protect the natural and cultural resources on which they rely.... In addition, under the public trust doctrine the federal government has a responsibility to*

> *protect the environment, including the atmosphere, for all people.... Under most federal environmental statues, including the Clean Air Act and Clean Water Act, American Indian Tribes are treated as states. Thus, Tribes like states are recognized as co-regulators of their environment and have the authority and funding available through Congressional acts to administer their own environmental program.... Numerous other federal legislative acts recognize the rights of Tribes to access, manage, and protect natural and cultural resources, including the National Indian Forest Resources Management Act of 1990."*

Despite these trust responsibilities, the federal government is still struggling with how to respond to climate change at the national level. In some political spheres, including the top level of the United Stated federal administration, climate change is not accepted as an important issue so financial resources are not being put towards regional planning and mitigation efforts. This has placed Tribes at the forefront of local or regional efforts to manage for culturally important species for future generations. This has been challenging since they are unable to manage adjacent lands that are fire prone if they are privately owned or on public lands. MacKendrick [32] wrote:

> *"Individuals described how the wilderness area adjacent to the reservation is susceptible to fire and areas to the east and northwest where a lot of blowdown has occurred contain a lot of ground fuels and could be sensitive to fire. One individual described, Forest fuels are high on Forest Service and BLM lands surrounding the reservation. The Tribe is trying to treat the boundaries of the reservation to protect it from surrounding fuels."*

Since most of the Tribal lands are not separated from other ownerships but are impacted by what occurs on these other lands, it makes it extremely difficult for Tribes to manage disturbances occurring outside their borders. The federal government is not meeting its trust responsibilities and Tribes do not have sufficient funds to address all these climate change impacts. Despite these challenges almost every Tribe has or is developing a management plan for their lands and waters.

5.2 Tribes, Tribal Resources and Forest Losses

Forests are a good lens to look at Tribes because many have timber lands managed to generate revenue for the Tribe. To understand forests and their role for

Tribes that manage timber, it is important to first explore how Tribes originally lost their forestlands. This will be followed by what cultural values Tribes have for their forests and how forest diversity is impacted by climate change. This section will conclude with how Tribal forest management is viewed to be better than what occurs on adjacent non-Tribal lands. This is a story of forest loss but the continued use of cultural factors to develop a sustainable forest management system on smaller pieces of land.

5.2.1 Historical Loss: Manifest Destiny and Loss of Forests*

At the time of its conception and declaration of sovereignty, the United States of America consisted of 13 colonies that ran along the East Coast, going no further West than the Appalachian Mountains. Today, the United States is composed of 50 federated states, the District of Columbia, 16 insular territories, and 304 federal Indian reservations. Over the course of two centuries (1816—present), the United States managed to grow from 13 colonies east of the Appalachians into 50 federal states sprawling all the way to the West Coast, along with insular islands off the continental mainland. This far-reaching expansion of territory during the 18th to the mid-19th century was primarily justified with the ideology of Manifest Destiny and financial gain due to the abundance of resources in the West. Incentivized by these factors, pioneers swarmed all the way to the West Coast to settle the land, and in the process, forever altered the ecosystem, and Indigenous people who called the Western Territories their home.

Prior to European colonization, many natural wonders of the West had been sustainably managed by Indigenous peoples for centuries. These lands were left untouched by the mark of civilization. However, once the pioneers settled the West, many of the structures and systems of the forests started to get treated unsustainably by the settlers. The U.S. Westward Expansion had a significant role in implicitly and explicitly changing the forests of the Western Territories by introducing different forest management practices and by relocating the Indigenous people who had managed and lived on those lands for centuries.

One of the biggest ideologies that promoted Westward Expansion during the 18th to the 19th century was Manifest Destiny. According to the general belief of Manifest Destiny, the United States, a country of great potential greatness and "human progress", had a duty to spread its greatness to the western "untrodden spaces" [79]. Essentially, it was their God-given duty to spread democracy and civilization to the supposedly untamed Western Territories, without taking in account the Indigenous people who had occupied those lands for centuries.

---

* By Jaskaran (Jesse) Singh.

The idea of Westward Expansion had been brewing ever since the conception of the United States, when it only consisted of 13 colonies bounded by the Appalachian Mountains and the Eastern Seaboard. As early as 1751, major figures in establishing and growing the United States (e.g., Franklin, Jefferson, Monroe, and Adams) had expressed desires for the U.S. to settle the west, as there were many benefits from this expansion. Benjamin Franklin believed that until the West was fully settled, commerce and economic growth in the United States would be stagnant due to lack of fertile land and resources for new businesses to prosper [80]. Franklin believed that the vast amount of resources and land would allow settlers to have their own businesses and no longer be a "Labourer for others", while also promoting more commerce due to the creation of trading hubs along new settlements across the West to East Coast [81].

Manifest Destiny, an abundance of resources, and vast amounts of land were the driving forces for Westward Expansion and incentivized many pioneers to set out to settle the West. However, it would not be until 1862 that the U.S. Government officially promotes Westward Expansion with the *Homestead Acts*. The *Homestead Act* of 1862 was passed by Congress and stated that anyone can claim up to 160 acres of land in federally surveyed lands in the west if they improved the land in 5 years. This Act was the final push for many, as it gave away free land to many immigrants and workers. Followed by the construction of the Transcontinental Railroad in 1869, expansion deep into the West was made even easier, allowing goods to flow from the East to the West. This promoted trade and commerce as Benjamin Franklin believed it would. As the government endorsed settling of the West, along with the potential financial gains that lied in the West, settling the West was inevitable.

First, it is important to realize that the forestland and their ecosystems were connected to the Indigenous people who occupied those lands for centuries before the Europeans colonized North America and before the United States' Westward Expansion. The Native Americans were, and are still, holistically managing vast tracts of forestland in accordance with the Tribe's forest management philosophy. There are three principles which most Native American Tribes uphold when making decisions relating to forests; the forest must be sustainable for future generations, must be cared for properly to provide for the needs of the people such as cultural foods or medicines, and all the pieces of the forest must be kept maintaining the diversity of species [82].

Tribes would only take what they need from the forest, as survival of the Tribe was a top priority. For example, one such Native American Tribe are the Menominee people, who have lived in Wisconsin land for over 4,000 years. However, before European colonization and U.S. Expansionism, their Tribal land once spanned across 9,500,000 acres (3,844,514 hectares), from the Great Lakes to the Mississippi River [83]. Like many other Tribes spanning across the frontier, they had developed a holistic approach to managing the forests they depend on for day-to-day life. They hunted, fished, and gathered all the various forest materials they needed to make homes, canoes, baskets, bows, spears, blankets, etc. For them, holistic management was required for survival, hence it was not influenced by any market forces or commerce demands. Any materials they required that were not present in their Tribal lands, they could trade for with other Tribes. The Menominee people practiced inter-Tribe trading with various Tribes; they imported catlinite (type of rock used to make pipe bows) from the Dakota Tribe, and copper from the Algonquians on Lake Superior [83].

Like many other Tribes, the Menominee people managed to survive for millenniums thanks to their forest management philosophy where they didn't view forests as financial resources, but as survival resources. This point is validated by the fact that most European cultural fallacies about the American western frontier described it as untamed and wild. This was due to the fact that in Europe and in the States, an agrarian society was the norm and forests were culturally defined to be uncivilized, but fertile farmland was considered civilized [84].

This cultural divide was that European/U.S. society was primarily agrarian and the majority of Native Americans practiced farming, however it was not a staple of their identity as a society. This is one factor that affected forestland during the wave of Westward Expansion by pioneers. They cleared forestland for timber and farming to sustain their new settlements. Although, it may have not been mass deforestation, it did affect the ecology of the forests since wildlife lost an important structure that was their habitat. This management shift was cleared for financial profits from farming.

Westward Expansion was driven by the goal of acquiring more resources, which would eventually translate to building new settlements and financial gain for many, but not for all. Often, settlers would encroach upon Tribal lands to gather resources such as timber, ores, and pelts as these were profitable and important resources for an expanding country. In most cases, the Indigenous people would be evicted from their land, as the pioneers and settlers would gather the resources on the land. For example, this excerpt from Simon Schama, an English historian and professor of History at Columbia University, describes one glimpse of life in the West, specifically focusing on settler—Native people's interactions. He describes a mining camp consisting of people from many different places and many different occupations. The mining camp was situated in the Mariposa Giant Sequoia grove in present-day Yosemite National Park. He describes in the year 1852 "violent crowd of Italians, Mexicans, and Germans... Hunters, loggers, ditch diggers, cooks, and whores, many of them practicing more than one trade..." [85]. The amount of diversity in nationality and trades, and the fact that certain groups of pioneers have already reached as far west as California in 1852, helps illustrate the drive for resources that compelled many people to venture westwards.

Simon Schama goes on to describe how pioneers occupied these areas being mined. Further he describes how a military battalion accompanied them and drove out the Ahwahneechee people who had occupied those lands. It is described how the Ahwahneechee used to repeatedly set-fire to the area, which cleared out the brush and opened space for grazing. They would also occasionally hunt, however all these activities halted once they were driven away by the settling groups of miners. The Ahwahneechee would occasionally raid the camp to "get some of their birthright back" [85], along with liquor and guns. However, the few of the Ahwahneechee people that survived the dislocation and dispossession of their land would call their torments Yochemate, roughly translated to "some among them are killers" [85]. This excerpt essentially describes a mining camp in Yosemite National Park and specifically focuses on the people working in the camp and how their presence has affected the Ahwahneechee tribe that had occupied that land.

John Muir, a nationalist/nationalists and early advocate for preservation of wilderness in the United States, assumed that the land was untouched and "Edenic", equal in splendor to the Garden of Eden. Most Europeans assumed that it was impossible for the natives to have handled the land to make it "Edenic". This helps to show the fallacy that was implanted in many settlers and Europeans minds regarding Native Americans and the vast wilderness in the west: The people were savages in need of finer, western-culture and that nature was wild and untamed. The mining camp and accompanied military battalion settled in the area had imposed control in the area thanks to their superior technology (firearms). This drove the Ahwahneechee away from their food sources. This explains why they desperately raided the camp to get the food they can no longer get, along with other resources; weapons, tools, liquor. Essentially, they took away the Ahwahneechee's food sources and land due to the mentality that land is free for the taking and unclaimed by any one person. Also, the government explicitly supported them through acts like the *Homestead Act* of 1862. The rights to the land and its resources that the Indigenous people, like the Ahwahneechee, had sustainably managed were disregarded and not considered to be valid. The, people like John Muir, a well-known advocate for conserving nature, were fooled to think that the land was untouched.

This scenario played out repeatedly in the mining camps, logging operations and towns which were established thanks to the railroad and *Homestead Act*. Increased Native American-Settler conflicts escalated due to settlers stealing Indigenous people's land and resources thanks to Congress's disregard for Native American land rights. This was done in the name of Manifest Destiny/expansionism and economic interest. For example, bison are an incredibly vital part of the culture of the Plains Indians as a source of food and hunting material to make clothing and tools (Fig. 5.1). Settlers in the Plains region were quick to realize the importance of the bison; however, they mainly saw the eco-

Fig. 5.1 Buffalo hunt painted by George Catlin[1].

nomic value as a trade commodity to the Eastern U.S. or Europe where there were no bison. During the winter of 1872–1873, more than 1.5 million buffalos were placed on trains and moved eastwards, for commercial gain as their leather was used to make clothing, boots, parts for machinery, etc. [86]. Relocating 1.5 million buffalos during the course of one season strongly affects the ecosystem and forests in the Plains. Furthermore, commercial hunting groups were at fault for decimating the bison population, and making them almost extinct, since perfect bison leather was a high-value commodity.

Hunting and relocation of the bison by settlers were mainly fueled by commercial and trade profits, but they were also hunted down as a military strategy to hurt the Indian's food sources, shatter their autonomy, and simply as an act to weaken the enemy (Plains Indians were often at war with United States, due to settlers settling their Indigenous lands). For instance, a high-ranking military officer (Lieutenant Colonel) in the U.S. Army, Richard Irving Dodge, once said [87]:

> *"There's no two ways about it, either the buffalo or the Indian must go. Only when the Indian becomes dependent on us for his every need, will we be able to handle him. He's too independent with the buffalo. But if we kill the buffalo we conquer the Indian. It seems a more humane thing to kill the buffalo than the Indian, so the buffalo must go."*

Furthermore, General Winfield Scott Hancock, a decorated General who served for the Union in the Civil War, Mexican-American War, and was a Democratic candidate for president for the 1880 elections, is quoted to have explicitly express the idea of dependency on the U.S. Government to several Arapaho chiefs [87]:

---

1 Accessed on June 17, 2018 at https://en.wikipedia.org/wiki/Bison_hunting

> "You know well that the game is getting very scarce and that you must soon have some other means of living; you should therefore cultivate the friendship of the white man, so that when the game is all gone, they may take care of you if necessary."

These two quotes show how strategically hunting down certain food sources, such as bison, was a valid military strategy to decimate the Indian's staple food source. This forced Indians to either shift to a different food source or lose their autonomy and become dependent on the U.S. Government. Either Indians would move to reservations where they can hunt their own bison without settlers, or to trade/receive rations. These tactics worked well, as most Tribes were often pushed out of the way of settlement, into reservations where they were promised certain provisions (despite most of the promises never being fulfilled in their entirety).

To summarize, settling the West resulted in unavoidable encroachment of the Indigenous people's rights to their land and resources. The Native Americans were the managers of the forests structures and animals that lived in these forests. This was all lost when they were relocated to make way for pioneers who cleared out forests for lumber or simply for land to farm. Although forests were not drastically altered, this shift in forest management and a shift in the occupant's cultural value and understanding towards nature, this affected forests for the upcoming decades. It set the tone towards forest management. In cases like the Mariposa Mining camp, the settlers managed the lands without having any experience understanding the many tree species that grew in the forest (such as the sequoia).

Many settlers and even the U.S. Department of Agriculture Forest Service didn't see the value in a timely and controlled forest fire management. In contrast, this was a practice which the various Indigenous people had been doing for centuries to manage the forest. This shift in forest management, from the Native Americans to the pioneers, helps illustrate exactly why forest and the Native Americans who occupied them are connected. The forests were originally managed using Native American's approaches of holistic management and through their cultural ideals. Later, once the settlements started occupying land, they practiced their own form of forest management based on their shared European cultural views towards forests.

Land management and transactions are important to analyze, since land transactions in the Western U.S. heavily affected the landscape that had previously been managed in a balanced manner. During the 1800s, using land as a payment method was a practice used by the government for many different transactions; military grants payed to veterans, *Homestead Act(s)*, payments for cash sales, but most importantly to provide the railroad companies with land to build the

Transcontinental Railroad (along with cash payment depending on terrain per mile).

Just to realize the scale of land transfers that occurred for the ultimate goal of Westward expansion, the total number of acres that were distributed by the *Homestead Acts* was 270,000,000 acres (109,265,123 hectares) during the peak of Homestead claims in 1913 [88]. On the other hand, a total of 142,883,372 acres (57,822,849 hectares) of land were granted for the Transatlantic Railroad's construction: 94,000,000 acres (38,040,450 hectares) directly to the corporations and 48,883,372 acres (19,782,399 hectares) via grants to states [89] [90]. Most of those acres pass through wilderness, especially the 270 million acres (109 million hectares) claimed by settlers in the West. The environment and ecosystem services provided by these lands were often damaged by the Euro-American settlers who were tasked with making the land more habitable (e.g., clearing forest land, logging for economic gain, hunting animals such as the American bison as previously discussed).

5.2.2 Fire Cause Loss of Forests, Cultural Resources and Timber from a Shrinking Land Base*

From left to right. Dr. Daniel Vogt, Dr. Kristiina Vogt, Jonathan Tallman, and Mark Petruncio at Ivars Salmon House in Seattle (2017).
Photo Source: Daniel Vogt.

The Yakama Nation is one of 29 federally recognized Tribes in Washington State. The 1.37 million-acre-Yakama Nation Indian Reservation includes 650,000 acres of forest and woodlands (Fig. 5.2). Yakama Tribal people have lived intimately on

---

* By Jonathan Tallman, Mark Petruncio, and Dr. Kristiina Vogt.

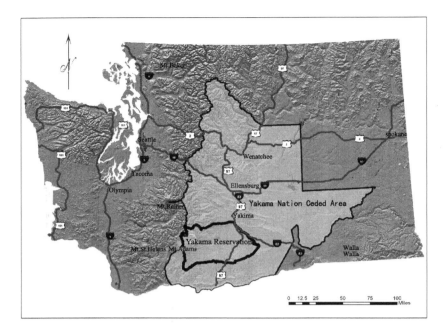

Fig. 5.2 The location of the Yakama Nation Indian Reservation in the State of Washington. The current boundaries of the Yakama Nation are outlined in black and the lighter gray area are Yakama Nation ceded lands.

Source: Yakama Nation Forestry 2017.

the land and have depended on the forests for millennia. Their wise stewardship of these forests has been acknowledged in the 2013 Indian Forest Management Assessment Team (IFMAT) Report [91]. The Bureau of Indian Affairs (BIA) manages the forest resource in trust for the Yakama Nation. Yakamas rely on the forest resource as a dependable source of spiritual renewal, clean water, food and medicinal plants, revenue, and employment. The Yakama Nation manages their significant forest resources to achieve multiple benefits including timber. The Yakama Nation manages their forests *"in accordance with Tribal visions; management priorities are shifting towards protection and commodity production receiving less emphasis"* [91].

Tribal management of forests is complicated by the multiple products and benefits that Tribes want from their forestlands and now the impacts of climate change on their forests. This has made it is more difficult to make choices on how to manage lands to meet multiple objectives, especially as the forest health declines in unmanaged areas of the forest. The IFMAT III Report [91] summarized well the economic and cultural benefits Tribes obtain from their forests and what they would like to continue to obtain from them:

> "*Diverse forest types provide irreplaceable economic and cultural benefits.* Forests encompass about a third of the total Indian trust lands, and sustain Tribal economies, cultures, religions, and spiritual practices. Forests are closely linked to community and cultural vitality in Indian Country. Forests store and filter the water and purify the air. They sustain habitats for the fish and wildlife that provide sustenance for the people. They produce foods, medicines, fuel, and materials for shelter, transportation, and artistic expression. Forests provide revenues for many Tribal governments, sometimes the principal source of revenue, and sorely-needed employment for Indian people and rural communities."

Tribal forest management is very challenging because forests need to provide economic, environmental and cultural benefits simultaneously [92–93]. It is not just getting the silvicultural practices correct but other resource issues have to be balanced with the need to generate revenue for the Tribe and employment opportunities for Tribal members. Tribes have been active forest managers before the arrival of European colonizers and there is ample evidence that they deliberately managed their ecosystems [63]. For example, there are many types of medicinal remedies that Indigenous people use in the Pacific Northwest that are collected from forests. Red alder (*Alnus rubra*) contains salicylic acid in the leaves and bark that is used as a medicinal for pain. Extracts from Douglas-fir (*Pseudotsuga menziesii*) and western red cedar (*Thuja plicata*) are used for cold remedies [94].

Maintaining Tribal forests and the multiple resources they provide is challenging today due to past land-uses and the occurrence of extreme climatic events that have altered forests and their health. Extreme drought events are decreasing forest health as forests are impacted by more frequent wildfires, disease and insects [91]. Extreme climatic events will not only increase the frequency of disturbances but will have significant impacts on available supplies of Tribal resources.

In response to these increasing disturbance events and their impacts, Tribes have been actively planning their response to these problems even though they are not responsible for them. For example, the Yakama Nation is developing their climate adaptation plan, i.e. "Climate Adaptation Plan for the Territories of the Yakama Nation" [95]. This plan summarizes the Tribe's approach to addressing climate change impacts. Following are a few excerpts from this report that address some of the issues that are relevant to know:

> "... We expect that both summer and winter temperatures will continue to increase and snowpack in the mountains will diminish. We have observed that many of these changes are already being realized today.

> "Climate change is real and, unfortunately, the effects appear to be in motion. We are witnessing changes in the seasons. Our roots and berries must be gathered sooner, and salmon returns are less predictable. Our people notice less snow in the mountains now, and there is less cool water during the summer when it was once abundant. The changes we see may not bode well for our future. Over the years to come, we may lose natural resources that are important to our culture and our heritage. Some of these losses may be irreversible."

> "Changes in climate can affect forest pest and disease responses in two fundamental ways: First, by lengthening or shortening the seasons when pests reproduce and complete their life cycles, and second, by weakening the host trees and leaving them more susceptible to insect attacks and plant disease. Specific effects in our region are difficult to predict at this time. For example, in some places in the Pacific Northwest, mountain pine beetles are projected to decline, and in other places they are expected to become more prevalent."

Tribes need to make their forests climate adaptive while managing their resources. The Yakama Nation is taking a holistic approach to their forest management that includes collecting multiple values from forests as follows [95]:

> "The Yakama Nation has taken an active role in managing and improving its forested lands, using an ecosystem approach, by implementing a complex and comprehensive Forest Management Plan (FMP). Implementation of the FMP is intended to enhance and maintain a diversity of forest conditions, maintain sustainable production of commercial and noncommercial resources, and thereby maintain the forest resource as a dependable source of spiritual renewal, food and medicinal plants, revenue, and employment for the Yakama people. The FMP is a collaborative effort of the Yakama Nation and BIA natural resources programs, including Archaeology and Cultural Resources, Environmental Quality, Fisheries, Forestry, Range, Roads, Soil, Vegetation, Water Code, Water Resources, and Wildlife. The main topics within the FMP include, but are not limited to, big-game habitat, forest health, old growth, revenue and employment, threatened and endangered species, and water quality."

There is strong evidence from climate models that extreme climatic changes in temperatures and precipitation (especially drought, less precipitation in the spring and summer months) will continue to decrease the health of Washington's forests. The models predict climate change will directly cause tree mortality in

Washington State and increase the outbreaks of secondarily pests and pathogens on already stressed trees. Several models have predicted the continuing impact of climate change on forests in Washington State and the occurrence of extreme weather events. According to the WA DNR Forest Health Program [96],

> "Drought conditions and warm, dry spring weather tend to increase tree stress and insect success, driving acres of damage up in both the current and following year.... According to the US Drought Monitor, all of Washington was in an abnormally dry condition during summer 2016, with the southwest and southeast areas of the state experiencing moderate drought at times.... Approximately 2.4 million trees were recorded as recently killed in 2016."

A recent study provided evidence that the high mortality rates are occurring in the upper-elevation forests and growth is not balancing the tree volume losses due to tree mortality. The high tree mortality rates are supported by the much lower precipitation levels in Washington State, and also on the Yakama Reservation, between 1981 and 2010. Based on the data provided in the WA DNR Forest Health Program [96], in 2016 the Yakama Reservation experienced from 110% to 200% departure or lower precipitation levels from the 1981–2010 normal precipitation levels in 2016.

According to the WA DNR Forest Health Program 2017, drought conditions were not as serious in 2016 as in 2015 (Fig. 5.3). Despite 2016 not being as bad of a drought year, the low spring rainfall during the spring of 2016 caused

> "the entire state being abnormally dry with moderate drought conditions". They further reported: "A severe drought in 2015 resulted in wide-spread damage evident by late summer. Many affected conifers remained green for months as the weather cooled over winter. Then with record-breaking heat in spring 2016, delayed symptoms became more noticeable and widespread.... These trees likely had a delayed response to the previous year's drought conditions. The 2016 aerial survey showed increases in ponderosa pine killed by western pine beetle and grand fir killed by fir engraver. Attacks by these bark beetle species often increase following drought events." [96]

This report by the 2017 WA DNR Forest Health Program supports the idea that these mortality events are not just a one-time occurrence as stressed trees are attacked by secondary disturbance agents. This further supports the idea that these unhealthy forests will not recover if no management is implemented on the stands.

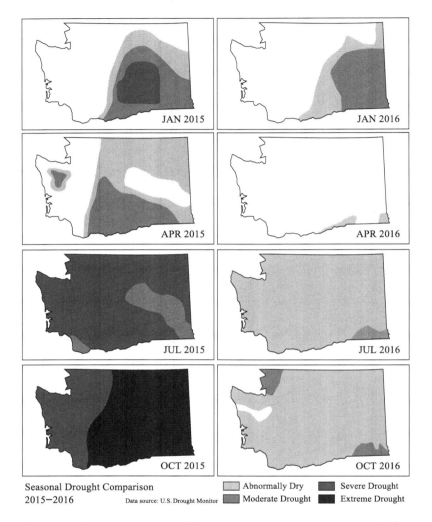

Fig. 5.3 Drought severity in Washington State in 2015 and 2016 [96].

The UW Climate Impacts Group has predicted many tree species will have very reduced ranges and suggested areas climatically suitable for Douglas-fir, pine species and subalpine forests will all decline[1]. The UW Climate Impacts Group have also predicted:

> "Warmer temperatures will result in more winter precipitation falling as rain rather than snow throughout much of the Pacific Northwest, particularly in mid-elevation basins where average winter temperatures are near freezing." [97]

---

1 http://cses.washington.edu/db/pdf/snoveretalsok2013sec7.pdf

This means that it is not an option to allow forests to recover on their own without the intervention of silvicultural practices. If forests are allowed to recover on their own, the diversity of the forests will be lower, the quality of the wood will be poorer, and the many medicinal plants and cultural foods that the forests provide will be less available.

Another cultural food — salmon — can also be impacted by not managing forests to restore their volume growth and diversity. The presence of forests and water flow/temperatures in rivers are linked to maintaining healthy forests which are essential for salmon survival. The National Research Council reported in 2008 that two-thirds of U.S. fresh drinking water originates from forestlands (Accessed on April 15, 2016 at www.sciencedaily.com/releases/2008/07/080714162600.htm). This fresh water is also important for salmon since the waters originating in healthy forests are cooler.

The UW Climate Impacts Group used models to predict that by the 2040s that the rise in river temperature may become fatal for salmon (Fig. 5.4). A decrease in density of trees and a potentially lower snowpack would also further exacerbate the low flow of the water from forests to the lowlands and further impact the viability of salmon populations. Therefore, not only are forests impacted by climate change but cultural foods such as salmon are also impacted

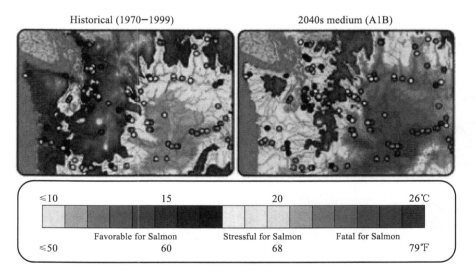

Fig. 5.4 Predicted changes in air and stream temperatures compared to historical levels. [Left grey colors are favorable for salmon, white-to-light grey are stressful for salmon, and right grey-to-black are fatal for salmon] (Excerpted from The Washington Climate Change Impacts Assessment, University of Washington, Climate Impacts Group, June 2009).[1]

---

[1] Accessed on July 16, 2016 at http://www.cses.washington.edu/db/pdf/wacciaexecsummary638.pdf

by what happens in forests and, therefore, forest management needs tools to mitigate climate change impacts. The physiological characteristics of true fir species provide an assemblage of overlapping cover for high-elevation watersheds essential to conservation of water quality [98].

5.3 Today Better Forest Management on Tribal Lands Compared to Their Neighbors

It is important to shed a light on how well Tribes manage their forest lands and why should they be considered a model for good forest stewardship. This highlights how Tribes obtain multiple values from their forests while continuing to obtain economic benefits from the same forests. Their goal is not to optimize profits. This was apparent from the interview of John Gordon, Chair of three IFMAT assessments of Tribal forestry and former Dean of Yale's School of Forestry and Environmental Studies. It is also demonstrated well why Tribal forestland management is a growing force in the Pacific Northwest U.S. (*see* Section 5.3.2).

5.3.1 Good Tribal Forestry under Federally Mandated Assessments (IFMAT): Interview of Dr. John Gordon*

Dr. John Gordon (left), Dr. Daniel Vogt and Dr. Kristiina Vogt (right).
Sources: John Gordon, Daniel Vogt.

Kristiina: Can you start off by mentioning your experiences and backgrounds in Indian forestry and your involvement with the assessments conducted every 10 years of their forest practices?

---

* By Dr. Daniel Vogt and Dr. Kristiina Vogt.

John: *I have been involved with all three so far of the so called IFMAT (Indian Forest Management Assessment Team) that are mandated by the 1990 law —* the National Indian Forest Resource Management Act. *They are supposed to occur every 10 years and be done by an independent group that is non-Tribal, non-governmental people (this is not always true since some federal employees have been involved over the years). The team is put together with about 10 people. We visit reservations and look at all available data on Indian forests and forestry and answer a set of questions that are mandated by the law. There are eight questions which basically cover every aspect of Indian forestry. Because of when the law was passed, there was a heavy emphasis on timber, but it was supposed to cover all aspects of forest uses. The idea was to come up with a nationwide report card — excluding the Alaska Native claims corporations on the basis they are not trust lands. So the idea of the IFMAT was to come up with a report card on trust lands throughout the United States. Each time an IFMAT assessment occurred, we visited 20 or so reservations and had field trips gathering other information from a whole variety of sources. A lot of the information existed at the Bureau of Indian Affairs and other federal agencies, and private sources and nonprofits, for example, who worked with Tribes.*

Kristiina: How long did they usually take?

John: *They take about two years from the beginning to the end when a product was produced. The product intermediary every time was the Intertribal Timber Council (ITC). They provided the funding and oversight. The ITC had to accept the provisions of the report. When ITC made suggestion, the IFMAT team didn't have to accept them. There is some tension there because the findings are not always uniformly made in line with what the ITC would like to have said. It is a fairly complex and pretty intensive process. When the Intertribal Timber Council accepts it, they pass it on to Congress. Then it is up to Congress, since they asked for it, to do something or not.*

Kristiina: So where does it go in Congress? Do they just deposited it there or does someone in fact look at it?

John: *It goes to the Secretary of the Interior who is supposed to present it to the appropriate Congressional committees. It's his view of what those are. In every case there have been Congressional hearings but not much has happened from those as far as I can tell. Well that is not fair*

> to say. There are the trends and there is some impact. It is an interesting thing to me since it is the only aspect of federal forestry that does have a congressionally mandated independent assessment. The other federal lands — National parks, National forests, BLM lands, Fish and Wildlife Services lands — don't have a Congressionally mandated, independent assessment.

Kristiina: Do you think with the current issues we face on taking public lands and giving them to states, the IFMAT process could be a viable approach to address these issues? Should the IFMAT approach be utilized instead of having local level discussions?

> **John:** *You mean an independent assessment? Well I am an assessment fan and I have written about that in the past. I think science based assessments are useful if they are done right. They very often aren't. One way they aren't is when no one asks for them. When white papers are produced and passed around and no one asks for them, nobody does anything.*

Kristiina: So going back to the IFMAT process, the Intertribal Timber Council is an intermediary. What does that mean? Do you think this introduces a bias into the process? A Tribal member is the Chair or whatever his title is in the ITC?

> **John:** *He is called the President of the Intertribal Timber Council (ITC) Board of Directors. Well I will give an example from the first IFMAT to answer your question. When we were being briefed by the Intertribal Timber Council and the BIA, Gary Morishima said to the IFMAT team: "Well just tell us everything." (Gary Morishima who has been a prime mover in the Intertribal Timber Council and a consultant for the Quinault. He is not a Tribal member but Japanese American, but he is very deeply involved in Tribal matters.) His phrase was just* "tell us the good, the bad and the ugly and then tell us how to manage our forests better". *We said, Norm Johnson actually said,* "What are your objectives?" "Why do Tribes manage these forests?" *Both the BIA and the ITC said we know the answer to that,* "We want economic improvement and the Tribes want money." *We said* "Fine but how do you know that?" *They responded* "That is just how it is." *So, we said* "Well we won't do it unless you allow us to independently assess what the management goals are." *Joyce Berry actually did a nation-wide survey of Tribes asking what they thought was most important.*

Kristiina: Is that in the reports?

John: *Yes, all three reports are on the Intertribal Timber Council website and downloadable (Box 10). Economics came in a distant fourth. The number one value was forest preservation. Keeping the forests as forests. Number two was keeping it healthy. Number three was cultural support for medicinal plants, sacred sites. The fourth was economics. It wasn't that economics wasn't important but it was not the whole story. But yet for how long, almost a century as the BIA had done most of the management of the Tribal trust lands and assumed that the economic thing was paramount. In fact I remember when a class mate of mine working for the BIA explained "We are more like Weyerhaeuser and the Forest Service since we manage for maximum revenue." That was at Warm Springs. Later when the Warm Springs Council did their forest management plan, they had four things they wanted — wildlife, water, cultural values and money.*

Box 10: IFMAT I Indian Peoples' Visions from Forests

"IFMAT I conducted surveys and focus group discussions during site visits to Category I and II timber tribes. The question asked was: "What do you most value/want from your forest and why?"

The first IFMAT report revealed a significant divergence between tribal public values and the perception among BIA personnel of those values. Tribal members articulated a clear desire to place protection of forest resources foremost, with strong concern also for cultural uses and aesthetics. BIA personnel, especially non-tribal foresters, placed greater emphasis on income generation as a primary management value. Tribal natural resource staff also rated protection less highly than did the tribal public."[1]

Kristiina: So, the first report was done about 30 years ago?

John: *Right. Not quite 30 years ago. Maybe 25 years ago. They all came out in threes. The first came out and was distributed in "93", the second one in 2003, and the last one in 2013.*

Kristiina: What triggered the first one to be done?

John: *The law. John McCain and one of his senior congressional aides was the Congressional spark plug for the first one. Two years later when*

---

1 www.itcnet.org/file_download/d045d458-e406-4515-baab-52ba685dc71d

we testified to the Senate, McCain was the only senator that showed up. Others sent aides.*

Kristiina: That must mean he had some experience with Tribes in his day?

John: *Yes, Arizona is the state with the most Tribal voters.*

Kristiina: When you got involved with this, were you the Chair of the first IFMAT?

John: *I was the Chair of the first one. John Sessions was co-Chair. I was the Chair of the second one and John was the co-Chair. The third one they just called us co-Chairs.*

Kristiina: Did they give you rules about what you were supposed to do? Did they want you to come up with particular results?

John: *No.*

Kristiina: You had pretty much freedom to do what you needed to do?

John: *We pretty much insisted on it. When we delivered a finding that the Intertribal Timber Council didn't like, we argued about it and listened to them. At the end, we put down what we thought the data indicated. They were really good about it. I mentioned Gary Morishima who was a prime mover on all of this getting started and funded. He is also one of the best reviewers of documents that I have ever seen. He was a copy editor as well as a contractor. That was a tremendous help.*

Kristiina: It is interesting to think about whether a similar process would work in assessing whether Western states should be given federal public lands to manage. What would be a comparable organization similar to what happened with the IFMAT that could assess these land issues? What is the process that you would go through to see whether lands should change ownership? You would have to have very forward-thinking individuals, like Phil on the Intertribal Timber Council, who understands both sides.

John: *I don't know of anything organization that provides a comparable assessment similar to IFMAT. There are some attempts to address these issues. I gave a talk at a meeting that Bernard Bormann had invited me to a few months ago. It was with the Washington State DNR and a group of other experts. What they were looking at, if I under-*

> stood correctly, are regional assessments where they had expert panels look at state lands and what they were doing. They also had a hearing process where people in the region could say what they thought about state lands and what should happen to them.

Kristiina: I know the new DNR Lands Commissioner is trying to do a lot of things and make changes. But DNR has problems that everything changes depending on who is elected as the new Commissioner. It is a political appointment so the new person who comes in tend to change everything, especially since they ran against the previous Commissioner. The new Commissioner tend to have a different philosophy and need to show that they don't make the same mistakes made by the former Commissioner who is now out of office. Nothing becomes institutionalized as the new Commissioner replaces all people in the top executive positions who bring in their own ideas of how to manage state lands and waters. You still have people lower down in the institutional structure, but they have no voice and are not listened to. This creates conflict. When I was on the Board of Natural Resources in 2000, the Lands Commissioner was very environmental and did not recognize that tree cutting could be done sustainably to generate revenue. She also did not understand that locking up trees was not going to retain them in the landscape. This situation was very tough for the foresters working in DNR. But it also made it tough for the Lands Commissioner because they resisted the changes.

> **John:** *Interesting. This is an aside from Indians, but I think it would be a fascinating study. Is it true that the Washington State DNR does a way better job with their forest lands than the Department of Forestry in Oregon on their state lands? Is that true and if so why? It is an important question.*

Dan: So how would you prove if it is true?

> **John:** *Well I would look at the same kind of stuff as we did for the Tribal forestry assessment. "How much and what do they produce?" "How are they in terms of the environment, in terms of timber production, in terms of public satisfaction?" "Are their outcomes aligned with the general policies of the state?"*

Dan: California want to change their forestry practices and recognize that Washington State DNR is managing forests better. The perception is that WA DNR is doing a better job with their forests.

> **John:** *Over the years, my perception has evolved and there are two states that stand out in terms of their forestry practices. One is Wash-*

> ington and the other is Michigan. I think Oregon has always had competent field foresters, but they have never been able to manage the political front very well. That is the downside. The upside is having an elected commissioner. Oregon has a committee that oversees forestry — the Oregon Board of Forestry. They have three environmentalists and three timber beasts. They are supposed to guide the program. Having one elected head can have real benefits.

Kristiina: If you think about organizations that are set up to look at timber like the Intertribal Timber Council, do you think they are open enough to address many of the other issues or does it bias them? It seems it depends on who is on the committee or in the organization that determines what happens. In the IFMAT assessment you had some people who were really open and systems thinkers. How did this happen?

> **John:** Well that is what the law says that it is supposed to be: An independent assessment done by people who aren't stakeholders.

Kristiina: So, you were brought in. You were at Yale and you were a Dean of a Forestry and Environment Studies program.

> **John:** I am sure that had something to do with it. I had no knowledge of Indian Forestry what so ever when I started. It was something other than that. That was considered a virtue since they did not want people who already had a lot of policy-based opinions. I think there is an important point since for a science-based assessment process to work it has to be like that. What it has to include is members that as are disinterested in the results as possible. You can't have people who have clear conflicts of interest on assessment panels. Another thing is that it is an elitist process. It is a science-based assessment but it needs to be an evidence based and independent of anecdotal and political initiatives. In so far as possible.

Kristiina: In a way you have the President of the Intertribal Timber Council who still has to respond to other Yakama Tribal members on issues occurring on the res (reservation). He told me a story once about having other Tribal members drive by and look at some forest management practice that was implementing in a forest. This is not months later that someone had questions but on the same day that he did something in the forest. If they didn't like what he did, he finds out about it immediately. The challenge for him is being part of an IFMAT and supporting what his tribe wants him to do.

I was on a National Academy of Science committee that was driven by one person who wanted certain things done. He mostly focused on not wanting whole logs to be imported to the United States because they were responsible for introducing the Asian gypsy moth. The committees charge was broad and needed to assess the impacts on rural communities by less trees being cut. We had about 18 scientists on the committee that were involved in the assessment but their inability to work together on a common question really stood out to me. You had 10 people in the last IFMAT sounds like you could have run into the same problem but didn't?

> **John:** *In addition to our supporting cast, we had three Tribal students attached to our team. They went through the whole process with the idea that they would become the memory of this whole process.*

Kristiina: I am sort of jumping again but still thinking about leadership. When Dan and I were at Yale, we watched you being a master at getting faculty to agree on an issue during faculty meetings. Usually everyone had a different opinion on a topic. Somehow you seemed to be able to develop faculty consensus. It wasn't like you were sitting there thinking about how am I going to get agreement here but it naturally flowed from how you interacted with the faculty. You were great at throwing out a saying or story that totally threw the faculty off so they stopped and started thinking about the issue. They had to talk and that is what happened. It seems before you set up an assessment the group has to go through some training in leadership like some of the stuff you have written about. Because if you don't do that, everyone is going to get caught up in their own thing. So I think the success has a lot to do with you but finding people like you is not easy. I don't think there are a lot of people around like you. So you have to somehow get new leaders to some level of understanding about how you move this thing forward. You have to get it where they leave the politics behind and outside the door.

Are there some rules that you can set in place? I am trying to think about what we want in this book and how you get people to collaborate differently. If you have a process that is going on that is working, how can you in fact transfer it outside of the Tribal community because the whole idea is that we want others to start thinking about how to look at issues differently? I think it goes back to leadership. I am thinking that you can't just form a group if you don't have an understanding or have thought about how leadership works first.

> **John:** *There are a number of kinds of leadership. Leading an assessment group is different than leading a tribe or a college. If you are leading a forestry college my view is that you need a vision where it should*

> go. That ought to be robust enough so you can sell it to outside people. If you are leading an assessment — a science-based assessment — you should not be having a single vision. The vision should emerge from the process. That means people can say whatever they think but they have to be able to prove it. They have to have data of some sort or at least base it on an observation.

Kristiina: Do you think that is what you do? When you had faculty bringing up things, you expected them to be able to support how you would do it differently or do it better. The process for how you engage can make a huge difference.

> **John:** *People have to be allowed to play their tape. If you absolutely don't want to hear what they have to say, you have to get rid of them if you can. Everyone gets to say what they want but it just needs to be evidence based. Let's say in one of these IFMAT groups we had a very highly capable person in their discipline but who was known as a vocal advocate for Indians. You would have to talk that person out of their view and say that you are right but that is not what we are doing. We are trying to make a neutral assessment. If we don't, we will be seemed to be advocates of a particular point of view and then no one is going to pay attention to it. Ultimately assessments are good if they convince people who are not believers to begin with.*

Visitor: Can we go back to what you just said? When you are doing assessments, it is critical that you have objective parties involved in that assessment. You said the outcome has to be evidence based.

> **John:** *It has to be free from political position taking. That is why on IFMAT people would say* "Who is the representative of this group?" "Who was the environmentalist and who is representing the timber industry?" *They responded* "Nobody." *They are individual experts and they are here on their own recognizance and if they act like a representative, let's say of Weyerhaeuser, they are gone. You take them off.*

Visitor: That is a quote right there.

> **John:** *I think a good example is Washington where at least your commissioner is elected. Whether you like the commissioner or not there is some substance there. In Oregon they carefully appoint representatives of given points of view. They have delegates. The Wilderness Society has delegates. The timber association has delegates. I am amazed that they expect to get anywhere using that kind of approach. It is like recreating a state legislature but with 12 people. You can't do it.*

Kristiina: I remember you drawing once on a napkin a square and you would use it to categorize people by their leadership potential. You had four boxes in this matrix that you could categorize a person. You had someone who was not a facilitator, but I can't remember all the labels for each box.

> **John:** *I can tell you. You have a four-cell matrix and on one limb you have completer and non-completer (see Box 11). On the other one, you have makes my life easier or harder. Then you classify everyone in your organization or panel or whatever you have into those four boxes. Of course, the one on the top left is completer and makes my life easier. Those people you give them whatever you want. Let them do whatever they want, they are fine. In the lower left and upper right corner as non-completer but make my life easier or completer and make my life harder you need some kind of balance. You can't have all of one kind or the other or it will drive you nuts. But your big job is to get that lower right-hand cell — the non-completer and make my life harder — get rid of them. Get them out.*

Box 11: John Gordon's Matrix to Evaluate People's Potential to Lead (Gordon and Berry 2006)

	Completer	Non-Completer
Makes life easier	A KEEPER	BALANCE
Makes life harder	BALANCE	NOT KEEP

Kristiina: Is there a way to take that same concept to select a committee to do an assessment?

> **John:** *First you need to figure out what the questions are, and they should come from outside of the committee. The committee should not make up its own questions. In the IFMAT, the questions came right out of the law. The questions were right there and you need to do whatever is needed to answer those. If you have question where you need to answer whether the timber sales carried out rationally and without graft. You need to have someone who knows about timber sales. Are forests protected from insects and disease? Then you need to have someone*

who knows about that. The questions dictate who you need in terms of expertise. But my advice to people doing this is get it in writing. First of all there has to be a recipient — someone has to want the answers to some questions, and they are usually questions that are not phrased as scientific questions. It is like what are we going to do with these forests? You get that in writing before you pick the group.

Kristiina: Could we use the same matrix but have it as something that allows you to determine who to have on a committee? It seems that one limb should include people that are totally disciplinary and are linear thinkers. The other limb in the matrix should be web thinkers. I am trying to think what you would call these people, but you need those that are ranked as being credible by their peers (Box 12). They need an expertise, but they also need to be open to ideas.

Box 12: Matrix to Evaluate People's Selection to be on a Complex, Transdisciplinary Committee

	Web thinkers	Linear thinkers
Willing to listen	FIRST SELECT	BALANCE
Not willing to listen	BALANCE	NOT SELECT

John: That is an excellent point. You can't have disciplinary chauvinism anymore than you can have any political focus. You have wildlife biologists but they have to be willing to listen to people who are not wildlife biologists. Most people who come up through the sciences are like me because they know a little bit or tiny bit but think they know everything. Willingness to listen is critical in an assessment. There are a lot of science-based people who think their discipline is the only thing in the world and write a lot of books. They think everyone else's knowledge is not as important. You can't really work with them. That is an excellent point.

Kristiina: Would it work then to have a linear thinking process versus web or holistic thinker on side of the matrix and then the willingness to listen or not listen?

Visitor: It is linear vs non-linear and non-linear does not necessarily mean holistic? You could be a non-linear thinker, but you could still be a chauvinistic person. Someone could have fantastic ideas but be all over the place, so they are not holistic in a sense. They are not looking at it from the 50,000-foot level and seeing the whole picture.

> **John:** *That is a good point too. The contrast between those people who are sequential thinkers or think in reverse. The other thing is how well do people recognize the limits of their knowledge?*

Dan: Or being willing to admit it.

> **John:** *I will give you an example. You know John Sessions? He is one of the best forest engineers in the world, but he is willing to acknowledge when he doesn't know something. This makes him an easy person to work with. Not everyone is willing to admit when they don't know something.*

Kristiina: I always thought forestry was a very interdisciplinary field. I remember having discussions with the guys in Marsh Hall (FES Yale) and it was mentioned that forestry started losing its direction when the FORTRAN program started being used. All of a sudden you just had equations and you put in numbers and come up with a result. We used to joke about it that you lost the art which is the core of forestry. When you are a forester you are translating the environment. The art comes in by your ability to see what's in the environment, what's maybe not working within the forest, and recognizing that it is going to vary depending on where you are. It's not like each forest is a duplicate of another forest so you need to be able to translate each forest. Equations make you think that you can replicate each forest and manage it the same way and it is always going to follow a certain process.

> **Dan:** *When you say linear is another way of saying tunnel vision? Or are you talking about something other than tunnel vision?*

Kristiina: Yes, it is tunnel vision because you have blinders on. You see the problem but don't let anything else get in the way. Even if it is clear that you have the wrong pathway. Or you can't admit the fact that you made a mistake.

> **John:** *You are not defining someone who is linear but STUPID. You believe you are right even when you are proven wrong.*

Kristiina: It is almost like when you have a complex and transdisciplinary issues, you first need people to serve on a committee. But you select them following a test or do a survey (Box 12). When I think about Tribes, they manage a specific piece of land using practical knowledge they have collected for a long time. Maybe the diagram has to include are you theoretical or applied? If you are applied, you are more like the forestry professional. You will be more open to practice the art of forestry. When you are applied you know that there may be many pathways to where you are going but if you are theoretical you are probably going to be hung up on ideas or theories. The theory says such and such but is not based on practical experience. I hear all the time in the social sciences that students are forced to build bridges to theory. They must melt both parts together. This seems to have arisen in response to the social sciences being told for decades that they did not have theory. They were not happy about that perception. If you don't have theory, it suggests that you are not building on a body of evidence that has accumulated over time.

Visitor: Don't define people where they are but where you are trying to go.

[Kristiina to John]: You have theory but also applied knowledge. What do you call somebody who combines those? Maybe there is a "sweet part" in the center of that.

> **John:** *I think the more important point is whether you are doing good research or bad research and not whether you are doing theoretical or basic versus applied research.*
>
> *A pitfall in identifying what research to conduct are time lags since it takes a long time to do research. There was consensus that it was a minimum of ten years after you knew something was theoretically possible before it showed up in the practical world. My favorite illustration of that was in the North Central Forest Experimental Station which then had one of the few forest genetics program other than in the south. They went to the pulp and paper industry, which was big in the central region then, and said "What do you want?" Well they said, "We want more white spruce because that is what we make paper out of." "We have trouble with planting white spruce because of frost kills." "We need better white spruce." The geneticists finally solved it — it wasn't an elegant solution but it was moving white spruce further south so they did not get exposed to frost kill and grew a lot faster. It worked. Then the pulp and paper industry said, "Why are you working on spruce?" "We have shifted the whole industry to aspen." "If we want spruce, there is tons of it in Canada and can ship it down here*

on a train." "You stupid researchers." *They questioned why the geneticists were working on this even though they had 15 years earlier said this was their problem. Taking end-users too seriously can be a real problem. It takes a little higher vision which is an elitist view.*

Dan: Isn't that the importance of having a mix of basic and applied research? Applied science provides a small picture of the research problem while basic science has the big picture?

John: *The Soul of applied science is doing something better and making incremental adjustments to what you are doing now.*

Kristiina: We have not set the bounds where the question is coming from. Research science becomes very narrow very quickly.

John: *I have heard more than once that the* **"forest is our super market" "our food, our medicine, our baskets"**. *Except under very limited circumstances that doesn't work anymore. It can be taken as an ideal but the problem is that often Tribal timber management is messing up the supermarket.*

Kristiina: Plus, they are on a smaller land-base than what they used to have. They will not be able to collect cultural resources from this smaller land-base.

[Kristiina to John]: Why do you think Indian Forestry is good forestry? You have been quoted as saying it is better forest practices than some of the other groups managing forests. Why have you said that and what is the reason for saying this?

John: *Of course, better is a loaded term. The evidence that we thought we saw from three iterations of IFMAT is that Tribal forestry is very diverse, very different in different places, some places it is good because of its environmental benefits. Tribes do a better job of balancing the different outcomes of forestry and natural resource management. They have an openly stated bias for wildlife and fish. So their timber programs are more likely to take that into account or have historically. So that is one thing. The other thing though is that for many timber Tribes selling wood or logs is a major source of income and many don't have other major options. So as the number of Tribal members is increasing, they realize they have to include revenue, usually from timber, in their equation for management. They avoid the solely biodiversity driven management that we see on a lot of public lands where all the constraints are in favor of fish and wildlife. The Tribes have a*

> bias in that direction but also have a need to produce revenue. If you look at the Self-determination Act, they all have at their center striving for economic self-sufficiency. They do forestry for both environmental goals and economic goals. That is why it is better.
>
> Plus, they also have a big advantage on reservations: They have to live with all the consequences of their management decisions. I can live here in Portland and say I don't want any trees cut because I like trees. It doesn't affect me at all. If Tribes do not cut trees, they do not have money. Similarly, I could say we need to do prescribed burning in all forests in Oregon because of the fire problems. When Tribes do that, they need to breathe the smoke. So they have to think about it. Not as a one-sided issue but as trade-offs.

Kristiina: Is there a way to have it where the agencies themselves, like the Forest Service, have to be responsible for the repercussions of their decisions? I guess today it has been estimated that half of the Forest Service budget goes into fighting fires. They won't move these costs into another agency that have a mandate to help communities respond to disturbances like hurricanes. Could it come out of the agency and then make it where people have to become more responsible for their decisions? I realize this is idealistic. Is there a way that politicians can be made responsible for the decisions they make? Kaleb — an investment trader who wrote a book entitled *Anti-fragile* — suggests that businesses need to be responsible for the decisions they make. He mentioned that we have more people today who have no responsibility for the actions they take. So the environment and people are volatile and we need to factor these in while being responsible. But today decisions are made. Then people leave their offices and are not impacted by their decisions since too much time has passed and they can't be linked back to the decision.

[Kristiina to John]: Do you have any examples of when the Tribes did not move forward on some business venture because of its impact on their cultural resources? Like selling timber because there were other factors that were more important for them. Are there any examples or is it just something they practice but it is not documented? Are there any examples in forestry?

> **John:** *You see on almost every Tribal forest they sacrifice some potential revenue, timber yield, or other goals. We talked about Warm Springs and they have some ~300,000 to 400,000 acres of what the Forest Service would call commercial timber lands. They say their commercial forest land is only about 125,000 acres. They have typed out whole areas that they don't cut because they are protecting fish. One*

Dr. Daniel Vogt.
Source: Daniel Vogt.

> of their goals is to have old ponderosa pine forests that look like the classic ones. They don't harvest any timber unless there is a fire or something like it in their pine forests. You will see that on the Colville and certainly on the Yakama. They don't cut some places because they like it that way.

Dan: Are you aware of how Tribes manage their forests because the decisions you make today may not be seen until five decades in the future? Do you think that Tribes are doing anything different compared to if you were Weyerhaeuser or some other company?

> **John:** *I do. Data shows that they tend to have longer rotations for timber, and they tend toward more equal consideration of fish. They talk a lot about seven generations, so their decisions should be good for seven generations. I think it comes from them thinking they are going to be there forever. They are going to stay there on the land forever. So instead of saying what are we giving up now, they ask what we have to do so that what we want will be there in the future. It is a different mindset. I am not saying it is universal and perfect. The constant thing the Chairman of the Coquille would say was, "Well just be sure it is a long-term thing, so we are not scarifying something in the short-term that we would wish we had in the long-term."*

Kristiina: What is happening with the Coquille and the interests of local people to have them manage BLM lands? I remember that local people felt that the tribe could manage better and get more return from managing the lands.

> **John:** *These were the Coos Bay Wagon Road lands. Their lawyer wrote a good article on this in the* Natural Resources and Environment Law Review *in 2007 that the American Bar Association puts out. He points out why it might be a good thing to have Tribes manage more federal forest land.*

Kristiina: I will be interesting to see the article. Many people seem to think that if you do all these things on the land like what the Tribes want, you won't make any money. This is a very common perception held by non-Tribal people. What it sounds like is the non-Tribal people felt Coquille could get more return from managing these lands. Did this happen?

> **John:** *No, it never happened on BLM managed lands. There is another good example in the Coquille where they abutted land managed by a company that was on a 39-year rotation and where they cut big trees. They find markets for the big trees that buyers pay more for them than little trees. The BLM manages solely for little trees now. So little ratty logs instead of beautiful big logs and you wonder why you aren't going to make any money. How can the Coquille do it while having their environmental ducks in a row? They have way more people per acre on the land. They don't have total more people, but they have more people out on the land per acre than BLM or their neighbors. They also monitor everything. This has puzzled me since the Northwest Forest plan days when it was all about monitoring, monitoring, monitoring. But federal agencies still don't do it. Its 25 years ago and they still don't do it. But the Coquille monitor stream oxygen, temperature and everything before the timber sale and after the timber sale. So if someone said they screwed up the fish, well here is the data. It's not rocket science. They also had people who knew regulations and how to manage. It is hard for me to see why that couldn't be done on BLM land.*

Kristiina: They almost don't seem to have enough people to do what the Coquille did? JD Tovey has talked the same thing for the Umatilla where they have many more foresters than what you find on federal forest lands.

> **John:** *The Forest Service used to have four times more people working in forests than BLM if you took the whole agency and divided the number of people working for the Forest Service. They need to get more people out in the woods and out of the offices.*

Kristiina: There is one book that Dr. Mike talks about called *"The Art of War"* that had an enormous impact on him (*see* Section 2.2). It was written by a Chinese military strategist in 512 BC. He talks about how you don't burn all the houses of the people you attack during a war. You win a war by not having a war. You figure out how you are going to do that. It has helped Mike to think about how you approach problems. Mike loves history and it is almost like we need to get more people reading history. Maybe history should be required teaching. If you are a resource manager, you better have a history class. It's almost like time starts just a year ago and how can you plan for the future if your time frame is so short. It doesn't make sense to reduce the time period in which you think about a problem.

5.3.2 Tribal Forestland Management: A Growing Force in the PNW U.S.*

Last year, we published a brief article on the potential of Tribal forestry [99]. This is a more complete version of that article. It is adapted particularly to the potential for Tribal forestry and for natural resource organizations to manage off-reservation forests, and particularly federal lands. We draw on a variety of sources to do this, but particularly the three Indian Forest Management Assessment Team reports (IFMATs I, II, III) [100–102]. These congressionally-mandated reports were done in consequence of the *National Indian Forest Resource Management Act* (NIFRMA) of 1990. Each assessment was done by an independent team of experts and submitted to congress and the BIA without Federal intervention. They can be accessed at www.itcnet.org. Our major theme is that Indian forestry organizations are of a quality and philosophical orientation to play a larger role in the management of Pacific Northwest forests.

The largest acreages of Tribal forest are in the West (Fig. 5.5), especially in the Pacific Northwest, with an aggregate national total of more than 18 million acres (about 1/10 the area of the National Forests). As far as volume, the Tribal forests in the Pacific Northwest (excluding Alaskan Native Corporation lands, which are not held in trust by the US government) contain 21 out of the 42 billion board feet of standing timber nationally.

Here are some findings that we believe are important when considering Tribes as off-reservation forest managers of federal forests.

---

* By Dr. John Sessions, Dr. John Gordon, Philip Rigdon, Donald Motanic and Vincent Corrao.

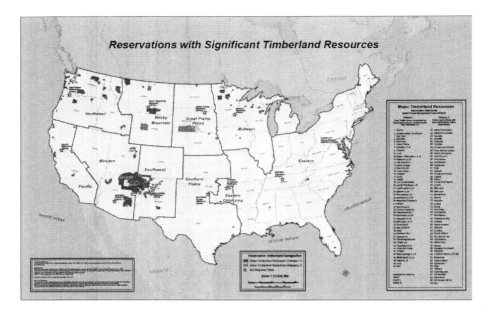

Fig. 5.5 Reservations with significant timberland resources (Courtesy: Bureau of Indian Affairs, Branch of Forest Resource Planning). [Dark grey colored are major timberland resource owners, light grey are minor timberland resource owners]
Source: Philip Rigdon.

1. SWOT analysis: Tribes have both strengths and opportunities

To examine this theme, we adopted the framework of a strengths, weaknesses, opportunities and threats (SWOT) analysis as a useful and concise method to appraise the potential of Tribal forestry for this larger role. **Strengths** of Indian forestry are a long term vision, the knowledge that Indians must directly live with the consequences of all their decisions and actions, progressive federal forest legislation (the *National Indian Forest Resource Management Act*, 1990) that specifies sustainable forest management, a trust land base that cannot be sold, an increasing land base assisted by gaming revenues as well as forestry receipts, more predictable and balanced forest policy, consistent production of ecosystem services and harvests that are significant, but often substantially below their allowable harvest levels. **Weaknesses** include unrealized potential gains through exploitation of economics of scale, the lingering effects of past federal allotment policy, unfunded environmental and consultative mandates, federal funding that lags behind federal support for other lands, challenges of workforce development including an aging and undercompensated workforce and the inability to use trust land for collateral. **Opportunities** include continued recovery of ancestral homelands, integrated landscape planning with their neighbors, shared management of adjacent federal lands, and marketing of ecosystem services. Threats include sub-optimal investment, particularly in fuels management on neighbor lands, and loss of processing capacity near to Indian forests.

Timber harvest levels (Fig. 5.6) and timber revenues have steadily dropped in the past three decades causing negative economic consequences on forested reservations. The "Great Recession" and poor timber prices from 2008–2013 were contributing factors along with shifting Tribal goals and the effects of historical federal practices. Many Tribes have put increased emphasis on fish and wildlife management operations, as well as on cultural subsistence and ecosystem service goals for forests (e.g., carbon sequestration, watershed protection, recreation, first foods). The trend is in a direction similar to many public forests, but much less pronounced than the shift on federal lands to management based on rigid biodiversity-based rules. The recent Annual Allowable Cuts (AACs) for Tribal commercial timberland totaled 751 million board feet [103]. A variety of factors contribute to the inability to reach the AACs, including lack of funding, lack of staffing, reduced processing infrastructure, and BIA administration difficulties (hiring freezes, failure to fill vacant positions when freezes are lifted, failure to get the funding that is available to the field through direct service or contracts/compacts with tribe, BIA procedures). The recent large fires will further impact AACs for some Tribes.

As forest management goals evolve, Tribes are also acquiring new trust and fee lands, so the total Tribal land base is slowly increasing. For example, the Coquille Indian Tribe of the south coast of Oregon just increased their land base by over 60% through the purchase of private forest land. Additionally, the Muckleshoot Tribe southeast of Seattle has purchased almost 96,000 acres of

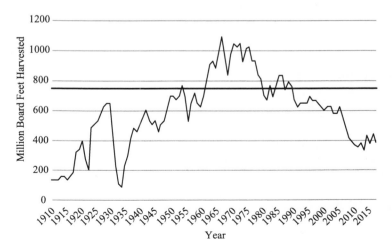

Fig. 5.6 Harvest from Indian trust lands from 1910 to 2017. The straight line shows current AACs.

private commercial forest land increasing their land base almost 25-fold. Between 1990 and 2015, Tribes added more than 2 million acres through purchases and transfers. Fee lands are managed under state forest regulations. Fee lands are eligible to be moved into trust but limit Tribes in their ability to repay costs of acquisition. Once in trust, the lands cannot be used for collateral for future loans.

2. They manage relatively well with minimal resources

Tribal forests and forest management present valuable models for sustainable forest management. The IFMAT reports cited innovative silviculture, better integration of timber production with environmental services, and low management costs in comparison with adjacent federal forests. Tribes are place-bound, must live directly with their decisions, and share strong, long term visions about the importance of forests for jobs, revenue, subsistence, and spiritual values. Although Tribes are sovereign nations, they have a trust relationship with the federal government that both threatens Tribal forest sustainability and has created opportunities for regional and national leadership. Indian led initiatives such as the Anchor Forests initiative provide a pathway forward for collaborative decision making for landscape management across multi-jurisdictional ownerships. Indian leadership resulted in successful passage of the *Tribal Forest Protection Act* (TFPA) in 2004 that recognizes federal responsibility for passive federal action on federal lands adjoining Indian forests. The passage of the *Indian Trust Asset Reform Act* (ITARA 2016) provides an opportunity for Tribes to move further toward self-determination and self-sufficiency by gaining firmer control of their forest assets though Tribal laws and management. Several Tribes (for example, the Cow Creek Band of Umpqua Indians) are now proposing pilot programs under which they would have firmer control of their own assets, including forests.

These ideas could be more broadly applied for the sustainability of all forest lands. Past federal policies with Indians have resulted in weaknesses that we recommend be corrected including more investment and restructuring of the trust oversight relationship. A new approach to oversight was first suggested in IFMAT I in 1993, which foreshadowed ITARA's provisions in 2016 (*see* below). Increased recognition of the broader potential of the Tribal integrated, locally-focused, traditional- knowledge- plus -science approach, supports the creation enhanced legal, trust-based mechanisms for Indian management and co-management of public lands.

Kenney [104] suggests management of the Coquille Forest as a model for Tribal management of federal forestland. This would in turn require scaling up of Tribal resource management capability, creating better mechanisms for intertribal cooperation and entrepreneurship, and development of venture capital sources for scaling up new and better cooperation mechanisms among Tribes. For example, recently, at the request of the Oregon State Land Board, a private timber company and a Tribe, with potential involvement with other Tribes, proposed a detailed plan for the purchase and management of the Elliott State Forest. Although the state did not follow through, the effort showed that Tribes and the private sector have the potential to work closely together on forest purchase and management opportunities.

3. Their management strength is a combination of focus and integration

Indian forestry benefits from both traditional knowledge and modern science. One of the most important aspects of traditional knowledge is a sense of place. In this view, there are no theoretical forests but only the one currently seen and alive in memory. Each place is unique and has its own whole set of attributes, all interconnected. For example, the non-Indian view is that organisms and species can be viewed as marbles in a bag, to be counted and thus managed and pronounced sufficient or insufficient to a given rule or purpose. In the Indian view, it is the relationship among them, as beads tied together on a string. This string tells a story, in the context of place and time that is important in aggregate to guiding decisions [105]. Thus, the Indian view of an "interdisciplinary team" is not a gathering of "ologists" each advocating for a specific "resource" but rather a true team all positively contributing to the achievement of an integrated, whole set of goals. This integrated "whole" outcome theoretically balances current Tribal needs and the long-term health of the land. Although they are not always successful in doing so they often practically demonstrate the closest approximation we know of to the definition of "sustainability" as offered by the Brundtland Report, "*Our Common Future*": Providing for present needs with minimum compromise of future options. Perhaps the best condensed version of this is the "seventh generation" principle attributed to the Iroquois which says that all decisions and actions should be taken with at least seven future generations in mind. An important element in the integrated Tribal approach to forests is the role of traditional story tellers, particularly when a change in management direction is required.

They help Tribal people discover the truth based on seeking the right questions. If the right questions are asked, the journey of transformation can be easier. **These stories then become a rough guide to understanding how transformation occurs. The story itself does not give answers but lead publics and managers to them** [106].

The long view that results from Tribes' intended tenure on the same land fits well with the long times involved in forest production. Conventional economics has had a difficult time dealing with the perceived drawbacks of long production times. (New Yorker cartoon, grandfather to grandson while walking in the woods, *"Yes, it is good to know about trees, just remember nobody ever made big money knowing about trees."*) **While foresters talk about 20 years as a short rotation, most corporate planners, even in companies eager to declare themselves immortal, rarely talk about planning periods longer than five years. Tribes solve this problem by turning it on its head. They say, *"We intend to be here into the far future; what does that demand of our current behavior?"* This way of looking at things automatically demands the "triple bottom line" approach, and often leads to innovative methods. In Indian forests, the triple bottom line (economic, social and ecological benefits in balance), unlike on federal lands, must include economic return because the Tribes need money to pursue self-sufficiency**. Thus, most Tribes are developing approaches to timber production that are economically sound and serve both community and environmental purposes.

Necessities born of scarce resources, respect for the sense of place, and the need to listen to Tribal citizens have produced innovative approaches to silviculture and integrated management. Tribes spend less per acre to manage forests than is usual on other lands and often do, from an external point-of-view, as well or better than the costlier comparators. For example, in 2011, BIA allocated 2.82 dollars per acre to Tribal forests, compared with 8.57 dollars for National Forests. What is different is the nearly universal sense that *"We are all in this together"*, and therefore must work to solve, not just proclaim, problems. The principles of a shared enterprise, applied together, allow adaptation to a variety of specific forest characteristics and human needs. As such they produce a mosaic of location-specific approaches to insuring long term forest values that are inevitably superior to approaches dependent on a thicket of rules promulgated from afar and intended to fit all forests and conditions. Lacking a rigid dogma, Tribes tend to politically support a variety of management outcomes, typically tending toward emphasis on landscape preservation, wildlife and fish, but also, given their dependence on forests as a major source of trust asset income, with an interest in the financial attractions of sustainable timber management not found in federal agencies.

Tribal management of so-called "treaty rights lands" in the federal forest estate makes sense in that Tribes are used to balancing competing ecological, cultural, and financial interests. Tribal objectives will be naturally built in to management actions from planning to execution, and Tribal managers often have long experience with similar, nearby lands.

To enhance Tribal management capabilities a number of initiatives could be undertaken. Some we think most important are listed below.

Restructuring of the trust oversight process and participants is both an urgent task and an issue that goes far beyond forests. Many schemes for accomplishing this have been suggested over the years, and we still think the one suggested in IFMAT I and echoed in IFMAT II and IFMAT III would work. Bringing together all the key messages about Indian forest IFMAT III introduced FIT (fire, investment, and transformation). The themes of fire, investment, and transformation embody the progress that Indian forestry has made over the period of the IFMAT assessments, as well as the opportunities and challenges the future holds. **Fire** and related forest health issues jeopardize the economic and ecological stability of Indian forest. Strategic **investment** is needed to achieve Tribal forest visions and plans and **transformation** of Tribes to self-governance, and toward the emergence of Indian forestry as a model for landscape stewardship. In any event, the current situation constricts Tribal leadership and is a legacy of a paternalistic past, long outdated.

Recognition of the broader potential of the Tribal integrated, locally-focused, traditional knowledge plus science model driven by a true "triple bottom line" approach should be the subject of coordinated publicity campaign launched by the Intertribal Timber Council and the Bureau of Indian Affairs. *The Indian Trust Asset Reform Act of 2016* may indicate progress in this regard. The *Indian Trust Asset Reform Act* was signed on June 22, 2016 and reaffirms the responsibility of the United States to Indian Tribes, authorizes a demonstration project for Tribes to voluntarily negotiate with the Secretary of Interior to manage their own assets, creates the option for the Secretary to establish an Under Secretary for Indian Affairs, and sets up a process to terminate the Office of the Special Trustee. The Act includes three specific titles with Title I reaffirming the responsibility of the United States to Indian Tribes, Title II directing the Secretary of the Interior to establish and carry out an Indian trust asset demonstration project and an Indian trust asset management plan, and Title III addressing improving efficiency and streamlining processes in managing trust resources. ITARA provides streamlining management of trust assets through the proposed plan and provisions that identify the functions or activities that are being performed by the Tribe under Tribal contracts and other agreements under the Act.

Enhanced legal, trust-based mechanisms for Indian management and comanagement of public lands, including but not limited to the Anchor Forests Initiative and an expansion of the *Tribal Forest Protection Act* initiative, should be designed and tested with state and federal partners.

Scaling up of Tribal resource management capability should continue, fueled by a five-year process to accomplish the funding and staffing goals put forward in IFMAT III. Indian forestry operations are understaffed compared to other public and private forest management organizations. Relatively low salaries, remote locations, and small organization lead to poor career ladders. The majority of current employees are over 50 years old. Exacerbating the problem are the large number of long-term employees eligible for retirement which threaten the loss of institutional knowledge and leadership.

Better mechanisms for intertribal cooperation and entrepreneurship should arise from pairs or groups of Tribes ready to contemplate formal mechanisms for joint management of existing and newly acquired lands. These should be informed by, but not be copies of, existing Timber Investment Management Organizations (TIMOs). Economies of scale are particularly important in forestry. Smaller forest units cannot afford the larger staffs commonly necessary for integrated forest management and even large forest units may not have the continuous workflow to support current staff. Analysis of BIA management cost allocations as a function of Tribal forest size indicated much larger per acre management costs for smaller management units.

Venture capital sources for scaling up new and better cooperation mechanisms should be vigorously sought with the assistance of investment managers and development professionals. Because trust lands cannot be used for capital, Tribes must carefully consider whether to maintain newly acquired Tribal lands in a fee status or to apply for trust status.

What might a "fully enabled" future for Indian forestry look like if the proposed steps for addressing the weaknesses outlined above are implemented?

- Tribes are directly managing a major fraction of the federal forests within their ancestral homelands resulting in more effective fire protection for all lands and more fiscally efficient management.
- Tribes can establish managed fire on the landscape to enhance cultural plant and wildlife integral to their knowledge of the ecosystem and bring fire back onto the landscape as it was historically.

- Through Tribal-industry-federal partnerships, Tribes have created a new wave of wood processing infrastructure based on an assured, high quality wood supply.
- Multi-tribe management organizations, with Tribal and other investors, are acquiring and managing private forestlands within their ancestral homelands, and in other places where sound investment opportunities exist.
- Partnering with universities and other research locations, Tribes are leading a renaissance in forest productivity and management research.
- Sustainable forest income is contributing effectively to Tribal economic self-sufficiency as other forms of revenue are stable or declining.
- The general public image of Tribes is of them as highly skilled and effective land managers.

Can Tribes **work together** with each other and their neighbors to make **real landscape-level management** work to the benefit of all? We think so.

5.4 Realities in Developing Resources on Reservations

Because of the abundant fossil and non-fossil resources found on reservations, it suggests that developing business opportunities for these resources would be an easy planning process for Tribes to pursue. But the historical customary lands were much larger in extent than what is found on reservations today. Tribes may only have 10% or less of their original customary lands [13]. Further, Tribes do not have the luxury to just move to other lands if they make bad decisions while implementing their resource management plans. Therefore, they have to live with the results of their decisions — good or bad. The decisions immediately impact the tribe's ability to pursue their economic self-sufficiency and sovereignty. Further Tribes are more significantly impacted by climate change and the changing fire regimes are common occurrences today (*see* Section 5.1, 5.2.2). So not only do they need to manage a smaller land-base to sustain and to acquire their natural and fossil resources, but they are the most impacted by changes in disturbance regimes. Tribes also balance the protection of wildlife, water, cultural values with the need to make money for the tribe (*see* Section 5.4.1).

This contrasts non-Tribal organizations and companies who are less concerned about balancing several values from harvesting or collecting resources. They tend to optimize the economic return from their decisions because their time frame of risk assessment is short, 5–10 years (*see* Section 5.3.1). Tribes have constraints that impact their management decisions since they still have to satisfy all federal government environmental regulations similar to non-Tribal companies.

Tribes are approached frequently by non-Tribal business or venture capitalists who want to access Tribal resources. There are tax benefits for non-Tribal entrepreneurs to establish businesses on Tribal lands (*see* Section 5.4.2) which is why so many try to develop business enterprises on reservations. These businesses do not understand that just because they can promise a lot of economic benefits, Tribes may not decide to move forward on a business venture. Most of these businesses do not respect Tribal values and do not comprehend Tribal respect for the wildlife, land and water. They focus on revenue generation quickly and then moving on elsewhere. This is not how Tribes make business decisions and the trade-offs of what business ventures to develop.

For any tribe, all economic decisions cannot impact their sovereignty, their cultural resources, or their ability to decide what land-use decisions to implement [13]. If any of these are impacted, Tribes will not choose to move forward on new business enterprises. Cal and Dr. Mike Marchand, who have managed many Tribal businesses, will explore the challenges faced by Tribes and the decisions they need to make (*see* Section 5.4.1). In Section 5.4.2, Dr. Mike Tulee will explore the many challenges tribe face when wanting to develop their energy economies.

5.4.1 Making Business Decisions: Interview of Cal Mukumoto and Dr. Mike Marchand*

Kristiina: Cal, why don't you introduce yourself and mention why we should be talking to you about how Tribes make business decisions?

> **Cal:** *So, the topic is Tribal economic development, right? Well my first experiences with Tribes started before 1980. My initial involvement with Tribes was only as a student at Humboldt State. We used to do things with the Native American group that was there, fundraisers and stuff. I took some classes in history and politics in the Native American Studies program. I took one class on Indian policy where a big argument arose between the professor and the Hupa Tribal members in class. I just used to sit between them turning my head back and forth listening. I had no idea what they were talking about half the time. But my real experience happened when I started working for the Makah Indian tribe in 1980. That started my career in Indian Country.*

---

* By Alexa Schreier and Dr. Kristiina Vogt.

Photo of Cal Mukumoto (top left), Dr. Mike Marchand (top right), Alexa (bottom left), Dr. Kristiina Vogt (bottom right).

Photo Source: Cal Mukumoto, Mike Marchand, Alexa Schreier, and Kristiina Vogt.

Kristiina: Did you already have your Master's degree in Business when you started working for the Makah Indian tribe (*see* Box 13)?

Box 13: Makah Tribe History

"On January 31, 1855, the Makah villagers, represented by 42 Makah dignitaries negotiated and signed a treaty between the United States and the Makah Indians." ... "The Makah tribal forefathers knew it was up to them to protect their peoples' whaling, sealing, fishing and village land rights from elimination. Certain rights were specifically outlined in articles in the Treaty to insure that the importance of continuing these traditional practices was clearly understood by both the United States government and future generations of Makah. In order to retain whaling rights, and to protect the health, education and welfare of their people, the Makah ceded title to 300,000 acres of tribal land to the U.S. In 1859 Congress ratified the treaty, ushering in the beginning of radical cultural changes imposed on the Makah by the federal government and those who implemented them."[1]

---

1 http://makah.com/makah-tribal-info/

Cal: *I had my bachelor's in forestry, and I went out there as a forester. I did pre-sale development and I started their greenhouse and a little business surrounding their greenhouse. I worked my way up to forest manager. I was working for the tribe, but I ended up being put on loan to the BIA for about four years. I was the Neah Bay Field Station Manager and basically managed the activities and represented the superintendent on the Makah, La Push, Jamestown Klallam, and the Hoh tribe. I also did all the public domain homesteads and other similar activities. But most of my work was on Makah lands, that was why I was there. Eight years later I ended up being, for a short while, the general manager of the Makah tribe. Before my position as a general manager, I got my MBA at U.W. and drove back and forth between the tribe and taking classes at U.W. During that time, we started a forest products business, actually a log brokerage company, at Makah. I also got involved with other ventures, like fisheries, at that time. As the general manager for the tribe, I was involved with all their economic development activities, as well as the social programs, the police department, and everything the tribe was involved with like the health services.*

Kristiina: How difficult was it for you to work for the Makah, because you weren't a Tribal member?

Cal: *It wasn't that difficult at first, because when I was working for the Makah as a forester, no one bothered me even though they would get upset about different issues. But it was alright for me to be a forester there and then eventually to become the forest manager. When I became general manager, no one bothered me at first because the tribe was insolvent. I noticed that a month or two before I got offered the job, I had vendors calling me complaining that they weren't getting paid. Then all of a sudden, one day, I was called down to the council and the whole council was there and they said, "Hi, Cal" and I go, "Hi"— it's only five people and I knew them, it's a small town. They go, "We don't think we're going to make payroll in the next two weeks." And I went, "Oh, really?" and they said, "And we'd like you to be in charge." So, they smiled, and I left. The next thing I know I'm working on a deal with the Chairman to take over management of the tribe.*

I was not called the General Manager, because traditionally a Tribal member holds this position. I was called the Operations Manager, with general management responsibilities, but I reported directly to the Tribal Council. They didn't want to give me the title of General Manager, but they said, "We're going to call you the Operations Manager." It's all semantics. Anyway, it took six months to turn it around, when I left, we were putting money in the bank.

It wasn't until the last couple months of being General Manager that people started complaining about having a non-Makah in charge of the government there. When the wolf wasn't at the door, people started complaining more, so that's when I left. After about eight years at the Makah, I thought it was time to leave. So, yeah, there is that issue if you're a non-member.

The other side of the issue though, is that when you're a non-member, you don't have all the ties with the relations, like you don't have your cousin working for you, or your Uncle Bill working for you. But because of the threat, the wolf at the door of being unable to pay vendors, I was able to go through and do a lot of changes within the organization that actually people wanted to do. These people didn't have the political "silver bullets" to be able to do it.

Kristiina: You were there at the right time then?

Cal: *Yeah, it was a good time to make the place more streamlined. So I did. I ruffled some feathers, but that was okay. There was one I didn't want to do. It's kind of an interesting story. The Chairman comes into my office and says* "I want you to put him in charge of fisheries", *even though he didn't have a fisheries degree or had never studied fisheries. I said,* "Well, the biologist, he's been running this program for years and he's doing a good job." *He says,* "No, I want this guy in charge." *I ask why and he goes,* "When his dad was on his deathbed, he said, 'Take care of him, make sure he's taken care of.' And I promised him, so you're going to make sure that happens." *I said* "Okay" *and I made that switch. I talked to the fisheries biologist, I didn't give him any lower pay, I said* "You're still in charge of this, but let him be the head." *That kind of ruffled some feathers but after six months, everyone seemed to be like* "Okay, things haven't really changed."

[Cal to Mike] So you've got a lot of power, you could probably pull that off?

Mike: *That's pretty universal though. When things are going bankrupt, that's when I got to be the president of ATNI. Everyone's leaving, and they said,* "Mike can be in charge." *That's the same thing that happened at Daybreak Star.*

Cal: *That's what happened. I found I was able to make a lot of changes. When things started turning around and cash flow was there, things looked pretty bright and cheery, then I started getting the letters coming in, the complaints. I could feel the tide turning as far as political support, so I left.*

Mike: *They said,* "That job looks easy, we could do that."

Cal: *That is not a real easy job, being in charge of the government. You get involved with everything. One of the things I did start though, that I'm fairly proud of, is the Dog Tag program. I brought my dog down there and I got the very first dog tag. It says #1 on it, and I think we might have sold three more. It was pretty funny, everybody laughed at that. We had dogs running around everywhere, in fact my dog was one of those dogs that probably would've been running around if I didn't take it in.* "Isn't that what happens? You live on the rez, you end up with a rez (*reservation*) dog, right?"

Mike: "Yeah."

Kristiina: But your dog's aren't rez dogs?

Mike: *No, I live between two towns, when people want to get rid of their dogs, they just drop them off on the highway.*

Cal: *Yeah and a lot of them are really friendly. You start scratching them on the head, and the next thing you know, they're in your bathtub and you're giving them a flee bath.*

Alexa: Working for the Makah, were there certain adjustments you felt like you had to make in your perspective. Or did you make decisions differently than you would have if you had worked at a different organization?

Cal: *Well it was very much consensus driven, a lot of things were done by consensus. Like, when we were trying to get policy and things made, I was working on the government side then, but we would probably go down and have coffee. Back then this is what happened. I think it's different now. I can't remember how many family groups there were, maybe four or five different family groups. I ended up in people's kitchens drinking coffee with the heads of different families talking about things they thought we might want to do in forestry and things like that.*

Then we had other groups that we would meet with like the timber committee. When I was first there, we would buy them breakfast and that would bring everybody in to eat the breakfast. So that's how you got the consensus, you'd get it through these little community meetings. You did more work at night, other than like grant work where you actually wrote grants and stuff. You did more real work at night, talking to people. It was a small community, so I would hop on my bicycle and I'd go down after work to get my mail. Just being in the post office, you'd run into 4 or 5 people you might want to stop and talk to. I'd have other people drop by my house. I'd be in the shower and hear a banging on my door. I'd get out of the shower, get dried off, pulling my pants on and there would be a guy there. He would say, "Hey, I want to talk to you about that thinning unit." and come sit on your couch.

[Cal To Mike] Have you ever had that happen? Something like that?

Mike: "No, it's more for me when I am in Walmart, it takes me an hour to get through Walmart."

Cal: "Yeah, we didn't have Walmart. We had the Washburn General Store." *You try to go shopping and you're just there to pick up a couple of things, yet you've talked to about five people. You can't get out of the place and they all have their thing they want to talk to you about.*

Mike: *If you try to reserve a place and set up a meeting, no one comes. But if you go to community events, everybody is there.*

Cal: *Yeah, if you don't do that, you can't manage effectively on a reservation. You have to go hang out where people are and that's how you get information. They'll do things like say,* "I don't know about you, but I think I saw someone leaving with a truck full of cedar from that landing up there." *And you go,* "What? They're not supposed to be cutting up there." *You go up there, and someone's cutting up there. This works because the Makah wasn't a big rez.*

[Cal To Mike] I imagine on a big rez like Colville, you really have to do that because you can hardly keep track of everything?

Mike: "Yeah, we've had a lot of theft."

Alexa: So it sounds like it's a lot more relationship based?

Cal: *Yeah. Another thing is the concept of time and how often decisions are made. When decisions happen, it's not in a meeting room.* "I think it's more like that now though, isn't it, Mike?"

Mike: *I think it can happen, but consensus is still a big driver and consensus is slow.*

Cal: "Yeah, it's much slower than you think."

Mike: *It frustrates the business people because they're used to things happening quickly. I tell them* "Well...." *They'll say,* "We could make this save you millions of millions of dollars and do this and this and this." *I tell them,* "Slow it up, this could take like five years!"

Cal: *Yeah, it could take that long. Like someone says,* "I want you to do this." *and then like another two months later someone could call you and say* "Well, so and so said I should talk to you about this." *Then another month goes by and they say,* "Well, we're going to go talk to the committee about this." *and then something might happen. It's just a different perspective of time.*

When I was working at Makah long ago, one of the recruiting attributes that we had for foresters was if they had spent time in the Peace Corps overseas. This was a high point for us. We actually had two or three Peace Corps volunteers that were working for us. They weren't surprised when we talked about this different sense of time and urgency and the idea that you had to have buy in from a larger group than just the people in the office.

Kristiina: It's interesting about the Peace Corps, because it makes me think about Mark Petruncio working for the forestry group at Yakima Nation (*see* Section 5.2.2). He's not a Tribal member but he was also in the Peace Corps. I wonder if people who have been in the Peace Corps are more in tune with Tribal culture and decision process. They've interacted with communities, so they have a different perspective of time and how everything gets done. They're probably more sensitive and they respect more of what's going on in Tribal communities.

Cal: *I think people in the Peace Corps, at least the ones I have worked with, had many experiences and they weren't too surprised when things didn't happen real quickly. We had foresters that were hired through the government working for the tribe, and they were really frustrated by some of the things that went on. But if you stay there long enough, you just start saying* "Ehhh, we'll get to it."

Kristiina: So what you're both saying really is that because it takes a while to go through and talk about something, you're not really going to try to run something by for a fast decision. It's not like what's going on today in the federal government where bills are introduced that have scribbles next to the margins of the bill have not been sufficiently vetted. If there's something that the community doesn't want, you're going to know about that way ahead of time. You're not even going to do introduce it if you have not gotten feedback on it. You're going to hear all the different sides first. So, the chances of you introducing something that everybody's going to go thumbs down on, isn't probably going to happen.

Mike: *Most things are slower than any other place. Not always though, every once in a while, everyone wants to do something real fast. But on average, things are slower I think. Also, the whole concept of money is way different too. For example, we have a mountain full of molybdenum and we've known about it since probably the 1950s. In 1978 a company came in seriously wanting to develop it, and they were going to spend millions of dollars. They drilled holes every so many feet and they got it all mapped out and they knew exactly what's there. So, on average, that mountain is worth about 2 billion dollars, over a 40-year plan. After 40 years it would be worth 2 billion dollars in gross mineral output, but it costs a billion to mine it. Two minus one, you would still get a billion dollars over 40 years. There's only like 9,000 Tribal members, so that could be a pretty big income for each one of them.*

The company went through the process of planning it and held community meetings. The impacts included blowing up the mountain and leaving a big hole and a mess. But the impact that really scared people was the idea of bringing in thousands of construction people, miners, and outside people. This would change the community from a hundred people to a thousand people. That terrified people, the possibility of changing the culture. So, they rejected it and the company got kicked out. The company had already paid our tribe millions of dollars, and they still got kicked out. They were mad.

Then about twenty years later, there was another big push by some elders in the tribe to look at it again. But I said let's put it up for a Tribal vote. They were all mad at me for wanting to do that, but I wanted a Tribal vote. Sixty percent of our tribe voted against it. I was kind of proud of them for that. These are low-income people that could be getting big checks every month, but they turned it down for the same reason they didn't want to blow up the mountain and they didn't want to change the lifestyle. The other thing was that it's like money in the bank by having the mountain still there, even in a thousand years the mountain will still be there. We might need it someday, so there's no rush. Maybe someone in twenty generations from now will find a better way to mine it. There's the idea that our tribe will always be there, they aren't going anywhere. So it's just as good if they do it ten generations from now as if we would do it now — we're the same tribe.

Cal: *Yeah, that perspective does make them unique.*

Kristiina: Well, that's like installing wind power in the Colville, right? You guys were looking at a company that was going to set up wind to generate electricity, right? That went on for two or three years. But the best place for wind power was the best prime hunting grounds. Thus, this was ultimately rejected?

Mike: *Well the windiest spots are the higher spots. These are special hunting grounds up there, so the hunters didn't like having wind generation in the same place. But then you see elk standing under the windmills, so they didn't appear to be too bothered by the windmills. I also think we were being used. I think this windmill company just wanted to lock up our land. I was opposed to it from the start.*

I worked as an environmental technician when I was younger and built weather stations, so I know it's not that hard to build. We could've put up our own towers and collected our own data. But people fought me on that, I don't know why.

Kristiina: Mike, do you think the company thought you guys wouldn't be aware of all the factors and issues associated with generating wind power? Sometimes I wonder if they think *"Oh, it's a reservation, and what do they know?"* Then they can come in and just sort of wing it by saying *"Oh look at this great deal we are giving you."* They toss a little bit of money in and then are hoping you guys will say *"Sure build it here."*

Mike: *I'm thinking they were just kind of like land speculators in the 1900s. They could tie up a bunch of sites at a relatively small cost. I think they just wanted to tie up all these sites. I don't think they were really planning to do things everywhere. If things got better, they would probably develop the top ten percent of their sites. I think we were at the margin since we have some sites that are just too far from power plants.*

Cal: *I've been approached by a lot of people who have worked with Tribes who have definitely thought they could cut a fat hog going out there. This is something that they couldn't get done on non-Tribal lands. They think that Tribes will just allow you to come in and do that sort of stuff.*

Mike: *I had this guy call me up from waste management, he said they'd give us a million dollars and they wanted to build a Northwest dump basically on the reservation. It was years ago. I said, "Well what am I supposed to do with this million?"* They said, "We don't care, stick it in your pocket, we just want the permits." *I said, "Well let me tell our council, but they're not going to like it."*

Cal: Yeah, *I had the same guys call and say the same thing to me. A guy comes out from Seattle. I take him on a tour and we're driving down this one road and he says, "Stop, stop!" We're looking over a valley which has some of the best old growth forests that the tribe wants to preserve. The guy says, "Yeah, we could put a dam on the end of that, fill this up, you'd be really wealthy." I'm thinking, no way, that's not going to happen. I knew I wasn't even going to bring it to the Council, I knew they didn't want to fill the valley. But you get people like that. I had people come up who wanted to take every stick of timber off the reservation. They offered to give a lump sum for it all. But it's not going to fly with the Tribes.*

Mike: *Sometimes we get proposals that sound kind of good. We had the guy from Sweden or somewhere and he wanted to do a log cabin business. He wanted information on our inventory, he seemed serious, seemed like a nice guy. We talked to the Bureau (BIA), and they said, "Well, we're pretty booked up schedule wise for the next three years, come talk to us in year four." The guy just kind of shock his head and gave up.*

Cal: *I've had the other side too. The frustrating thing is if I find a market, but you can't get the product or whatever is produced to sell it in the market. They're not driven by this idea of making, doing production, getting stuff done like that. I had one where we actually made this canoe out of Cedar veneer. This is an example from the Makah. We had this local artist design this canoe so that it looked like a normal northwest Tribal canoe with the neck and everything. We took it down to the wooden boat show in Port Townsend. We got some press coverage, and then I got a call from REI who wanted me to bring one*

> in for them. I got a call from Eddie Bauer, they wanted me to bring one to them. I got called by a trading company, one of the representatives also wanted a canoe. The Tribe said if we start a company, that guy is going to be in charge — the artist. I said okay let's get a couple more canoes put together and then we can start taking orders. It was all set up for manufacturing to mass produce them. Then the artist got involved in making a canoe for some Hollywood director, he was making a gigantic one. I said, "We don't need to be making a gigantic one, we should be making these short ones that REI would like to buy a few and Eddie Bauer wants a few." and he's replies, "Oh, yeah, when I get done with this one." Then a few months go by, he finally delivers that one to the Hollywood director. Then he found that the Hollywood director referred him so he was carving a totem for another Hollywood director. So it kind of went downhill then. It's the whole perception of what's important and the artist was having a good time.

Kristiina: Right. Then he's no longer an artist if he's making many canoes for REI.

> **Cal:** *The Tribal council said he was the one that was going to be in charge of this if it happens. You know, I couldn't go out and find an alternative for him. So, it's just a whole different perception.*

> **Mike:** *We get the same thing, but with smaller things like bead work or moccasins or clothing. They'll do it on their own, for their own family. But the idea of doing it for production, they don't want to do that. Or they might try it and then decide they don't like it, it takes all the fun out of it.*

> **Cal:** *A large biomass project was another one that got killed on a rumor that we had dispelled years and years ago. I was gone by this point, but they brought a person in who would totally finance a 30-megawatt biomass plant. It would've created a market for all the small stuff coming out of the woods. But somehow, they got on the wrong side of the Chair at the time and the Chair started talking about all the smoke this biomass plant would be spewing out. Right at the last minute, it went out the door.*

Alexa: Cal, what kind of companies or business models do you think the Tribes are more willing to work with?

> **Cal:** *Most of them are very traditional. They're usually charted through the tribe via two ways, one is constitutional, and the other is Section*

seventeen in the IRA, that's a federal corporation. They usually have a board or sometimes a business council that will manage all the different businesses they own. It just depends on how they want to do it. I've seen them own business under state charters, where the Tribe wanted the business to be ready for loans or partnerships. Most of the ones you run into are just traditional where there's a council, a board and then there's a manager and then there's the company underneath. It could be chartered under the tribe's Tribal government or it can be chartered under the federal government.

[Cal to Mike] Mike, do you see any others?

Mike: *That's probably the main thing, there are also some individual Tribal businesses. It's a pretty small sector, but in the future it might be bigger. Like a lot of smaller stuff, the tribe isn't doing that stuff, but it might be good for a family. It's kind of awkward for the tribe because we're big and can be bureaucratic. We tend not to do smaller things. But stuff like a big casino, the Tribes have been successful at that, a big sawmill, the Tribes have been successful at that.*

Cal: *The Tribal government drives most of the development on the reservation. You'll have individual entrepreneurs and they'll be a LLC or whatever. A lot of them are both on and off the reservation.*

Mike: *Like our new three-story office building, one section was set aside for a café/deli side and that's just leased out. It seems to be doing great, it's kind of like a family business.*

Cal: *I think in the future, Tribes might try to push more individual development. Although when someone gets too successful everyone points to them and it's not a good thing. In fact, I know a Tribal logger, this was a while back, but he's still doing fairly decent. Even though he ran a good business and was making money, he never bought a new truck. He would always just buy these old trucks, fix them up, and then drive them around because he didn't want to look too wealthy. In fact, at one point his wife was driving around in an Audi and he was getting complaints about his wife. But he was driving around in a 1966 Chevy pickup that he kept wrenching on.*

Kristiina: It probably worked better than some other car.

Cal: *Oh, yeah. It was a nice truck. He was a mechanic, he went to diesel mechanic school. He was good at it, he liked wrenching.*

[**Mike to Cal**] Remember that old councilman that used to drive his Beetle to work?" *He'd park his new European sedan back behind the trees somewhere, so no one saw it.*

Cal: *Yeah, and when I was working on the Oregon Coast for a Tribe, I was driving around in a Camry. What's a more standard American car than like a Camry, just a 4-door sedan? I had a Tribal member who's now the Chief of the Tribe come up to me and say, "What kind of fancy executive automobile is this?" His wife goes, "That's a Toyota Camry, I have one." He goes, "Really?" They just kind of put this thing on you when you're in different positions. If you're a consultant, they expect you're probably driving around in a BMW or something like that because consultants come in with leather bags and nice cars.*

Mike: *I lost an election over my Hummer. I bought it over the summer, it was old and cheap with a lot of miles on it, but it still looked nice. They thought I was like a millionaire driving around in this Hummer. I lost my election.*

Kristiina: That's right, you've had a fancy motorcycle, you owned a Hummer.

Mike: *Motorcycles didn't seem to bother them.*

Cal: *Yeah, motorcycles don't seem to be any big issue.*

Mike: *Even though it probably cost more than my Hummer. But the general public doesn't know that.*

Dr. Mike Marchand on his motorcycle.
Source: Mike Marchand.

Kristiina: Didn't you have a little sports car at one point too?

> **Mike:** *Yeah, a little Mazda sports car. That didn't seem to bother anyone either. They couldn't figure out whether it was a luxury car or an economy car, but it was a fun car. It was like my third midlife crisis red sports car. It was good for commuting.*
>
> **Cal:** "It was a Mazda Miata, right?"
>
> **Mike:** *Yeah.*
>
> **Cal:** *See, those aren't big though. If you had a big red sports car, then you'd get in trouble.*
>
> **Mike:** *Like a Corvette, my brother has a Corvette.*
>
> **Cal:** "Remember those brothers from Flathead?" *One was a forester. They all drove Corvettes and I used to think they were going to get into trouble for that. They were government employees.*

Kristiina: Can we talk about business, and the kind of businesses you would have. When we were meeting the last time at the Colville, there was a lot of discussion about gaming and that Tribes like games. So having a casino is probably not unusual.

> **Mike:** *Well it's a big problem that I don't questions. While we have some people with PhDs and college people, but on average, a lot of our people are undereducated. So, when we've tried to recruit manufacturing type businesses, they survey our work force and say you guys don't have enough people with basic math, basic English, or stuff like that. I think it is a problem. We get a lot of dropouts. But not everyone's like that.*

Kristiina: There's a different way of thinking too. A University is very different if you're more of a visual learning type. I think your learning style does have an impact on what and how well you can learn. A lot of the classes, even at U.W., you're being taught stuff. You need to say, *"Why do we need to know that? What are we going to do with that?"* Sometimes I think there's a disconnect between what's being taught and what people really need to learn.

> **Cal:** *I know when I used to live out in Makah, I used to make a point to go out to the high school to get people interested in natural resources.*

Mike: *That's a big problem too. They aren't exposed to what's possible out there. When I was kid, I was ostracized for going to college. People said, "Why are you going to college?" "You could stay here where the big parties are at." I just laughed and said "There's actually bigger parties at the college, if you want to think about it that way."*

Cal: *One of the things I remember, a lot of those kids were thinking they were going to become fishermen. This is back when the fish industry was bigger. I remember one time someone asked me, "How much does the tribe pay you?" So, I told them and the whole class laughed, they said, "We could make more fishing."*

Mike: *That's true though. We had CAT drivers, like logging contractors. It could make a decent job. Just drop out of high school and drive a CAT, make some good money.*

Cal: *"But I think with every generation it gets a little better, doesn't it, Mike? As far as kids going to college?"*

Mike: *We're probably getting more but we started at a large deficit.*

Cal: *There was a big deficit but it's getting better with each generation.*

Mike: *A lot of our people who worked in the mills were high school dropouts and they've been floundering for ten years.*

Cal: *That's one thing I know that from a Tribal standpoint, at least from all the Tribes I've worked for, they like businesses that can provide entry level jobs. Places where people can come in, train, and learn, even though it might not make the best economic sense. At one Tribal job, the chair really wanted me to build gas stations. It just didn't make any economic sense. The reasoning, I could tell, was that you didn't need much of an education to work at the gas station.*

Mike: *We're getting a lot of that now, in Washington State. We're getting a tax rebate now. For every gallon we get 29 cents.*

Cal: *We didn't have that advantage down in Oregon, plus the market was oversaturated with gas stations.*

Mike: *Yeah, normally it's a real low return business.*

Cal: *But that was the concept. Now I know tribe's trying to build a coffee stand. You don't need a high education to become a coffee stand operator.*

Mike: *A lot of our people like manufacturing, assembling stuff, like what they were doing in 1960s, maybe 1970s. But all that stuff has moved overseas or moved to Mexico or somewhere else. There's nothing in the United States like that now.*

Cal: *Another Tribe I worked for had a clothes manufacturing business and they made these very unique items they were selling in outlets like Coldwater Creek. What killed that was the fact that it employed a lot of people. But when you make vests (they made vests, like Pendleton vests that were embroidered beautiful garments), they were having problems because Coldwater Creek was making complaints. I went through the racks and you could visually see that a woman's medium were different sizes depending on who made them. Every one of them was different. I said we need to get consistent on quality here. But they had been making so many of these things that they had a big giant warehouse filled with all these clothes that were totally uneven in size. The buyers didn't want them because they don't want that type of variability in the sizing of the clothes. I ended up renting a closed-out store in the Bend River Mall and I got advertising on the radio that said, "Big Sale!" We sold like three quarters of the stuff.*

Kristiina: I remember my great aunt who lived in Turku, Finland telling me a story. People would come to a clothing store from the country or farms. They would come into the store to try on a wedding dress or a suit to get married. The sales clerks would stand behind them and put this oversized thing on them. They would then pull the jacket from the back, standing behind them. Of course, these people would go back to the country to get married and they would put on their suit and it was way too big. Apparently, it was a very common thing to do and the sales clerks knew the people probably wouldn't come back or ask about it.

But anyway, I was wondering about the location of businesses. When you look at Tulalip and you have the Quil Ceda Village, it's right on the highway so it's very easy to get to. This must be a huge barrier for setting up businesses when you are not close to the highway. What kind of businesses can you get established? Most reservations are not located close to highways. Look at Colville, how long it would take you to get there if you had to buy something. Then you have all those other things you need to think about. Is that a big barrier still? Or is that not an issue today?

Cal: *No, that is an issue. Some Tribes are lucky, they're right along rail lines or right along a major highway.*

Mike: *Nothing they do at Tulalip is anything we could do at the Colville Tribe. We don't have the customer base. The outlet mall, giant casino, we couldn't do that. We have a casino but it's like 1/10 the size of theirs.*

Cal: *You really have to work with the market that you have. There's a tribe in Wisconsin that's been very successful and after visiting it people would come back and say, "Well why can't you do that here?" And I'd say, "Because we are on the Oregon Coast." They think of that as an excuse, but some people have the mentality of "build it and they will come," and you see that in casinos. There's a casino down in New Mexico and they got into a joint venture with a major hotel chain. They built this gigantic casino that's really had some issues. You just don't have the traffic moving through it. You have to plan for the local market. There's not a one-size fits all. That's just basic business. Sometimes, what you have on the Council are people who are not experienced in business. They all come with their different narratives on what a business is and what they want to build.*

Mike: *They might be housewives or ranchers, they might be businessmen, but it's whoever people elect. It's like your mayor or city council. Maybe they could run a business, maybe they can't.*

Cal: *That's one thing about successful people. I've seen successful Tribal businesses, they really have a good slice of board members with business experience to local experience. You need them both on the Tribal Enterprise board. You can't have all local experience people, that doesn't work because they don't make good business decisions sometimes. But then you can't have all business people on the board because then they make decisions that don't connect to the local communities. The ones that have been successful have had a cross section of people and the most valued people are ones who are both Tribal members and business people.*

Mike: *We've hired board members or CEOs who on paper look like they're like blue-chip, corporate-type, successful people, and maybe who started their own corporations. Then they get out to the reservation and they get chewed up by government politics. They can't survive.*

Cal: *I'm working for the tribe on this log merchandising enterprise and there's a three-person board. One's the CEO of the Tribe, one's the Tribal natural resources director, and then there is one who actually ran an international large forest land-owner company. He told me, "I don't want to be involved in anything with the local politics." So, I use him that way. I consult with the CEO and the local natural resource person, he's a Tribal member. Then when I want to know just about forest industries issues, I call him up and he tells me, "Oh yeah, that looks about right." I keep it kind of isolated that way and so you kind of have the best of both worlds. It's different with these larger boards.*

I was lucky when I was at the coastal Tribe, I was the only non-Tribal member on the board. But because the Tribe is a restored tribe, when they were terminated everybody went out and they had to make their living. When they got re-recognized in the late 1980s, you have this diverse board with lots of experience. I had a vice president from a major retailer on my board, a Tribal member. I had a high level financial person on my board, also a Tribal member. I had a former college president, also a Tribal member, and I had a local guy, a construction guy, who was a Tribal member. I loved having this person as a board member because he had been making his living running a construction company. He was actually the most practical guy when it came down to business decisions.

Kristiina: It sounds like people that have a foot in both are much more effective.

Cal: *I think that's what you're seeing in Tribes in other places. You're seeing members who have either gone off and worked outside of the reservation or have gotten highly educated and now can start serving on these boards for the tribe. I think that's the best combination. The other thing is, are they respected? A lot of times, the council, even though you have this acumen of business people, because they are a Tribal member and they're in the same crab pot, they don't listen to them. The worst critics that you have, if you're a Tribal member and you're on your own res, you could be fantastic on the outside, but you're nothing sometimes on the inside of the reservation. I've seen that happen, it doesn't matter.*

Developing business opportunities on reservations is challenging even though Tribes have many opportunities because of the political status as a government. Tribal sovereignty is especially challenging since Tribes can't just form business ventures with anyone who is interested in such endeavors. This is discussed in Section 5.4.2 by Dr. Mike Tulee using an energy perspective.

5.4.2 Challenges: A Boom? or a Dis-economy of Scale for Tribes?*

Tribes all over the United States have a great opportunity to develop renewable energy but are facing major challenges [107]. Even though many Tribes are located in areas specifically suited for renewable energy generation, there exists very little, or no viable infrastructures that are able to usher in harvestable actions on a major scale. The implementation of renewable energy systems on Indian Reservations is a monumental challenge because of multiple and oftentimes, complex interest clashes stemming from both inside and outside of Indian Country. Even more, generational inheritance of passed down histories of betrayal of the "Indian" has resulted in a legacy of "distrust" of the non-Indian world.

As Tribes seek to expand business activities, the need to further develop robust legal business environments and sound governance becomes critical. Underdeveloped legal, or outdated governance structures, insular Tribal policies, and politicized business management are barriers to business and economic development in many Native American communities [108].

Sovereignty

Also, there is a strong determination from virtually all federally recognized Tribal members to retain what is known as "Tribal sovereignty". Because of aboriginal political and territorial status, Indian Tribes possess certain pre-existing sovereignty that is subject to diminution or elimination by the United States, but not by the individual states [109]. Tribal sovereignty is a concept on which Tribes base their very existence as a nation, separate from neighboring states and the federal government. All federal treaty Tribes made agreements with the United States government in years past that even today, guarantees Tribal "sovereign" rights on their lands.

Tribes guard their sovereign rights with vigilance and any perceived weakening of their sovereign rights has never been acceptable, especially in relation to state governments within the United States (Fig. 5.7). State incentives for renewable-energy development has proved challenging to tribes because oftentimes the state require Tribes to create a state-chartered organization. Tribal governments are guarded against relinquishing sovereign immunity to individual states, subjecting themselves to state laws and reporting requirements [110]. Therefore, any kind of outside interest or actions being introduced in Indian country is carefully vetted before any kind of allowances are made by Tribes.

Tribal leaders carefully listen to not only elders, but to "the will" of their Tribal people. Cultural values passed down from elders to the next generation

---

* By Dr. Mike Tulee.

Fig. 5.7 Depiction of Tribal governments guarding their sovereignty rights against infringement by State laws and regulations [111].

include retaining "naturalness" of Mother Earth. The earth, trees, mountains, rivers and animals are all spiritually considered sacred to Native Americans on their lands. Even when animals are hunted and killed, it is a common practice for Native people to give reverence and thanks to the creator. Native people on Indian reservations can play critical roles in answering a looming national energy crisis but first, social and environmental obstacles must be overcome.

Lack of funding for renewable energy projects

Although federal legislation and funding approved during the George W. Bush administration enabled enhanced renewable energy production within the US, progress has been slow in coming, and the current Donald J. trump Administration has not made renewable energy production a priority. In fact, the amount of U.S. dollars invested into renewable energy research and development has dwindled in the last 5 years from over $9 billion to a little over $7 billion. In 2009, Tribes sought $52 million for renewable energy projects but only $6 million in funding was made available. In addition, some Tribes are deterred from participating due to the Department of Energy's cost share requirements [112]. What this means is that federal funds made available to Tribes will also require a set percentage of Tribal dollars being made available to the funded project. Most Tribes will not have the necessary assets on hand to contribute to the proposed project.

In today's world, and in times past for that matter, Tribes have had to compete for investment dollars to be leveraged on their Tribal lands. To be successful,

Tribes must offer outside investors the opportunity to earn economic returns that are commensurate with the returns they might earn elsewhere [113]. In fact, it is challenging just to acquire capital for business start-ups and economic expansion for Tribal business enterprises and independent Indian-owned businesses located in Native American communities [108]. The structure of business transactions on Tribal lands have traditionally not been in Tribes' favor. In 1934, Commissioner of Indian Affairs John Collier (under the Roosevelt Administration) pushed through federal legislation that enabled Tribal self-governance and eliminated land allotments that effectively halted the onrush of land grabs by non-Indians. This legislative act became popularly known as the *Wheeler-Howard Act* [114]. Also included within this new Tribal governance system was the establishment of a revolving loan program for Tribal development and Tribal business charters. This effectively marked a "new" beginning for Tribal business development on Indian reservations.

Bureaucratic process

Bureaucracy in multiple forms can and does prove to be quite challenging when getting energy programs off the ground. For instance, in order for a Tribe to even acquire a permit requires approval from four different federal agencies, whereas for non-Tribal land interests, only four "steps" in totality are required [115]. One needs to look no further than the Navajo Nation to understand the difficulties in setting up desired business ventures on Indian land. The Navajos require conducting an archaeological survey, obtaining a letter of support from the Tribe's president as well as a dozen additional steps that slow the process [116].

According to a report conducted by the Government Accountability Office (GAO), the Department of Interior Bureau of Indian Affairs (BIA) was overloaded with internal breakdowns which resulted in delays and caused Tribes to miss out on opportunities to generate revenue. The report also said that Native American reservations could have produced much more solar and wind energy, but the BIA had actually hindered energy development on Tribal lands. An existing antiquated 1961 regulation resulted in the BIA taking a "one-size fits all" approach to processing all surface leases. This system, lacked a defined process or deadlines, making it common for a simple mortgage application to languish for several years waiting for approval from the federal government [117]. Let alone being slow to identify land and resource ownership, the BIA lacked adequate staffing to review energy-related documents [118].

In 2012, the Department of Interior announced a streamlined approval process for renewable energy projects. Included in the recent policy was improved efficiency and transparency in the land leasing approval process that resulted

Chapter V Forestry Lens: Culture-based Planning and... — 245

in Tribes gaining more information in negotiating with developers [119]. These changes were intended to increase the efficiency and transparency of the BIA approval process for the residential, business, wind energy evaluation, and wind and solar resource leasing of Indian land [120].

A great majority of Indian Tribes throughout the US are struggling to help their members make economic gains in our society. Although the introduction of gaming casinos has proven to be an economic boom for a number of Tribes, a great majority of Tribes however are still in a state of great peril.

Need to work in concert with U.S. federal tax policies

There are federal incentives available to businessmen/investors that Tribes can utilize, one being the "New Markets Tax Credits (NMTC)". The NMTC is a mandate established by US Congress in 2000 that enables individuals or corporations to invest in Tribal lands that will create jobs or material improvements on Indian reservations [121]. However, there are multiple and complex challenges to overcome in investing in Indian country through NMTC. One of the major challenges Tribes face in securing NMTC investment dollars is the recovery of monetary investment for the lenders in the event of borrower default [122].

Needless to say, the solutions for Native Americans are complicated because of the special legal obligations owed to them by the federal government as a result of long-standing treaties and federal Indian policy [123]. For instance, when first written, the intent of the *Indian Tribal Energy Development and Self-Determination Act* (ITEDSA) of 2005 was to provide Tribes with a framework for developing renewable energy infrastructure. However, certain complications still hinder the ITEDSA from helping Tribes fully realize energy prosperity. It is claimed that the ITEDSA is flawed for failing to correct for misplaced financial incentives for renewable energy development by Tribes [124]. Also, the law continues federal policies that ensure that non-Indians and non-Tribal business entities often reap far greater economic rewards than Tribes or members of Tribes. Thus, it can be seen that Tribes have great opportunities to be providers of renewable energy, but it is currently a major challenge to establish these enterprises [125].

Historically, federal tax policies did not favor Native Tribes on renewable energy development. The guarantee of tax-free status of Tribes on Indian reservations guaranteed that Tribes could not utilize tax credits granted by the federal government [126]. However, both wind and solar energy investment rules changed. As recently as 2013, tax codes prevented Tribes from effectively utilizing what are known as production tax credits. Production tax credits did not enable economic profit for Tribes, nor did it afford them decision-making powers while working with investing partners [127].

Recent alterations to IRS (Internal Revenue Service) policy now allows broader economic production on Tribal lands. The IRS also announced through "private letter", that an American Indian Tribe can elect to "pass through" a thirty-percent credit to a third-party lessee. This means that investing entities can now share economic benefit of a tax credit with partnering Tribes [128]. Tribes have the ability to partner with and pass through to a 3rd party equity investor a 30% federal tax credit [129]. This particular federal tax arrangement provides Tribes a unique opportunity as renewable energy developers because they will be the only entities that are not subject to federal income tax. Tribes will have a unique advantage especially in the solar development field as they are not subject to federal income tax but may pass through tax credits to tax equity investors.

Also, through the *American Recovery and Reinvestment Act* in 2009, the US government made available to Tribes, a $2 billion bond issuance known as the Tribal Economic Development (TED) bonds [130]. New federal guidelines meant that Tribes could apply to IRS for "TED" bonds that could be issued as a tax exempt or as a tax credit bond. Either way, it enables Tribes to borrow at a lower rate in concert with local governments or states on Tribal lands.

Problematic to the TED program however is that many of the Tribes that are seeking dollars to fund projects (such as renewable energy) must have the ability to pay back the borrowed dollars based on a challenging formula [131]. Multiple factors such as repayment period, number of households, and per-capita income on an Indian reservation determines the ability of a tribe to repay a loan. The per-capita allocation statistics are high. I suspect that the financial analysts and the policymakers who designed the TED bond program believe most Native American Tribes receiving a TED allocation would not be able to repay the TED loan.

The arrival of Community Development Financial Institution (CDFI) is proving to be instrumental in breaking Tribal economic barriers. The origin of the Native Initiatives dates back to September 1994 when Congress passed the *Riegle Community Development and Regulatory Improvement Act*. This Act created the CDFI Fund and mandated that it conduct a study of lending and investment practices on Indian reservations and other lands held in trust by the United States (CCH). The CDFI Fund is a $195 million initiative that is specifically designed to promote financial services in Native Communities. These initiatives are leading to increased credit, capital and financial services in Native Communities. Specifically, the study recognized barriers to private financing, identified the impact of such barriers on access to capital and to credit for Native people, and provided options to address these barriers. Through the CDFI initiative, strategic planning of renewable energy programming can be established on Indian reservations.

The politics behind policy change must take effect. For instance, just too even establish siting for transmission infrastructure, it is a difficult obstacle for the US government to overcome. It is quite common for states to question and challenge a federal authority's jurisdiction to site land for transmission development [132].

Renewable energy still has not taken hold in the United States to the extent it has in other countries. In fact, European countries have now increased utilizing non-fossil fuel energy to power their needs at a 15% rate, while the United States is hovering right at 7%, with no support from the current Trump Administration.

Although the numbers of renewable energy system startups nationwide will increase significantly for the next few decades in the form of wind turbines and solar panels, fossil fuels are still overwhelmingly proving to be "king", i.e., dominating the energy market [133]. This is because the overall market of renewable energy is still not "cost-competitive" in relation to fossil fuels, and there is no support from the current Administration. For instance, additional bastions of natural gas have been discovered within the United States and harvest has been accelerated because of the practice of "fracking". This increase in natural gas harvest is causing natural gas prices selling on the energy market to drop, thus "undercutting" both wind and solar powered energy systems.

Even so, renewable energy is becoming more cost competitive. According to a study by the International Renewable Energy Agency, costs associated with extracting power from solar panels have fallen as much as 60 percent in just the past few years (IRENA). Also 118 countries now have at least some type of renewable energy policy or goals by the early part of 2011, suggesting that renewable resources are likely to become the major source of energy worldwide in the coming years [134].

Chapter VI
Tribes, State and Federal Agencies: Leadership and Knowledge Sharing Dynamics

There is a reason that we include the leadership and knowledge sharing dynamics between Tribes, state and federal agencies. The policy process followed by states and the federal government directly impact cultural resources that Tribes have rights to through treaties. There are success stories but also destruction of cultural resources due to the activities of the state and federal government. Rodney further introduces this below and specifically focuses on what he experienced while he was the Tribal liaison for the Washington State Department of Natural Resources (WA DNR).

The present school of thought believes that the destruction of a cultural resource such as an archaeological or historic artifact may be a case of irreversible loss if the item is unique and irreplaceable or non-renewable. Native people view natural and cultural resources much the same and have the same belief about all of these resources. They don't want to see the inanimate and animate resources threatened or endangered or become extinct. Thus all of these resources are important to support and maintain the quality of the land, air and water. Cultural resources which have been inherited from past generations can be seen to have something in common with natural resources, which have also been provided to us as an endowment. Both impose a duty of care on the present generation. Additionally, a similarity can been seen between the function of a natural ecosystem in supporting and maintaining the natural balance and the function of what might be referred to as a cultural ecosystem in supporting and maintaining the cultural life and vitality of Native culture.

> "We're the advocates for the salmon, the animals, the birds, the water. We're the advocates for the food chain. We're an advocate for all of society. Tell them ... how they're our neighbors. And how you have to respect your neighbors and work with your neighbors {Billy Frank}."

Tribes use certain rocks to heat up for their sweat lodges and certain rocks for pit cooking foods. They use certain trees for carving, for weaving, for medicines, and for tipi poles. Traditional and religious leaders have certain places they will gather fresh water to serve to their guests at winter dances. For all of these reasons and many more, Tribes feel that natural resources are cultural resources.

Since the coming of non-Indian people to the Pacific Northwest, natural or cultural resources have been impacted in many ways:

> The economy of the pre-White settlement Northwest has been described by Jim Lichatowich as a gift "economy", which had evolved over 1,500 years. In this economy, one attained social position not by accumulation of wealth, but through the size of one's gifts. Gifts were the basic form of exchange and commerce. Natural resources, like salmon harvests and fishing sites, were gifts from nature, not for individual ownership and exclusive possession, but to be shared with others and passed on to succeeding generations. Salmon, which had a conscious spirit, would remain abundant if treated with respect. The Natural World was filled with such spirits, which humans needed to cultivate to ensure a continuous food supply and other necessities.
>
> The new White economy introduced by White settlers was fundamentally different. Salmon and other natural resources were commodities to be captured and sold for profit [136].
>
> Harvests declined 50 percent between 1884 and 1889, rebounded briefly as harvesters switched from chinook to coho and sockeye then fell again. Thirty-nine canneries in 1887 became twenty-one by 1889. By the turn of the century, harvests in both Alaska and British Columbia exceeded Columbia River harvests. This "free-for-all" was over within just thirty years. Never again would Columbia Basin salmon be thought of as a limitless resource [137].

These examples and those to follow will all demonstrate how natural resource decisions have impacted cultural resources that are also highly valued by the Tribes. Rodney Cawston interviewed Francis Charles to describe the Tse-whit-zen Village DOT Construction project and how it impacted the Lower Elwha Klallam Tribe (Cawston R. & Charles F. Oral History Interview 2008) [138].

Tse-whit-zen Village DOT Construction project[1] (*see* Section 6.2.2)

> *"My name is Frances Charles; I'm the Tribal chairwoman for the Lower Elwha Klallam Tribe.... I'd like to take this opportunity to express... what we are dealing with spiritually, mentally and physically*

---

[1] http://leg.wa.gov/jlarc/AuditAndStudyReports/Documents/06-8_Digest.pdf

Chapter VI Tribes, State and Federal Agencies: Leadership and... — 251

for what is known the Tse-whit-zen Village site in the Port Angeles harbor area.... This was a reconstruction project with the Washington State Department of Transportation (DOT) and the Port Angeles (PA) to build pontoons and anchors for the Hood Canal Bridge. It was an opportunity to bring jobs to PA.... What really brought the project to a halt, was the partial skull and jaw that was found in one of the areas.... It's landmarked on the maps of the coast guard as well as the maps of the historical museum of PA.... We had several elders who could not confide to us... because it brought back... really bad memories for them. We were trying to interview them to recall the cemetery, it was really hard for them to open up and expose what they knew because of how our people were treated back then, like animals.... West Shore Heritage was contracted and hired by WS DOT and when we asked them this, they had done some testing cores and surveys.... It was in the winter time and the weather conditions were not the greatest.... Whenever they had dug holes the test pits would always fill with water. So it was very hard for them to really evaluate what was taking place on the ground.... The report stated a clause identifying that there were no significant findings at that point in time.

We had called in all of the agencies to come forward.... We started having staff and a couple of our lawyers start taking a look at what the Tribes rights were and what the responsibilities of not only the tribe but the agencies responsibilities.... We did ask the agencies to stop the project, to slow down so we could take an evaluation of what was happening on the site. There were a few remains that were exposed that were fully intact individuals that were exposed... to our amazement we had unearthed over 850 etched rocks. In these etched rocks not one of them were duplicated but told a story about the village... through this process it was really hard because we had witnessed our ancestors to be used as back fill... since the early 1900's mills have been built down here time and time again, and they had disturbed soils, and had mixed it up. But after we went down six feet below the surface, that's where we hit our ancestors and that's where we hit the village site... The size of this village site itself remains unknown... we know that our longhouses are bigger than we have imagined, 50 to 100 ft. or even more than that... We have found all kinds of trade materials from ivory to Chinese coins...

... We have seen and witnessed pilings that are used for holding up the mills. These poles that have been driven right through our ancestors' remains and have split them in half. We have found mothers and children that were embraced together. We have found husbands and wives that were lying next to each other... We had agreed that we would not

> be sending our ancestors to be carbon dated one way or the other... we used materials around them which was identified as early as 200 years old and as late as 2700 years old... where we felt that is was really important for the tribe to maintain our ancestral remains which to this day we still have at an undisclosed location 335 cedar boxes... we have over 100,000 artifacts that are maintained at the Burke Museum... we still were being bullied by the agencies stating that because there was an agreement and that they had this million dollar project they had to continue on... 103 burials came out of the one area that were fully intact individuals and that's where we started working frantically on trying to halt the project itself with the agencies...
>
> ...We started outreach to our delegates, our senators and our congressional representatives and also the Governor to stop this project... it was always the value of the project, there was a lot of money involved and they did not want to see this project come to a halt... we were able to go to the Centennial Accord meeting with Governor Locke... but there is a lot of prejudice yet, even though people don't like saying that, prejudice, racism, that still hurts because we still deal with that with our youth, when they're in school... to me, it wasn't about the money, it was about the ancestors and how they were being disturbed... it was also seeing the gifts of what came out of the ground and seeing the pride of our youth and our community they wanted to be enriched with the culture...."

The Tse-whit-zen Village DOT Construction project was curtailed after the Centennial Accord meeting with Governor Locke in 2004. WSDOT's internal auditor identified $86.8 million-dollar loss in expenditures related to the Port Angeles graving dock project. This total includes $60.5 million for construction at the now abandoned site, and $26.3 million of inefficiencies to the bridge project caused by the shutdown of the site [139]. This is one of the largest nationally and inter-nationally known cultural resource disasters. The WSDOT construction project at Port Angeles has served as a training example in colleges and cultural resources conferences all across the United States.

There were many reported mistakes in the JLARC audit report:
- For the Hood Canal Bridge project, WSDOT's project management and project development schedules were inadequate, and a fast-tracked project schedule reduced the time available for analysis of alternative graving dock sites and options.
- The legislatively-mandated Transportation Permit Efficiency and Accountability Committee's (TPEAC) inter-disciplinary team process for permit streamlining entered the project late. Also, the compressed project time

schedule limited the ability of permitting agencies to fully consider proposed site alternatives for the graving dock.
- WSDOT did not follow a consistent documented protocol for addressing compliance with cultural resources assessment and consultation requirements of Section 106 of the *National Historic Preservation Act*.
- WSDOT's consultation for the graving dock, with the State Historic Preservation Officer and the Lower Elwha Klallam Tribe as required by Section 106 of the *National Historic Preservation Act*, began late in the Port Angeles graving dock project site selection process [139].

When WSDOT first attempted to contact the impacted Tribal government, they sent letters to Makah, Jamestown S'Klallam, and Port Gamble S'Klallam, but not to Lower Elwha whose Tribal headquarters is less than 15 miles from this site. There were many recommendations regarding all of the issues with this construction project, but it would take too much space to present this information and beyond the scope of this book. With an approximately $90-million-dollar cost to the State of Washington, cultural resources management is taken very seriously with land management agencies today.

6.1 Tribal/Federal/State Cultural Resource Policy*

Rodney Cawston.
Source: Colleen Cawston.

The *Archaeological and Historic Preservation Act* of 1974, commonly known as the *Moss-Bennett Act*, helped to fuel the creation of Cultural Resources Management, which also created growth in archaeological jobs in the federal government,

---

* By Rodney Cawston.

academia, and with Tribal governments. Today, Tribal governments are some of the largest employers of cultural resources staff, including field archaeologists. Federal legislation was passed earlier in 1906 under the *Antiquities Act*, but it was not until the 1970s when the term "cultural resources" was coined by the National Park Services. This term came into more popular usage after two meetings in 1974: the Cultural Resource Management conference and the Airlie House conference. Following these conferences, the National Park Service (NPS) defined cultural resources in the Cultural Resource Management Guidelines as being [27]:

> "Those tangible and intangible aspects of cultural systems, both living and dead, that are valued by or representative of a given culture or that contain information about a culture.... [They] include but are not limited to sites, structures, districts, objects, and historic documents associated with or representative of peoples, cultures, and human activities and events, either in the present or in the past. Cultural resources also can include primary written and verbal data for interpretation and understanding of those tangible resources."

Federal regulations for cultural resources are governed primarily by Section 106 of the *National Historic Preservation Act* of 1966. Implementing regulations for the Section 106 review process are found in 36 CFR 800. The goal of this process is to offer a measure of protection to sites which are determined eligible for listing on the National Register of Historic Places. Federal regulations only come into play when a project requires a federal permit or if it uses federal money.

In Washington State there are numerous agencies, departments, divisions, bureaus and commissions that have been established in the Executive Branch and this can be very confusing and frustrating to Tribal leaders, especially those who have been newly elected. There are many agencies, programs, and boards and commissions that are not accountable to Tribal governments. Washington's complex organizational chart has such a large reporting structure that it is difficult for Tribes to focus on strategically important information and initiatives important to them. Much of this same story exists in natural resources. Here, the organizational problems are two fold. First, the state has no central oversight over all-natural resource departments. The second problem is that coordination to provide effective policy development of interest to Tribes is nearly impossible. Natural resource policy-making is a maze. Washington has a Board of Natural Resources which passes policy for the Department of Natural Resources; a Fish and Wildlife Commission has oversight with the Department of Fish and Wildlife and its director; and the governor has oversight of some of the natural resource agencies such as Parks and Recreation and Ecology.

Chapter VI Tribes, State and Federal Agencies: Leadership and... — 255

In Washington State, dozens of boards and commissions are responsible for regulating particular natural resource issues. Although the governor has an annual meeting with federally-recognized Tribes, this meeting does not include all of the state agencies and departments. The Commissioner of Public Lands is a state-wide elected official and does not report to the governor. The Director of Fish and Game reports to the Fish and Wildlife Commission, which is appointed by the governor and has no Tribal representation. Both of these departments are not present at the Centennial Accord which had a goal to improve relationships between federally-recognized tribes and the State of Washington[1]. The Department of Fish and Wildlife does not have a Tribal liaison that directly reports to the department director and this is required by Washington State Senate Bill 6175: "Designate a Tribal liaison who reports directly to the head of the state agency." This bill further states:

> *"In establishing a government-to-government relationship with Indian Tribes, state agencies must: (1) Make reasonable efforts to collaborate with Indian Tribes in the development of policies, agreements, and program implementation that directly affect Indian Tribes and develop a consultation process that is used by the agency for issues involving specific Indian Tribes."* [140]

Most private development projects in Washington have to consider cultural resources under the *Washington State Environmental Policy Act* (WSEPA) whenever county or city permits are required. In such cases, SEPA mandates that the project's potential environmental effects, including impact upon cultural resources, be reviewed. If the review indicates that no impacts to environmental or cultural resources would occur from the proposed project, the project may proceed without further consideration. When cultural resources are involved, the project cannot proceed without first evaluating the importance of the cultural resources.

Many federal and state agencies, cities and counties have passed their own local management plans or ordinances regarding cultural resources. Many of these ordinances have specific provisions to protect local cultural resources that are deemed important to the local Tribes but which may not be sufficiently protected by state and or federal regulations.

Tribal Historic Preservation Officers (THPO) are officially designated by a federally-recognized Indian Tribe to direct a program approved by the National Park Service and the THPO must have assumed some or all of the functions of State Historic Preservation Officers on Tribal lands. This program was made possible by the provisions of Section 101(d) (2) of the *National Historic Preservation Act* (National Organization of Tribal Historic Preservation Officers website).

---

1 Accessed on Jun 18, 2018 at http://goia.wa.gov/Relations/Relations.html

Not all Tribes have a THPO. If they don't and they are a federally-recognized Tribal government, this function is usually handled through the Bureau of Indian Affairs. To comply with the *National Historic Preservation Act*, a federally-recognized tribe must submit a formal plan to the National Park Service describing the functions of the Tribal Historic Preservation Officer. Many Tribes have Tribal Historic Preservation Offices or programs that have responsibilities for managing cultural resources both on and off of their reservations. Many of these programs include oral history components, archives, language preservation and archaeological repositories. Many Tribes have also established Cultural Resource or Elders Committees or Spiritual Leaders who are consulted for their special knowledge of Tribal place names, archaeological sites, traditional cultural properties, and places that are important for protecting and possibly inclusion into the National Register of Historic Places. In Washington State, Tribes are asked to identify sacred sites to be included onto a database at the State Historic Preservation Office. However, many Tribes are reluctant to share this knowledge for fear of exploitation by professional and amateur archaeologists that "pot hunt" known archaeological sites for priceless artifacts that are often sold in artifact auctions and elsewhere.

The leadership and knowledge sharing dynamics of non-profit organizations and state and federal agencies tells us a lot about how holistic management can occur on our lands and waters. When culture is included in natural resource management, it tells you that place-based knowledge is being included in management. When it is not part of organizational management, culture is less integrated into resource management and science at the airplane level is mostly used to make land and water use decisions. It has been challenging for state and federal agencies to adopt and build on Indigenous knowledge in making nature decisions. State and federal agencies frequently are reactive when society doesn't agree with their decisions. This means that agencies are trying to respond to multiple stakeholders who are all shouting to get attention but are not using the tools that would allow them to respond to local stakeholders. When these agencies are unable to proactively respond to these conflicts and do not address place-based knowledge, the agencies lose the trust of the public and Tribes.

The next series of stories tell how Tribal members and non-Tribal members collaborating with Tribes have forged new approaches to address the continuing issues Tribes face today. It shows how Tribes are still not receiving the treaty guaranteed rights from state and federal agencies. It also highlights how non-profits and private and public partnerships may be an approach that can contribute towards moving us from the current stalemate where Tribes are not accessing resources guaranteed in treaties (*see* Section 6.3). These stories are important to read so these practices do not continue but are addressed.

6.2 Tribes and Washington State

Our attempts to provide insights to leadership and knowledge sharing dynamics between the state of Washington and Tribal members is possible because we have the stories of two Tribal members who successfully balanced having a foot in non-Tribal agencies or legislature and with their Tribe. Both are authors on this book. Our first story focuses on John McCoy who holds a position as a Senator in the Washington State legislature. He has been voted into this position by the citizens of his district because he has worked tirelessly for their benefit and to solve their problems. He was not voted into office because he is a Tribal member. The other story focuses on Rodney Cawston. He was the Tribal Liaison for the WA Department of Natural Resources, has a PhC degree from the U.W. and is now on a Chair of the Colville Tribal Council.

Both John McCoy and Rodney Cawston are able to speak on the challenges facing Tribes in Washington State. They are Tribal people who have managed to bridge Tribal culture and American culture. They have worked in both cultures. We can't think of better individuals to describe leadership and knowledge sharing dynamics occurring in the State of Washington. Their stories support how abysmal is the knowledge held by non-Tribal Washingtonians on Tribes and their treaty guaranteed rights. Therefore many Washingtonians look at the lands and resources as theirs and not that Tribes have customary rights encoded in treaties. In response to this lack of knowledge of Tribal law and rights by non-Tribal Washingtonians, John McCoy as a Washington Senator — with the help of many others — was able to get a bill passed that would teach K-12 students about Tribal law.

Chapter VI will end by describing how Tribes are successfully collaborating on natural resource planning through inter-Tribal collaborations (*see* Section 6.5) and through the policy process (*see* Section 6.6).

6.2.1 Washington State Policy Process and Tribes: Interview of John McCoy[*]

Kristiina: Why did you decide to go into the Washington State Legislature? Why did you not feel that you could contribute to Tribal issues as a Tribal member?

> **John:** *Well, I get asked this question all the time. My answer has always been, I did not wake up one day and say I'm going to run for*

---

[*] By Alexa Schreier and Dr. Kristiina Vogt.

John McCoy.
Source: John McCoy.

the Legislature. It was a process. When I was managing the team that was putting Quil Ceda Village together, we discovered that there were state and federal laws that made it difficult for us to accomplish what we needed. So, I got permission from the board do create a governmental affairs department and I had three ladies in that department. So they were in Olympia and Washington D.C., taking care of those issues that needed addressing. Because of that and because I also needed to be out and about doing government relations, I also had to do community relations. I was out in the community a lot telling them what Tulalip Tribes was doing and that what we're doing was in no way was going to change or make their life worse. If anything, it was it was going to make it better. So, I told them exactly what Tulalip was going to do and Tulalip did it. So, we gained the reputation of people of our word, because we advertised that we were going to do something, and we did it. So, no matter if someone told us, "No you can't do that because..." and I would always respond, "Okay, where in law does it tell me I can't do something?" I'd always challenge them back.

So, over the years and then in 2000, there were people talking to me about running for political office. I did make an attempt at a run, but Tulalip was then in a Legislative district that did not fit their need. Tulalip is located next to Everett and Marysville, so that's where all our politics and our economic influences were. For the tenth legislative district, everything happened in Oak Harbor and that was an hour and a half drive away. So why were we up there, you know we had no business up there? There were already two Tribes on the outskirts to those city limits and that was their territory, not ours. Well, I ran in that District. I didn't win, but I impressed a lot of people on how well I did.

Kristiina: Was that difficult for you, because it's a very different group of people you had to convince to vote for you? So, it wasn't just the tribe, but you also had to think about the people that were living in that district and satisfying what they wanted. Or, were the things that Tulalip wanted similar to the things that the non-Tribal community wanted? So it wasn't that difficult to make that adjustment?

John: *No, working in North Island County was difficult, because the cultures were different than what we experienced. The culture of the county government was a lot different than our culture. They were a group of people that were extremely conservative, and Tulalip was, as you know, it's got it conservative streak, but it's very liberal- progressive. There was that clash, if you will. So, I found pockets of democrats that liked what I had to say and there were a few republicans that liked what I had to say and so, like I said, I did extremely well. I impressed the Democratic Party on how well I did. Because of that experience of running in that district, I went back to the Tribal council and said,* "Tulalip's got no business there, we don't have the same interests, so we should be over here in the 38th legislative district." *That was in 2000, and I convinced them to allow me to lobby the redistricting committee to put Tulalip in the 38th legislative district. So I did that.*

Well, the 38th had three very good democrats. So, I wasn't going to run against any of them because you know we were getting what we wanted from them. So, I decided my short political career was over and I'd just stick to doing economic development for Tulalip. Well, the next year, the sitting senator called me and said, "If I retire, will you run in my place?" *I said,* "Well if the seat's going to be open, yeah I'll run." *But one of the sitting Representatives wanted that Senate seat. Because there is a pecking order, no matter where you go there's a pecking order, I said,* "Okay, you run for the senate and I'll run for your open seat." *Well, he hand-picked an individual to run for his seat, so I wound up having two democratic opponents, and because I had the better name recognition, I handled them quite well. I won the seat in 2002 and the rest, as you can say, is history. So, it was just a number of event that just led up to it and then I decided okay, I'll run.*

Kristiina: So it sounds like the Tulalip tribe never had a problem with people from the Tribe running for the Washington legislature. Also, it sounds like it wasn't like it wasn't a problem or concern with non-Tribal people. Sometimes you hear people saying, "Oh we need Tribes to stay in their place." But it sounds like you never really ran into that. Maybe it's because of how you deal with people, you had a balanced view, and you were talking to people?

> **John:** *No, I had that problem. The majority of the board supported me in running but there were a couple of individuals that didn't like the idea at all. They'd cause an issue on occasion, but nothing we couldn't overcome. So, I continued to work for the tribe when I was not doing legislative work. But then as my experience grew in the legislature, you might say I started climbing the ladder. So, in August of 2010 I decided to retire from the tribe, because I was having conflict between the work I was doing with the tribe and my legislative work. I had to make a decision and so I decided that because the Tribes knew the direction it needed to go, that all they needed was a strong manager to continue that movement. They could survive. Then I'd go ahead and turn my attention to legislative work. So that's what I did and I'm quite happy that I did that.*

Kristiina: One of the things that I find interesting is that you were able to go between two totally different worlds and you were able to balance and think about what everybody wanted. Do you think anybody else from the Tulalip Tribe, at the time, could have done what you did? Because you seem very effective at bridging, you're able to think about both sides and what's needed and sort of make the choices or the options. I just wondering if this is something unique, because you're really the only one that's doing this? Is this something where you have an ability to better balance between two worlds and it's not just something that just anybody could do? This is the impression I have, am I correct?

> **John:** *Yeah, but there are some individuals I know that could do the work, but they're quite happy just sticking to the Tribal work. But I had come to the determination that we needed to work in both worlds in order to accomplish what we needed. So, I have become very adept at doing that. I've had to learn to communicate in about 4 different ways, because of who I have to work with. It gets to be a challenge at times, to understand how they communicate. For example, my legislative district is high-tech, so we have a lot of highly educated people. But then there's a legislative district in Pasco, Washington, where a GED is all you need because most your labor is manual agriculture work. We had to*

learn to communicate with both those two groups so that we understand one another. So, you know, a word that you use with one may have a totally different meaning with the other group, so those nuances needed to be addressed.

Kristiina: Let me just throw something else in. In the *Wall Street Journal* there was an article about somebody asking what makes a good business person and the answer was that people from different sides have to trust that you're going to make the right decision. So, the thing I was getting at is that this is one of the characteristics that makes someone a leader in business. Do you think that different groups of people you talk to have trust that you will make the right choice? I'm thinking that that's what's happening with you because if people didn't trust you to make the right choice, they would not vote for you.

John: *Yes, they do trust the right choice will be made. When I was developing Quil Ceda Village and going around to all these service organizations and local governments, I told them what Tulalip was going to do and how we were going to do it. We did it just like we had planned it out. So, one of the former County execs told me that the best part about me was that when I said I was going to do something, I did it. So, I do have that reputation.*

Kristiina: I think that this is an important point and something we definitely want to highlight here.

Alexa: Before we get into some of the other questions, are you the only Tribal person in the legislature?

John: *No, there's three of us. But I'm the only Washington State native. One's Alaskan and one's Bad River Chippewa from Minnesota.*

Alexa: How do you think the Tribal perspectives should be included or considered in the Washington State Legislature?

John: *Indian Law has to be a big part of it. So, I do a lot of educating. I bring in experts to help educate the legislature. I got a piece of legislation passed, in the beginning it was voluntary, that is now mandatory that the K-12 system must teach Washington state Tribal history and culture. In that way, the students are starting to receive the basics of Indian Law. For instance, yesterday I went to a high school, and I spent four and a half hours talking to the civics classes. I found it amazing that in each class, a student would ask a question they wanted an answer from another angle, another direction. I'd say, "Oh, now* *you're going to get basic Indian Law." So, I'd explain it and walk them through that and I'd the students would say, "Oh, I didn't know that." I would say I know because you weren't educated properly. I speak at the U.W. Law School and when I walk through there, students that know me will stop me and say, "You know, I'm paying big bucks for information I should have gotten free in the K-12." I say "I agree with you and I'm trying to fix that." So, education is huge, I think every state ought to have Tribal history taught — their Tribes history taught. Cherokees aren't everywhere, they may think they're everywhere, but they're not everywhere. Like Navajo, they cover a lot of territory, but they're not everywhere. So, each state should be teaching about their own Tribes.*

Kristiina: I think that's really an important point, because I have a class that I'm teaching, "Forests and Society", and it's pretty interesting how they really know very little about colonialism, the impacts of the colonialism, and the changes that have occurred in the American landscape. They don't know that Tribal people have been good business people way before the colonialists showed up or how they've had to adjust to living on a smaller land base and not always been able to access their customary lands for resources that they used to collect. It's amazing how they just don't know any of this. I just find it strange.

John: *That's what the white civilization wanted. They wanted us to blend in, to assimilate. They had an expression, "save the man, kill the Indian".*

Kristiina: How difficult was it to get people to pay attention to the education bill? Do you think it was easier than it might have been 10 years ago?

John: *No, it was difficult. It took me three sessions to pass the initial bill. So I had to do a lot of educating and I still educate to this day. So, no, times have not changed. Take a look at Trump—that's at its worst. As long as we have people around like that, racism will never go away.*

Alexa: Education is clearly a really important piece of this, both in schools, K-12, as well as in the legislature. Do you think that's the best method to really get people to change their perspective and consider Tribal issues more? Or do you think there's other things like policies, or other tools that need to be used other than just educating people?

John: *You have to educate people because they don't understand treaties. They don't understand that Tribes are sovereign nations. They have a* *hard time grappling with the idea that the state of Washington has 29 sovereign nations in it. So, they have difficulty wrapping their heads around it. Right now, there is a law on the books in the state of Washington, telling Congress to aggregate all the Indian treaties. I've tried to eliminate that law and I can't do it. I've tried a couple times, can't do it.*

Kristiina: But the state can't do anything anyway though, right?

John: *Correct. But it's just a law that's "a burl in the saddle". So, education is the heart of everything. I found that even the most hardcore person, I can get them to move a little bit once you thoroughly educate them. If you can get them to understand what the issue is and why it's that way, you can move them. So, if I can get them too neutral, I've won.*

Alexa: We're wondering a bit about how often you're approached by other members of the legislature looking for a Tribal perspective on issues that you might not even be directly involved in,. Do they want to hear your perspective and input?

John: *That happens every day.*

Alexa: Are there specific issues people come to you with, or is it across the board?

John: *It's across the board. Like yesterday, I had a student just ask, apparently he knew a little bit about the Boldt decision, and he asked, "Is the Boldt decision being adhered to?" I said, "Well it is, but there's still a lot of people that don't like it." But it is there, and it's being enforced. We're starting to get through but we still have a long ways to go. I mean, the United States Constitution was written with racism at its core. The opening statement "all men are created equal", that was a bunch of white men in a room. There were no people of color in the room when that statement was made.*

Alexa: Can you think of any specific examples where legislator came to you and asked for your perspective on an issue and because of that there was a change in what they would have done or what they proposed based on the information they got from you?

John: *It happens all the time. Those legislators that consciously want to help or improve the situation ask me. Then I will explain to them* how they should go about it so it doesn't infringe on the Tribal sovereignty and they'll make the changes.*

Alexa: Are any of the things they come to you generally surrounding the legality of Tribal sovereignty or is any of it more cultural and perspective oriented?

John: *More of a perspective orientation. They pretty much understand sovereignty now, not as in-depth as I'd like them, but at least they got the concept. They will generally ask the right questions and they know that we have to co-exist and co-manage. They are interested in making things work.*

Kristiina: John, if you were not there, who would they go to? Who would they ask those questions to?

John: *They wouldn't ask anybody, they'd just go off and do whatever they thought was right.*

Kristiina: That would suggest we need to figure out how to replicate you.

John: *I keep looking for someone.*

Kristiina: You're on the Natural Resources Committee, is that right?

John: *I'm on the Agriculture and Water Committee. Because the democrats are in control, we consolidated agriculture, water, natural resources, and parks. They're all in one committee. And yes, I sit on that committee.*

Kristiina: So a lot of times you may be getting questions related to each one of the areas covered by the committee. You've talked about the Skagit, and the wells, and if development can occur. Which still hasn't been solved from the sounds of it.

John: *No, not yet.*

Kristiina: People bought land and then they've been told that they can't dig wells on their land because there's not enough water for everyone. They're still trying to push through legislation so that people can dig wells on their lands to get drinking water.

John: *They want to destroy water laws and make it the Wild, Wild West again. Those with the most money get the water.*

Alexa: If you had to summarize the most important pieces of information that legislators need to know about Tribes to make well informed and conscious decisions, what do you think that those pieces of information would be?

> **John:** *Well, every legislator that has a tribe in their district or at least next door to them, which for the State of Washington, that's all of them. I tell them, why don't you pick up the phone and call them? Go out for a cup of coffee with them and understand what their needs are, because each tribe's needs are different. They all need money, but if they had money, they'd all spend it in a different way.*

Kristiina: For instance, the Tulalip tribe, you have the Quil Ceda Village, you've got the Casino there, but you're the only tribe that has something like. You're the only one that has a federal city, right?

> **John:** *Yeah, this idea emerged in the summer of 1994. I was talking to my Tribal attorney and beginning the planning process for our economic development. I said to him, we in Indian country know that we don't have a stable base of financial income. We don't have taxes; therefore, we can't borrow money in the conventional way that everybody else gets to borrow money to do their infrastructure work. So we have to be creative in how we finance these projects. So my question to him was if we were able to convince the state to step back from taxes, "How do we need to be structured to take advantage of using taxes to fund infrastructure development?" He said that was a pretty big challenge because we've been trying to figure it out for 500 years and we haven't yet. But somebody has got to ask the question, so how are we going to do that?*
>
> *I went off and started doing the same old thing of tracking money down and getting grants and loans to get our projects done. The Tribal attorney also made it a class project. So 3 or 4 students took it on as a project. They went all the way back to the first tax case in court and then worked their way up to present time. Then they provided a report on what we had to do, and it was quite simple. The operation had to be on Tribal trust land, it had to be owned by the tribe and it had to be managed by an Indian. It was that simple. So that's what we did. So now we are in federal court with the state of Washington suing over taxes.*

Kristiina: It seems like every time the state needs money they start trying to figure out how they can get money from Tribes. Even though you guys already pay taxes, which I don't think a lot of people realize. One of the things that is probably going to be your legacy is the impact that you've had on youth. You interact and talk to the youth all the time. These are going to be the future leaders. If they already start thinking about things since they've heard something different, they're not just going to act like their parents. They will already have developed a different view. So, ultimately your strategy probably is education, and connecting with a lot of the youth that are going to be making the decisions in the future. You don't want to focus on adults, because they're pretty set in their ways. You can kind of make them change their ways a little bit when they ask you questions, but as soon as you decide to retire they won't have you to come to. You will no longer be the filter they can use to understand Tribal issues. So you've got to have something else in the pipeline, something that's going to have an impact.

John: *Yeah, that's why I keep talking to the kids.*

Kristiina: That's a very smart way to focus on the future. You have to impact the current generation by going and talking to the youth, because they'll remember these things. I think it will make a difference.

John: *We're getting there. It's why I'm teaching at the MPA program at Evergreen. I'm starting to see those students in the leadership ranks of Tribes, so we're getting there. It takes time.*

Kristiina: One question I would like you to comment on is "How challenging it has it been for you to keep your feet in two different cultures, without losing your integrity or your values?" At some point you're going to have to make decisions and sometimes they're not going to work for or benefit the Tribes and sometimes it might be very negative. So, you're going to need to be able to somehow deal with the fact that you're in a particular position where you bridge two cultures and not everyone will always win. At times you are going to need to say, "Alright that was a loss, can't do anything about it." Or are there approaches you try to do knowing you will have to respond to such situations?

John: *There are times when a piece of legislation will be going through, and I try to change it and I'm not successful in changing it right then. But I cast the "no" vote and then the mere fact that I voted "no" causes change. People go back and look at it and then the next session there's a bill going through to fix that. So, sometimes I don't get it done in one shot, it takes a few shots. I keep telling them, I'm patient, I'm very patient, so they'll come around.*

6.2.2 Washington Department of Natural Resources and Tribes: Interview of Rodney Cawston*

Alexa: To contextually frame our conversation, could you talk about your job working with the Washington Department of Natural Resources (WA DNR). Could you also comment on the survey that you did with all of Washington's Tribes while you worked for the WA DNR as a Tribal liaison?

> **Rodney:** *I worked for DNR for a little over two years actually, under two different state land commissioners. I was the Tribal liaison with all of the Tribes in the State of Washington, and actually, even Tribes outside of the state that bordered Washington State. Many of these Tribes have natural resource interests or treaty rights that extended into the state.*

Kristiina: Oh, I didn't know that. You were also a WA DNR liaison even for the Confederated Tribes of Warm Springs even though it is mostly located in Oregon?

> **Rodney:** *Yeah, when I was there, I met with Warm Springs and the Nez Perce tribe of Idaho and Coeur d'Alene. So, when I worked for the WA DNR, one of the first things I did was actually meet with every tribe to see what they wanted from the WA DNR. I interviewed all the Tribes but do not have access to my notes on the responses of the Tribes to my questions. Some of the information is probably lost in my house right now since I moved back to the Colville reservation after leaving WA DNR. But I did ask each Washington Tribe what some of the most important natural resource issues that the Tribes faced at the time. One of the things that really struck me was the common issue of access to DNR lands for cultural resources that were promised by the treaties that Tribes signed. They wanted access to State resources and State lands and public lands for all kinds of reasons that they had been promised.*

Kristiina: Can you comment further on this access problem?

> **Rodney:** *Yes, Tribes should have access since most of them have treaty rights to collect from these lands. When they signed the treaties, they ceded millions of acres of land to the federal government. As part of this, they reserved certain rights and certain lands for their own use and occupancy within the "reservations". But they also reserved their*

---

* By Alexa Schreier and Dr. Kristiina Vogt.

rights to hunt, fish, and gather on those lands that they ceded to the federal government. So, this was probably the biggest problem Tribes mentioned. I think the next issue was probably conservation. They were always concerned about conservation of lands and land management policies and land management plans. They were concerned how forest actions or forest practices, or even agricultural practices would impact lands that are very important to them. This was especially important for those ceded areas and especially areas where they gather traditional foods or traditionally used plants or trees for weaving and carving.

Kristiina: Rodney, can you go back just a little bit, like why did you decided to do the survey? Was it very clear when you were there that no information existed on what you were asking about?

Rodney: *Actually, when I did talk to the Commissioner, there wasn't a lot of clear cut directions on the relationships between WA DNR and the Tribes. It seemed there wasn't a lot of communication between the agency and the Tribes. So I really wanted to open that up and let them know who I was, that I was hired, that I was there. I wanted to gain some familiarity with Tribal concerns related to WA DNR and for them to become more comfortable with me. But also, just to learn what their major concerns were related to DNR.*

The present Commissioner of Public Lands signed the "Commissioner's Order on Tribal Relations" in 2010 [141]:

This Order is an expression of the intent of the Commissioner of Public Lands to build strong relationships individually and collectively with the twenty-nine federally recognized Tribes of Washington State (Tribes) and the Washington State Department of Natural Resources (DNR).

"As the Commissioner of Public Lands (Commissioner), I recognize the importance that Tribal governments and their respective Tribal members place on perpetuating their culture, language, oral history and traditional teachings from one generation to the next. I also recognize that Native American culture is characterized by an intimate relationship with natural areas and resources. It is with an acknowledgement of our shared commitment to protecting natural resources that I seek to build intergovernmental relationships based on trust and mutual respect."

I wanted to report back to the Commissioner, which I did. I wrote trip reports of the concerns for every tribe I visited. The summarized information from these visits, led to the Commissioner's Order on Tribal Relations (Box 14). I wrote the initial draft of this Order which was followed by State agencies going through many reviews. But the order really was about how to establish a government-to-government relationship with Tribes. Tribes also wanted to have at least one annual meeting with the Commissioner so they would be able to discuss their issues. Also, I believe Tribes wanted quarterly meetings in some of the regional offices with the State Land Commissioner. All of this really did open up a lot of the communications with the Tribes and, for the two years I was there, we had really successful statewide Tribal leader's summits with the State Land Commissioner.

Box 14: Commissioner's Order on Tribal Relations

The following Guiding Principles will guide how the Department of Natural Resources will interact with tribal governments[1] :

Principle 1: Respect for Sovereignty
Principle 2: Interdependence
Principle 3: Sustainable Use
Principle 4: Sound Science
Principle 5: Transparency
Principle 6: Respect for Traditional Knowledge and Cultural Values

---

Alexa: What two years were you there?

Rodney: *I'll have to look that up, it's been a while.*

Alexa: You met with all of the different Tribes, and you said that their largest priorities were access and conservation?

Rodney: *These seemed to be the biggest issues that arose during the discussions with Tribes. In Western Washington, most of the Tribes are on small land bases. So, they need to access federal and state lands for their hunting, their fishing, for cultural purposes, for gathering. They really rely on working with state and federal governments and having access not only to the uplands but also aquatic lands. The Washington State government through the Department of Natural Resources man-*

---

[1] Signed by the Washington State Commissioner of Lands Peter Goldmark on September 10, 2010. http://file.dnr.wa.gov/publications/em_comm_tribalrelations_order_201029.pdf

ages both aquatic lands and state lands. Tribes are interested in access to fish and wildlife and also geoducks which are managed by DNR and a really important industry here as well.

DNR has a lot of problems on the state lands, especially forested areas. People use these lands as garbage dumps and body dumps. So almost all of these lands are gated to keep people out. For Tribes, it was really important to have access to lands behind these gates. A lot of elder Tribal people couldn't walk. State land managers would always say "Well, why don't you just walk around the gate?" But how practical is that especially for big game hunting or anything like that? So it was that was one of the biggest issues I faced was to look at how Tribes could have access behind those gates. Before I left DNR, we had worked on creating a Memorandum of Understanding between DNR and the respective Tribes to access to those areas that were of most interest to them. Tribes were going to receive a key to access these areas.

Alexa: Do you think DNR was doing an adequate job of making sure that Tribes' sovereignty and their treaty rights were being upheld?

Rodney: *Well it's always difficult because there are many special interest groups that really don't care for Tribes to be able to exercise their treaty rights. They would even come to the offices at DNR and protest or they would call and write and let you know their views. It was always a difficult issue. The state is also charged with raising revenue for the state from DNR lands because these lands fund many different schools and hospitals and counties throughout the state. It was always a challenge for DNR to try to find a balance between managing those lands for DNR's interests, for the public's interest, and also for the Tribes.*

I don't think that many times they really respected the tribe's treaty rights to accessing DNR lands for cultural resources. For instance, Tribes use old growth cedar to carve canoes which are found growing on DNR lands. Canoes are still used every year for their canoe journey and even for traveling the oceans and the rivers. Some of those canoes are even maybe 80–100 years old so that's how long they use them. Old growth cedar is really protected. Tribes felt that they had a treaty right to gather them, they didn't ever give up that right to be able to harvest those trees. When we worked through that issue, it was decided the trees could be purchased and gifted to the Tribes. This is not really quite the same as the Tribes having the right just to go out there and get, or harvest the trees themselves for their own purpose. Tribes could not go to an area and select a tree they wanted so in that respect I think that the state didn't really fully respect Tribes' treaty rights written in treaties.

Sometimes it would be a long drawn out process for a tribe to be able to acquire cedar trees. Tribes' were not familiar with some of the processes to get cedar trees or some of the management processes. Many times, the Tribes just didn't feel like they were really adequately addressed or maybe responded to in a timely manner when they requested cedar trees. So, there's just a lot of issues all over there.

The state has to also deal with 29 federally recognized Tribes in the State of Washington, as well as those other Tribes outside of the state whose rights extend into our state. WA DNR has to deal with a lot of governments within Washington who have treaty rights to resources. The multitude of cultures and languages makes it difficult to make decisions. The existence of 29 federally recognized Tribes caused constant issues that I had to resolve while working for the DNR.

Alexa: Besides yourself as the Tribal liaison, was there someone else who had a role in DNR to advocate for the Tribes?

Rodney: *No, I think I was pretty much the only position there. Not all agencies even have Tribal relations managers or Tribal liaisons. There was a recent law that was sponsored by Senator McCoy requiring agencies to have a Tribal liaison included in their budgets. But I don't even think today all the agencies still uphold that Law. It was always a challenge for one person to be able to deal with so much and so many issues. I mean, one day I was out on the Puget Sound dealing with geoduck issues. The next day I was in Eastern Washington looking at shrub steppe areas and traditional food plants, or agriculture. It was just always something different coming in.*

[Kristiina to Alexa]: What Rodney did was to make it where DNR's interactions with the Tribes was really strong. But he had to in fact translate and educate the Commissioner on each Tribe and what are their needs, what they want. This allowed the Commissioner to interact with Tribes and to address their concerns in a credible manner. You really can't have a non-Tribal person in this role. Rodney made the Commissioner look really good because of being able to make these connections to Washington Tribes.

> **Rodney:** *It was a lot of education that had to occur all the time. Not only to the Commissioner but even to the executive staff at DNR. I produced a lot of materials about each of the Tribes just to inform them who they are, what their interests were, what their membership was, where their land base was, and what Tribes lived in Washington. Most of the general public, or even professionals that come and work in DNR, really didn't have any experience at all working with Tribes. They didn't have a clue what it meant to work with Tribes. Like my own tribe, we're a confederation of twelve Tribes, in Northeastern Washington State, it's not just a single tribe. The traditional homelands of these Tribes go way up into Canada as well as down into Oregon. So just in one tribe, the cultures and languages, the political and legal histories are all very different and very unique. It's very complex even on one reservation, let alone looking at the state as a whole.*

Alexa: In any decision-making that happens on DNR lands or for land management, are the Tribes ever brought into or asked to inform those decisions?

> **Rodney:** *A lot of times this occurred because there are several natural resource agencies in the state, you have the Department of Ecology, Fish and Wildlife, and DNR. You have several agencies that deal in a particular area but with a different focus. At DNR, I always tried to get them to really interact with the Tribes as early as possible, even at the threshold of something just beginning. That's when the Tribes really deserve to be contacted and informed. But many times, many agencies don't. They will wait until the document is drafted and then they'll send it out for the review. Such a process has gotten the state into trouble a number of times with the Tribes basically filing lawsuits against them. So, I always recommended that they work with the Tribes as early as possible, because it's better to develop those documents together. Even once they're produced, there are people who will have questions about it. But at least they will know everything that's in there and also know that their own perspectives are honored or respected.*

Alexa: Do you think that there are challenges in that the Tribes likely have a different decision-making model than DNR or some of the local governments?

> **Rodney:** *I think that really is the case a lot of times. The Tribes really do vet the issues with their people, they'll listen to the people, and everybody can make comments. So, it's more like a bottom-up approach to really looking at decision-making. Whereas at DNR and some of the other federal agencies, it's more top-down. They'll make*

the decisions. They'll draft the documents. Then they will share them with the public and everyone else.

Alexa: Do you think that complicates the relationship between the governments and the Tribes and their ability to work together?

Rodney: *I think it really does at times. Like I said, I've met so many people who came to work for DNR that didn't have a clue about Tribes were really about. They had learned about management from their own education, no matter what their discipline is. So, I think that came into conflict a lot of times, or maybe it's just ignorance of the staff person. Some of the letters I remember reading that they were going to send, I couldn't believe. I had to put a stop to and just say, "No, no, no, we've got to redo this, we've got to rewrite this."*

There's always conflict because of looking at policies and their perspectives on the land management goals. DNR isn't all about money, but it seems like a major objective of theirs is to generate money for the trustees and all the schools and everyone else who is really demanding funds generated from these lands. Where the Tribes, their values are a lot of times different. They look at the wealth of the land, of what it has to offer, traditional foods and the materials that they gather and how they approach that and those very sacred areas. There are cultural resources out on the landscape, we have gravesites out on the landscape, which is very personal to us. Sometimes I just felt like it was always just something that had to be managed at DNR. But for Tribal people, they took that very seriously and still do.

I don't know how to convey these differences or how to explain them. We have ceremonies. All of the Tribes have ceremonies about their traditional foods, even about the way they gather these foods and the way we raise our young people to respect that. For instance, the canoe tree, they make take one canoe tree and use that one tree for maybe 50–80 years. But when the state harvests a cedar tree, it's just used for lumber or shingles. The purpose just isn't the same.

Another difference is bear grass. The Tribes use that to weave baskets that can last up to a hundred years or longer. Where most of the people in the commercial floral industry just use it for a one time in a floral arrangement and then that's it. I just couldn't believe the devastation when you see people who have harvested and taken literally everything. They just destroy areas while collecting for the commercial floral industry.

> *While Tribes have selective harvesting on just about everything. They really have a lot of respect for those areas and have gathered there for many generations at some of the same areas. So, when they come there and see someone who has just obliterated the place, and you'll see just a huge mess, it's just devastating.*

Alexa: Do you think some of that comes from both different perspectives of the land, like different types of respect? Or do you think some of it is ignorance of the ecosystem and the landscape or what kind of implications those actions will have? Whereas the Tribes seem to have a much better understanding of the land and the impact of their actions?

> **Rodney:** *I think it's probably a lot of everything and the above. I see people who basically are just trying to make a living and sometimes that's done legally, and sometimes that's done illegally. For whatever reason, they will strip an area. A lot of times they take everything without really thinking about the landscape and trying to harvest in a sustainable way. Whereas the Tribes, and I can't say this about every Tribal member, but in general Tribes always go out and pray before they harvest those things. Also, they will have their youth witness how they harvest. Harvesting cedar bark and things, Tribal members will pray with their young people who will then harvest the tree but they'll take care of it. They also believe that the things you don't use, you leave it there on the landscape. I don't know why that's always been a teaching with native people, I think maybe it's just a better understanding of having that respect for those ecosystems.*
>
> *Where on the other hand, I have gone out and seen where people will harvest different plants, especially for the floral industry. They collect a lot of different greens and things, especially from cedar trees and fir trees. They'll just strip everything as far as they can reach. It'll look real nice and pretty along the road, but they'll go just beyond the road and they'll take everything. Again, it's just for them to make money and a lot of times they're not thinking about that in a sustainable way. A lot of it is done illegally too, so they have to do it quickly and then get out of there.*

Kristiina: They lose money too, because they collect too much, and the floral industry doesn't accept a lot of it. I've heard about rejection rates of almost half of what was collected. So, they're just collecting everything, because they want to collect as quickly as possible and get as much in a short period of time. This creates huge problems. Salal (*Gaultheria shallon*) leaves and branches are wastefully collected because the floral industry pays good money to harvesters who collect them.

> **Rodney:** *There's dumps outside Olympia where you'll just see huge dumps of these plants and stuff that the industry didn't accept. They've just taken it all, and yet we're so regulated. The Tribes are regulated in what they can take. There have been times when the Park Service will follow Tribal members to make sure they're doing everything, harvesting things, in a sustainable way. But yet there's just no comparison at all how Tribes harvest, in comparison to how it is done for the commercial industry.*
>
> *Tribal weavers might take a bag or two of materials. We've got to fight for that in the legislature and with the agencies just to have the right to be able to go out there and do that. Yet they'll permit people to go out there to gather these things who won't do it legally. There are so few enforcement officers so it's just difficult to regulate. There's all kinds of poaching going on out there. That's why a lot of the gates are actually in place, even though these people find a way around them. People are even stealing the gates now for the iron. When I was there, they were having problems with that. The activities that go on out there when people get past the gate, it's just unbelievable.*

Kristiina: They've talked a lot about berries that are over-harvested by non-Tribal collectors so Tribal members no longer find berries when they go out to collect in their customary areas. I have heard how difficult it is to find Huckleberries to collect when Huckleberries are needed for special ceremonies. Non-Tribal collectors just come in, overharvest and then they're selling it to industries to make jams.

> **Rodney:** *Huckleberries are now an international market product. You can buy Huckleberry lotions, potions, jams, syrups, and everything else you can make out of them in international markets today. It's a big industry that generates a lot of money* (Box 15).

Box 15: In 2009, Price/Unit for Huckleberry Fruit[1]

Fruit, Raw $6-8/lb
Fruit, Frozen $13.33/lb

---

[1] Huckleberry Product Prices, November 2009; 1 lb = 453.6 grams. www.Opb.org/programs/oregonstory/harvest/market/page_2.html

The Tribes have this political forum, it's called the Affiliated Tribes of the Northwest Indians (ATNI) and they meet quarterly. The DNR State Land Commissioner and I attended one of the meetings together and he gave a speech at one of the conventions. Several ladies from Yakima and Colville actually got up and they were questioning him about the Huckleberry harvesting. A lot of the Tribes will have a Huckleberry feast. But before anybody in a tribe can go out to harvest Huckleberries, they will pray for those foods and give thanks for them. It's just something a lot of the Plateau Tribes do before harvesting. But they were saying to the Commissioner that by the time they got out there, there was nothing left to collect. They could not find any berries in the areas they usually go to.

There have actually been a couple of Huckleberry summits held in Washington State because of this problem. The University of British Columbia and the University of Idaho are the two Universities that were really researching this topic of overharvesting. There are people using rakes to harvest the bushes. If it gets too late in the day, there were people just pulling up bushes and throwing them into the back ends of their trucks to pick the berries off the bushes later or in a warehouse or something. They would even pull bushes with green berries and try to bring them where they could mature somewhere else.

So a lot of those Huckleberry sites were being devastated, so they tried to regulate it. For one, they outlawed the use of rakes in Huckleberry harvesting. The Forest Service actually started a Huckleberry season (see Box 16). They would try to communicate with the Tribes to give them like a two week notice ahead of time, so they would have time to get out there and get berries for themselves and for the feast. Since I haven't been at DNR for a few years, I don't know how that's working. We went up to see a Huckleberry warehouse. I can't remember statistically how many gallons or what weight of the berries are being taken out of the Cascades, but it's huge. There's a lot of commercial buyers that have warehouses set up in the mountains to manage their Huckleberry collection. We actually went into one of them and I've never seen so many huckleberries in my life.

Kristiina: What about the example you've talked about in the past where you had to set up a training program on Tribal cultural resources and Tribal treaty rights for the Department of Transportation?

> **Box 16: Washington Public Land Harvest Regulations for Wild Huckleberries**[1]
>
> - Every harvester of huckleberries is required to have a permit on the Gifford Pinchot National Forest.
> - If you are picking for personal consumption and under 1 gallon of berries per day (up to three gallons of berries per year), you can get a <u>free use permit online</u> to print and take with while harvesting.
> - If larger quantities are wanted or if you plan to sell your berries or berry products (jams, ice cream, fruit-leather or other items), you need a Charge Use (commercial) Permit-available at your local <u>Ranger District or Monument Headquarters</u>.
> - The Gifford Pinchot National Forest does not start issuing commercial huckleberry permits until mid-August.
> - Mechanical removal of berries is not allowed. (Rakes or other brush disturbing devices.) Harvesting of berries is allowed on the majority of lands on the Forest. Areas closed to harvest include Wildernesses, Research Natural Areas, Mount St. Helens National Volcanic Monument and the Mineral Block. A free, detailed map is available at your <u>local Ranger Districts</u>.
> - Please respect lands reserved as American Indian harvest areas.

Rodney: *We actually did set up training programs for state level government agencies and private landowners. I didn't do these training programs just by myself but pulled in other staff people from other agencies as well. We did training on cultural resources and management, and Tribal law and traditional cultural properties.*

We also did trainings with private landowners and private forest land owners on some of those issues as well. There's just so many people who don't have an understanding of Tribes and treaty rights. I think it really did help in a lot of the places we did these training programs because a lot of the private landowners had a lot of questions. They just didn't understand much about Tribes, their treaty rights, or anything else related to Tribes. We'd give them examples of what they (Native Americans) were harvesting, and what they were looking at. Afterwards, a lot of the private landowners even offered to have Tribes come and harvest cultural resources on their private lands. I think that was somewhat of a win-win situation. But the kind of formative edu-

---

[1] https://www.fs.usda.gov/detail/giffordpinchot/passespermits/forestproducts/?cid=fsbdev3_005071

cation is something that you almost need to do constantly to keep the public informed on Tribal issues. You need to go from one area of the state to the next all the time.

Kristiina: Do you think the problem also gets back into people's view of culture as something that is only a thing of the past. In Puerto Rico, if you found glass that was dated to be over 50 years of age, then it was considered a cultural site according to the U.S. Forest Service. This would then limit what you could do on that piece of land and you could not collect natural resources. So most people were incentivized to not report cultural artifacts on their lands. What we seem to have also institutionalized is that the culture is the past and not that it's a living, changing thing. You looked for artifacts from the past but not practices of a group of people that still continues today. I wonder how much of this view persists today for Tribes.

Rodney: *That was one of the key points I used to always try to make as a Tribal Liaison person in DNR. Tribes have gathered and visited these same sites since time immemorial and they're continuing to still use those sites and have handed down knowledge of those sites from one generation to the next. So, looking at cultural resources or archeological laws that protect those sites, it always aims to protect the artifacts or the graves or the funerary objects that are like 50 years old or older. Or even anything on the landscape that's 50 years old or older.*

But you can't really look at it in that perspective only because Tribes are still using those same sites today. It used to be where Tribes would leave mortars and pestles at those sites, but a lot of those things were all looted and are no longer there. They left them there, so they didn't have to carry them all the time. They would use them at those same sites all the time and they still do. I think about even my own family and we are still going back to the same sites that our grandmother brought us to because she made us aware of them. She used to tell us about those places all the time and how to take care of them. So, yeah, you can't look at those things as just in the past.

I think one other thing you brought up that's important too is that when I was at DNR we actually had delegations visiting DNR from foreign countries. I remember there was one from the Philippines and one from Mexico. China was another country sending a delegation to DNR. When they came here, they always wanted to meet with me because they wanted to know the legal/political relationships that Tribes had with the federal and state governments. They also really wanted to look at Tribal management of resources. I remember a delegate from Mexico, their national forester, saying they were basically having the same problems that we were. They were really fearful of identifying resources that were important to them and letting the public know about them because there was the risk that they would be exploited. So, they were really trying to find ways to protect those and institute policy or laws so that those areas could be protected. A lot of Indigenous Nations world views and their culture are almost parallel to ours. So, they were experiencing a lot of the same issues that we were, especially with resource extraction.

Kristiina: So, Rodney, just thinking about everything you're saying and the loss of collection rights on customary lands. The Tribes were put on smaller pieces of land but were guaranteed collection rights to customary resources. But these lands are now owned by private and public groups so accessing resources from these lands has become challenging. Tribes can't just go on what is now private land. If Tribes have to interact with Washington state agencies or even federal agencies, Tribes should have a greater chance of accessing customary resources. But collecting from these lands also have issues. You just mentioned about the overharvesting of Huckleberries on DNR lands. So even when you have access on public lands, you've already have others that are coming in and overharvesting in very unsustainable ways. So, you're losing access to these resources. It seems that the agencies, even the public agencies, have not really come up with good policies for how to facilitate Tribes being able to obtain their treaty rights for resources. This doesn't look very good for Tribes. Am I missing anything?

Rodney: *You bring up a few points again but there are security issues of continuing to collect on State trustee lands. A lot of the land management agencies buy, sell, or exchange land all the time. DNR does as well. DNR really wants to block up land for efficiency and use of resources and managing those areas instead of having scattered parcels across the landscape. Some of those scattered parcels can be pretty large parcels, several hundred acres or larger in size. Yet all those areas can have high interest to the Tribes. So, that's always been in contention with the state because when the state is looking to buy, sell, or exchange land, they may be selling some of the only lands Tribes have to gather on. I remember once they were selling these large parcels of land and the tribe met with us and said you're taking everything away, we have nowhere to go after this. We went through and looked at all the maps with them and it was pretty obvious what they were saying.*

The other thing is that when they do sell or exchange those lands, sometimes even to private ownership, new owners develop those lands and there's a lot of loss of resources in that way too. DNR is really charged with making revenue for the state. So if they think somebody else can make better use of the land, especially over in Eastern Washington, they will sell that land or trade it or exchange it.

Even lands that do not look like they have any value, like the shrub steppe areas, there's a lot of resources that Tribes harvest from these lands. You may look at that as a waste land but there's a lot of traditional foods, a lot of medicines that Tribes access on those lands. Some of the shrub steppe areas may even have more resources for Tribes than forested areas. So, when they sell those lands, especially if it goes into private ownership, Tribes can no longer enter onto those lands for anything, without the permission of the new land owner. Or if DNR develops the land themselves to lease it for commercial use, to make gravel or for agriculture, it changes the landscape as well and Tribal resource access.

Kristiina: When I was on the DNR Board for Natural Resources, forest lands were being exchanged for lands where warehouses could be built and leased for storage by commercial companies. What other land-use changes did you see that reduced Tribes from harvesting resources from DNR lands?

Rodney: *There are a lot of issues that impact Tribal harvesting on DNR lands. If DNR sell land to some of the major lumber industries, some of the companies put the land into questionable uses. They'll log the forest and then they'll sell it or they'll parcel it out and put it into homeownerships. This creates all kinds of problems. A lot of times, this is done without any permits for wildland fire management or emergency services after a landslide. When somebody has a home built, whether it's permitted or not permitted, and they have it on a landscape where something happens, it's always a priority to protect that land.*

Over in Eastern Washington this is a problem because homes or structures are the number one priority when fighting a wildland fire. Look at what's happening in our state now, the wildland fires are increasing unbelievably in size and in number. So, if you have a home here, those firefighters will all work to protect that home while all of this forest is burning.

Kristiina: A year and a half ago, there were several large fires in Eastern Washington. Mike drove us around the Colville reservation, and it was shocking the extent of the forest area that had burned. So the high intensity disturbances like fires also cause loss of resources and therefore access to them.

> **Rodney:** *If you noticed in that fire, all the firefighters were dedicated to saving homes while basically millions of acres of forest land just burned up.*

Alexa: Going back to when you were talking about DNR trading and selling land. You gave the one example of the Tribes that came and said that if you sold that parcel, it was everything they had. How often do you think DNR, when they're going into making a sale or trade, takes into consideration the Tribes use of those lands? Do you think they do at all?

> **Rodney:** *They inform the Tribes, most of the time. But it's not early in the process. That's a problem, that's a huge problem. Because usually Tribes find out when it's being put up for sale. Then it is too late to give meaningful comments. I've seen too many meetings with a lot of heads nodding saying "Yes, yes, we hear you, but the land sale goes through any way". That just happens a lot.*

Alexa: From the tribe's perspective, do you think a lot of them want a more active role in management of those lands with the government agencies? What do the Tribes want from DNR as a management agency? Do they want to be involved, do they want to be better informed and considered when decisions are being made?

> **Rodney:** *Yes, to all of the above. There were two cases, the Boldt decision where the Tribes sued the United States government on the right of the Tribes to harvest fish. They felt that that was their treaty right to harvest those fish. It was illegal prior to this decision and there were a lot of Tribal members thrown in jail for fishing. If you look up the fish wars, you'll see they were also harassing women and children and others who were fishing. This brought a lot of the Tribes together to file that suit. Basically, Judge Boldt said that the Tribes can harvest up to 50% of the harvestable catch, which was a really key decision in tribe's legal history.*
>
> *The other case was the Rafeedie decision. This extended the Boldt decision to shellfish, saying that Tribes can also harvest up to 50% of the harvestable shellfish.*

> Through both of those actions, Tribes were recognized as co-managers of the landscape primarily for fish and wildlife. It is actually a legal position for the Tribes. But even with co-management rights for fish and wildlife, the land is managed by DNR. That's also one of the things the Tribes have always had complaints about. When they want to deal with one issue they've got to go to the Departments of Ecology, to Fish and Wildlife, to DNR, to all these different agencies. But when anyone from an agency comes and meet with the Tribes, they can just meet with the Tribal leadership. It's just a one-stop-shop. But for the Tribes it's a much more difficult, because then they've got to find parody with each of those agencies to move an issue forward. So it's pretty difficult, especially that they have segregated management functions. To some extent, I think agencies will use that as an excuse, "This is not my area of management, go see them." I've seen that a lot of times when I was working there and it's very frustrating to Tribes.

Alexa: Even if Tribes can take advantage of their treaty rights and really use that to their ability, it's not a direct path to access these rights. Is it really caught up in the bureaucratic government systems and convoluted?

> **Rodney:** I think it can be very convoluted because the rights extend not only to the fish and the animal species itself, but it is their habitat and those things that impact their habitat, whether that is water or pesticide use. There's a lot of different issues but those are all handled in different agencies. For one issue they have to find agreement or work through a number of processes or regulations to just deal with one subject.

Kristiina: This gets back to what we've talked about where you have an agency with this piece of land, but you've got somebody else living next door. It could be another agency. These owners don't coordinate at all on addressing a problem that crosses both lands and they don't want to. There are good reasons for this lack of coordination. If it is an agency, they're doing it because they're fighting for their own survival as an agency. This is especially important when agencies are political and are impacted with whatever happens in washington D.C. The Forest Service has to worry about getting funding from Congress to do their job and they don't receive enough funds to do this.

> **Rodney:** You've got to remember too, I can't stress this enough, that the aquatic lands are just as important as the uplands. The Tribes are always at the table on those issues occurring on the aquatic lands.

Kristiina: Has the Boldt decision and its co-management requirement working better for Tribes on issues occurring in the aquatic lands?

> **Rodney:** *I think it's worked a lot better. It's given the Tribes a very strong legal foothold. But it doesn't mean there aren't issues still, there are still a lot of issues. For one, the Tribes are very regulated in their harvestable catch, which the state can control for the commercial catch, but they can't do as much with the sportsmen. They really have no idea how much fish are being caught by sportsmen. When I sat in meetings where they were really questioning the Tribes, and the Tribes they said if you can show me the harvestable catch of sportsmen and commercial, then you can come criticize us. Otherwise, get out of here. They couldn't provide this data but there are many boats and people out there on the river during the fishing season. I don't know if you've ever been out there to see this but it's pretty packed during this time.*

Alexa: Do you think that DNR is capable of sufficiently or effectively addressing the cultural needs of the Tribes or do you think there's something that would need to dramatically change?

> **Rodney:** *I think that there's always a possibility that they can, but you have to remember that the Commissioner is elected every 4 years. This makes a huge difference on their perspective, as well as the executive staff that they bring in when they win the election. When I was there, when the new commissioner came in, all the executives were let go and they brought in their own staff. You don't know who those people are, where they came from, what their backgrounds are, what kinds of experience they've had with Tribes, the decisions they'll make or the policies they'll come and change. It is a constant battle. It's difficult.*

Kristiina: It's funny because you have rules and regulations that have to be followed and it seems like every time you have a new Commissioner, everything changes. The Commissioner who ran DNR when I sat on the Board of Natural Resources, as the Dean of the College of Forest Resources, really didn't understood natural resources. I don't know what her background was, but I was not impressed that she really knew what was going on and because of that she couldn't really provide leadership. It's interesting, over the weekend (November 2017), I saw an article about 500 company leaders who were asked them about their leadership styles and what works for them. One thing that came up was a comment made by one person was that they can always "trust" that the leader will always make the right decision no matter what. But the leadership focus seems to be a predictable process, it's like you do 1, 2, 3 and you start off by asking everybody for their opinion.

But I remember when Dan and I were at Yale, the social ecologist would always start off saying *"That was a great job you did! But..."* I always waited for the "but" because I always knew there was a "but" that was going to follow. It was not a balanced approach that cared for your input but focused more on making a negative comment on what you had done. In fact, the Tribes cannot "trust" that the decisions being made are what should be occurring, based on the "science" and our understanding of it. If you could trust decision-makers, you would bring in some of the issues and knowledge that the Tribes have. This knowledge should be part of any discussion. But it's all this politics. For some reason that resonated when I saw that business article. Today, the Tribes cannot trust that the right decision will be made. My opinion.

> **Rodney:** *Well, for me, it was really an experience when I was one of two positions that was kept when a new Commissioner came in. The structural change was extensive since I saw everyone being let go, even their staff assistants, and a whole new group of people come in. They completely changed the organizational structure of DNR. They brought in all of their own political agendas, which maybe were or maybe were not congruent with the ones that were there before. Even though there were existing laws already in place, it was the interpretation of those laws and how they would approach that determined how they would direct. Everything changed.*

Kristiina: So, Rodney, why do you think that the Commissioner listened to you? Why did the Commissioner empower you to go out and talk to all the Tribes in Washington State?

> **Rodney:** *I think a lot of that was on my own suggestion. I worked for a time under the previous Commissioner and they really just wanted me to deal with the issues that came up. There was really no process to try to comprehensively look at how we could work with Tribes. I thought it was a good plan to talk to all the Tribes, just to put out fires everywhere. It would be better for us to meet with the Tribes, find out what those issues are. This really was an attempt to develop some form of policy or structure or communication for how we could actually improve things overall and not just dealing with issues as they came up.*

Kristiina: That's what you were doing? You were proactively looking at working with Tribes?

Rodney: *I was trying to be proactive. I suggested that, and the Commissioner just gave me the go-ahead and said that was a good idea. The Commissioner's Order on Tribal Relations is still there. They're still having the annual meetings with Tribes, so hopefully that did make some positive headway.*

Kristiina: That's an important point. If a tribe reacts or responds to a specific problem at the time it occurs, you can never win. This is a reactive approach. The news will have published the problem and most agencies will want to get rid of the problem, so people are not thinking about it. It's also going to make somebody look bad. I was just thinking about what you were saying and how to change this type of approach to problems. It means that the agencies have to become proactive and not reactive. But you've got to get the right people into an agency. So for DNR, there must be somebody in there as the Tribal relations person who you can trust to proactively present Tribal issues. But I never even hear about the Tribal relations person in DNR. Does that mean it's a paper thing, so it doesn't really matter?

Rodney: *Yeah, I don't really know the focus they're taking now. But anyhow, I liked the job. I did really like it there, it was a really busy place. But I wanted to pursue this degree too, which is why I took a job up here.*

Kristiina: How much did the legislature impact or make it difficult for the Land Commissioner to really move ahead on some ideas? Because they have the purse strings, they control the money that comes in, don't they? Is that another issue?

Rodney: *That is another huge issue, not only just for the Natural Resources Committee but for other committees as well, it could be Energy or any other committees. They're always looking at natural resource extraction at other agencies. But in the State of Legislature, I can recall us always working constantly with a lot with the Natural Resource Committee, also the Energy Committee. We needed to know the legislative perspective as well since they are passing those state laws. So in DNR, we would have legislative sessions once the legislatures started meeting. We would get together every week and look at all those laws that were proposed, or the code changes, or anything else. So, during that period, I was constantly reading those and trying to stay ahead of what was being proposed.*

It could come from just about anywhere too. I remember land sales were put into budget provisos that might be a little bit hidden. You wouldn't even see them if you didn't read the minutes of the committees to find out a lot of the things that were put into a bill. For me, I could never lobby, that would be the death of me and my position. But I could inform others on what was coming up. So, I was always constantly trying to come up with messages to the Tribes to inform them of some of the actions of the legislature that could actually impact their interests or policy.

Kristiina: Did any laws change when the Department of Transportation was digging up the old cemeteries? Or was it just educational?

Rodney: *It actually impacted a lot and much of it can be found in the Joint Legislative Audit and Review Committee (JLARC) Study where it's talking about the Lower Elwha tribe at Port Angeles [139]. There was an archaeological site there. It was just a classic example of ignoring this site that cost the Washington State millions of dollars. There was a gravesite where they were looking at as a construction site for Pontoons or something. They did a JLARC study that documented all of the impacts and cost the state some 80 million dollars or something like that. It was all on cultural resources and the Lower Elwha tribe. They contacted all of the Tribes around that area but the Lower Elwha. The State Department of Archaeology and Historic Preservation left it up to the consultant to contact the Tribes and not the agency themselves, so they were just ignorant. They said that they were ignorant to it.*

So, they had this area of potential impact or affect, and they did all the studies there. But then as time went on, they found just a few artifacts, so then they increased the construction site area. But they didn't do a review on that expanded site. The expanded site area included a village site and next to the village was a cemetery. They actually unearthed like one or two graves during the construction. They mitigated that. They were going to excavate it and move the graves. A lot of Tribes were screaming. The Lower Elwha tribe was screaming. The construction kept going. I don't remember how many graves they disturbed, but it was a lot before it went to a meeting with the governor and a lot of Tribes. The Tribes basically said "Enough is enough." You can't keep going. This city of Port Angeles was pushing it because they wanted the economic development, they wanted the jobs that it created. It was the Department of Transportation, it was a WSDOT

> *project, and they didn't want to stop because they'd already put millions of dollars into the project with all the architectural design and all the studies that they did. But they had to stop in the end. The Governor agreed with the Tribes and called it quits and told them to shut down the project. It cost the state millions.*
>
> *JLARC, they're kind of like an investigative arm of the state. They did a study, and it's a good study. It changed a lot of things — not only here in Washington State but even across the United States because it was so many classic mistakes that were made in that whole process. It was really difficult politically too because Francis Charles was the Chair of the Tribe and she was trying to do everything that she could to stop it. The people in Port Angeles were threatening her physically, cussing her out in town, she put up with a lot, but she held firm. In her interview, she said that she realized she had to take the issue out much more. They brought the story to the* Seattle Times *and internationally people weighed in and actually started coming to the site. With the pressure of the Tribes and pressure of even international groups, that's what brought it to a stop.*

Kristiina: But what you're really saying is that when a problem arises, and there is an agency, let's say the Department of Transportation, that in the past Tribes have not been able to interact and be heard by an agency. To get some resolution on what's happening and be at the table really, it sounds like an issue always had to be elevated to a higher level. You need to get the governor, or even international involved. People then write about it. So, it sounds like there still isn't that respect for Tribes. That's what I'm hearing, right? Because otherwise the only way they can get justice is to elevate it where others are talking about and then suddenly there's a response.

Alexa: It sounds like there's also not enough value of the Tribal knowledge of the land.

> **Rodney:** *If they would've just talked to the Tribes. In hindsight too, in that process they're supposed to research the subject. If they would've just researched it, it was a known traditional site and village and grave site. People knew, it was documented. But they just went ahead with it anyway. That did change everything. It changed a lot of things, money changes a lot of things.*

Alexa: Do you think that incident did change any of the agency's contact with the Tribes? To be more proactive?

Rodney: *At the time, we actually did a lot of trainings on cultural resources and I remember partnering with the Tribes, because the Skokomish tribe was one of the impacted tribes, as well as Lower Elwha. We had some of the Tribal Council members do some of the trainings and they then did some of the keynote speeches. The training was done for a while, but I think that's probably kind of withering again. Hopefully we won't run into something that catastrophic again. But the potential for that is always out there.*

Alexa: It also seems like if there's always new people coming into the agencies, then they're lacking that training or cultural awareness. You may have been able to help the current people but not those that arrive later.

Rodney: *There's a lot of issues that come up. You just never know where they're going to come up. There's a lot of construction going on, even look at when they were boring that tunnel through Seattle. They ran into artifacts, so they had to deal with that. A lot of the construction in these cities was done before archeological laws were passed. I'm sure there's all kinds of things under all of these buildings being developed here in Seattle. Unless these are torn down and new construction happens over the same site, then those issues will come up again. But it's not just Seattle, it's wherever they're doing development.*

Kristiina: I wonder if Mukilteo ever looked at the ferry landing area that they built. It is one of the stops of the *Tribal Canoe Journey* and Mukilteo gets its name from an Indian word "Muckl-Te-Oh" — translated as a "Good Camping Ground" [142]. So, you wonder, there's probably a lot of places that nobody thinks about but used to be important for meeting point for Indians and will have artifacts.

Rodney: *Well it's always difficult for the Tribes because the State Department of Archaeology and Historic Preservation maintains a database if they know of the sites and they're mapped. If somebody applies for any kind of a permit there, a red flag will come up if there's something there. They'll have to investigate it further before giving a permit. But a lot of the Tribes don't want to have stuff put on those databases because that has been exploited. People have broken into those databases and have used them to mine places where they can search for artifacts. Artifacts return a lot of money now. You see them auctioned on eBay, in Santa Fe and other places.*

I remember going to the Celilo Village along the Columbia River, which is one of the oldest fishing sites in the world and Native people have been there since time immemorial. You can go down there and there are holes dug everywhere where people went looking for artifacts. A lot of artifacts were taken out of there. Those artifacts have been there for maybe thousands of years — carved pestles, bowls, arrowheads, net weights, all kinds of things. It was very rich with artifacts. You can still go there and still see some of them almost right on the ground. They have to patrol all of the time. But they've been robbed of almost everything. You can google "Columbia River Artifacts" and you'll probably find things on auction today. I couldn't believe it when I went down there and took a look at some of those places and saw the potholes that were just everywhere from people looking for stuff.

There's actually another one over on Lake Roosevelt where an amateur archaeologist got into a database somewhere. He found out where a lot of those old village sites were along the Columbia River and he was digging things up and had them in his garage. He was wanting to store all these things and was trying to make one massive sale to make a lot of money. People found out about it and that was put to a stop. I don't know exactly what happened to him, but he had been doing it for quite a number of years, just going out there and taking all those artifacts off the landscape, out of graves, out of cultural sites. It was pretty disgusting to me that somebody would do that just for the dollar, but it is happening.

That's one reason a lot of Tribes won't put things into a database. They prefer to be contacted a lot of times when there's any type of groundbreaking activity taking place.

Kristiina: If you look at all we have talked about and knowing that agency knowledge of Tribes and cultural resources fluctuate a lot, in a nutshell what do you think are 2 or 3 things that would make a difference? Like, maybe having a really viable and empowered Tribal relations person in DNR? Every time you have a new lands commissioner in DNR, they have a Tribal person able to advise them but also with the authority to stop activities that reduce Tribal sovereignty? Should some of these agencies be forced to have a Tribal liaison person? Or, anytime there is going to be some activity, have a Tribal person on the committee from the region where some land-use activity is being proposed? Or, maybe ATNI should have a cultural experts committee that's able to respond and would know how to interact better on these issues. I'm just sort of wondering if you have thought about this and what do you think it would make a difference? We're seeing what the weaknesses are, every time you have a new political appointment, the memory is lost within an agency. Usually this hurts Tribes.

> **Rodney:** *I think that to have a Tribal relations manager is key and really having somebody in that position that understands Tribal governments, Tribal cultures, and the Tribal history of a specific area. Maybe even more than one person. That was one of the things I used to try to tell the Commissioner. I just said there are cultural differences from Western to Eastern Washington and there's enough work that we should've had at least two Tribal liaison people in DNR.*
>
> *Then I think there should be conferences with all of the natural resource agencies, not just DNR, but others and to just talk about specific areas. Say the Puget Sound because the Puget Sound is an issue for all of the Tribes here. Look at that management with everybody in the room sitting there at the same time and really trying to make some decisions of how they can really work together to improve the management of the Puget Sound. When you're doing this piecemeal or agency by agency, it just seems to take an awful lot of time and an awful lot of effort and it just seems like we could do a lot better if we could do some of this collectively. With the Tribes, this has to done collectively since the Tribes are independent sovereign entities as well. So to have them there together is best. It might take a while and there would probably be some really heated discussions. But I think we would go a lot further than just independently meeting with everybody all the time.*

Kristiina: Would you have to have somebody in the Governor's Office?

> **Rodney:** *Well there is the Governor's Office of Indian Affairs, and I agree with that. There's a Centennial Accord where the Governor meets with the Tribes every year. The Centennial Accord really just almost for show since how much can you really get done in a day?*
>
> *The Governor doesn't supervise all of the agencies in the state. DNR doesn't answer to the Governor and that person isn't at the Centennial Accord. The Department of Fish and Wildlife don't report to the Governor, but to a Commission. The structure of the state doesn't really quite work for the Tribes, I don't believe. That's why I think if it was just done by the landscape or aquatic lands and management area or management zones, everybody could weigh in and really work towards making some decisions.*

Kristiina: I look at the Tribal leadership meetings that happen at U.W., and it's mostly for show. You never see anything come out of it, nothing changes.

> **Rodney:** *I really think the key people have to be there, just to have a really good working situation. If it takes multiple days, then so be it. I think we would get so much further, because I don't think the Centennial Accord works. I just don't.*

Kristiina: The other thing that we have to address is that a lot of agencies have very good intentions and interact well with Tribes. But it doesn't carry any weight. It never impacts anything. It's just a way of saying, "*We're working with Tribes.*" I had a student that worked in Alaska and the EPA wanted to know how they could better work with the Native Alaska Corporations and communities. But the EPA is a regulatory agency. The Native people are brought in at the end of a decision process. So when somebody comes in and says I want to build a mine here, kind of like Bristol Bay, they assess the mines impacts. They first assess are there any environmental impacts, any chemical contamination issues? It ends up being a process where trade-offs are made. For Bristol Bay, the decision to stop the mine occurred because 48% of the Sockeye salmon caught globally came from this bay. Research by Professor Tom Quinn's lab at the University of Washington reported that the mine would eliminate the habitat for salmon and therefore would cause the loss of this rich salmon fishing site.

One of the Vogt lab graduate students was doing research in Alaska and commented that the salmon runs were really poor in 2017. The only place that had abundant salmon runs was Bristol Bay. But the 2017 political changes in Washington D.C. has triggered a reconsideration of the earlier decision to not allow a mine to be built in Bristol Bay. This has the potential to cause the loss of the rich salmon runs in Alaska during extreme weather events. Bristol Bay may be the hotspot source of salmon once Alaska returns to more normal weather conditions but not if a mine is built in this Bay. This has a federal agency making the trade-offs using a top down approach and where local communities are consulted but have little influence on the trade-off decisions made.

In contrast in Washington State, Tribes are co-managers of regional fisheries and are able to decide whether to fish or not depending on the sizes of the salmon runs. In 2017, the Swinomish, Upper Skagit and the Sauk-Suiattle decided not to open a commercial fishing season for Coho salmon [143]. As Yuasa wrote:

> "*Treaty Indian Tribes in Western Washington will restrict fisheries again this year — including culturally important ceremonial fisheries — to protect weak salmon runs caused largely by lost and damaged habitat.*

> "Our ability to catch salmon to supply food for our funerals and ceremonies is being constrained because of low returns, made worse by a lack of will to protect salmon habitat." said Lorraine Loomis, Chair of the Northwest Indian Fisheries Commission. "Fishing is at the heart of our cultures and economies. Cutting these fisheries is painful, but conservation must come first so that future generations will be able to exercise their treaty-reserved rights."

Do you want to comment on this?

> **Rodney:** One of the things I don't think that a lot of people in the Washington State realize is that Tribes are the ones who are conserving these areas to improve the habitat for salmon. Tribes are putting a lot of money into salmon habitat restoration.
>
> This also occurs in how much they take of wild game or big game. Their share of take is miniscule in comparison to the state. A lot of Tribes, like the Lower Elwha, they don't harvest antler-less doe or cows or elk. They just take the bucks or the bulls. Which to me, I don't know if you've ever eaten any wild game meat, but during rutting season, that meat is really tough to eat. It's really gamey, the best hunting are really the doe and the cows. But for them to make that decision for the whole tribe and still conserve the populations, I think that's a huge sacrifice for that tribe. The tribe may only let one animal be hunted per family. Even looking at that, Tribes have hunting and fishing regulations that a lot of people just don't realize that happens.

6.3 Alaska Natives, Conservation and Policy Process

Alaska is an interesting state to examine related to Indigenous people and how much of a voice they have on decisions made on their lands. Alaska Natives did not sign the treaties like what occurred in the lower 48 states. They are also Indigenous communities most impacted by climate change impacts. So they provide comparisons on how Native Alaska communities can partner with state and federal agencies to pursue conservation goals. Their goals are not necessarily compatible since in the Western world, conservation mainly means protecting wildlife and not seeing them as part of a group's food security. Further, the policy process follows a western approach and inadequately represents the lands and water that Native Alaskans need to manage holistically and across borders for their survival (*see* Section 6.3.1). Since most of the lands in Alaska that are not already in parks belong to the Native Corporations, it calls for some unique partnerships to expand the area in conservation (*see* Section 6.3.2).

6.3.1   Alaska Native Perspectives on the Governance of Wildlife Subsistence and Conservation Resources in the Arctic*

Photo of Victoria Buschman in Alaska.
Source: Victoria Buschman.

Current conflicts in subsistence and conservation

Engagement and motivation around natural resource conservation in the Arctic is growing alongside the need to mitigate and adapt to ecological and anthropogenic impacts across the landscape. Global interest in the State of the Arctic has compelled collaborative efforts to manage a variety of key natural resources including offshore oil, shipping waterways, and wildlife. Wildlife pose a unique challenge to natural resource management as many species are migratory, transcending ecological, jurisdictional, and political boundaries. They carry cultural, environmental, and economic significance that spurs conflicts of interest in management decisions. Challenges for global and national policies and management strategies are only becoming more complex through shifting environmental realities brought on by climate change, environmental degradation, resource exploitation, among other impacts that seem to characterize the rapid loss of biodiversity in the 21st century.

---

* By Victoria Buschman.

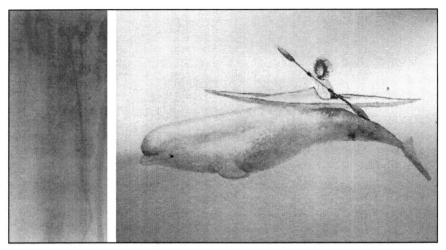

Water color by Victoria of an Alaska scene.
Source: Victoria Buschman.

Certainly many communities within the Arctic are concerned about these changes and their impacts to society. We as Alaska Natives have been deeply connected to the landscape for millennia and our approach to conservation and natural resource governance will be integral in finding solutions for the future's most challenging Arctic issues. Alaska Native participation in governing institutions and venues has increased in part through a legislative fight for regional natural resource rights and in part through the increased international attention to Indigenous sovereignty, an examination of the ethics of climate change impacts, and the recognition of increasing stress on the ecosystems on which Indigenous communities depend for their livelihoods [144–146]. Historically, Alaska Natives have been largely absent from discussions around natural resource use and subsistence management despite being arguably the most impacted by these policy decisions. Given the opportunity and inclusive policy processes, Alaska Natives are determined to be heard.

Indigenous peoples have a vested interest in environmental protections as our livelihoods depend directly on the health and provision of the ecosystems and landscapes. Alaska Native culture is inextricably linked to wildlife as a source of food, continuity of language, spirituality, community, and wellbeing. In many Arctic landscapes, the ability to subsist on a variety of plants and animals is the crux of Indigenous food security. Subsistence in these environments is an important way of life that continues to support the world's Arctic communities. Subsistence refers to the act of Indigenous peoples using certain species for personal, familial, or community use. Arctic peoples practice some number of five forms of subsistence: Hunting, fishing, gathering, herding, and farming. Alaska

Natives have traditionally practiced subsistence hunting, fishing, and gathering, and with the introduction of reindeer in some communities, subsistence herding as well.

The increase in natural and anthropogenic disasters across the landscape threaten subsistence and food security with new and dramatic losses to biodiversity and shifts in species dynamics [144]. Many governments and environmental organizations acknowledge the role of natural resource conservation in mitigating environmental impacts but some conservation practices that focus on sanctions burden Indigenous communities [23, 147]. The fundamental issue that has pitted contemporary wildlife conservation and subsistence practices against one another is the impossibility of both conserving and consuming a single unit of resource.

These days, much of wildlife management in the Arctic concerns the conservation of lands and species [144]. Oftentimes environmentalists and decision makers neglect the importance and input of Indigenous people such that some conservation practices have inequitable or detrimental effects on local communities [146]. Many Alaska Natives would consider unfair policies as representing "the burden of conservation" [23]. This burden is often placed on communities while the drivers of climate change, environmental degradation, and resource exploitation occur elsewhere [146]. For instance, Indigenous communities are often expected to change their behavior to improve immediate environmental conditions while the greatest contributors of biodiversity loss and other impacts are not penalized. International law protects the fishing industry, but not an Indigenous community's right to those fish as a source of food security. This disconnect, between who is responsible for the problem and who has to suffer the consequences, is a major source of tension [146].

Alaska Native communities have a sovereign right to be a part of decision-making and policy processes as they must live directly with wildlife management decisions. The 2007 United Nations Declaration on Indigenous people gives Indigenous community rights to their culture, language, identity, health, and environment, among other critical issues [145]. Articles 3, 4, and 5, respectively, give them the right *"to freely pursue their economic, social, and cultural development..."*, *"to autonomy or self-government in matters relating to their internal and local affairs..."*, and *"to maintain and strengthen their distinct political, legal, economic, social, and cultural institutions"* [145]. The United Nations also declares "Indigenous people have the right to the lands, territories, and resources which they have traditionally owned, occupied, or otherwise used or acquired." [145] Alaska Natives are interested in addressing this sovereignty:

> *"We will continue to strive for recognition and protection of our basic human rights to food security and self-determination, and to maintain our own unique cultures — rights that are collectively recognized and*

codified in the International Covenant on Civil and Political Rights, the Universal Declaration of Human Rights, adopted by the United Nations in 1948, and the United Declaration on the Rights of Indigenous peoples." [148]

These rights have been recognized in part because Alaska Natives and other Indigenous people tend to approach environmentalism, decision-making, and policy differently than the typical Western model of natural resource management. Unique perspectives on the drivers, mechanisms, and systems of ecosystem dynamics also set Alaska Natives apart from Western ideas of wildlife science and policy. This poses a major challenge to governing these resources, especially when governments, agencies, and environmental organizations are unwilling to accommodate these differences. Despite all this, the general sentiment of Alaska Native communities is that they should play a larger role in the fate their homeland. As one Indigenous knowledge holder said:

"All of the plants, all of the animals, the water, the air, the land is all of what we are.... It is who we are. This is our understanding. People making decisions have a different understanding." [23]

Identifying challenges to collaboration

The inclusion of Alaska Natives in collaborative policy processes is a momentous shift that, albeit moving slowly, is backed by an increasingly powerful force of Native voices. The work of many Alaska Native interest groups has opened numerous venues for participation in decision-making processes. However, inviting a group of people to the table is different from opening the discussion to truly inclusive management [149]. Oftentimes, Indigenous participation is treated as a formality without respect for how inclusive and engaging that collaborative process is in reality. In building a framework for the management and governance of Arctic wildlife, Western policy institutions must recognize their place in supporting diverse perspectives.

Fundamental differences in how Alaska Natives and Western policy institutions define, approach, collaborate around, and solve natural resources issues complicates the working relationships of these groups and serve as barriers to the policy process. This challenge also poses an incredible opportunity to pool all of our collective resources and strategies to combat the impacts of climate change and work towards responsible and sustainable natural resource solutions in a tumultuous time.

Modern environmentalism tends to pit conservation and subsistence agendas against one another despite the mutual benefit of melding the two practices into cohesive and equitable management policies [146]. Alaska Natives and conservationists have many of the same end goals, namely the continued proliferation of healthy, endemic species population dynamics, a healthy environment, and climate change prevention and mitigation. Why, then, are Alaska Natives and many conservation agencies and organizations still at odds with one another despite desiring many of the same environmental outcomes? In consulting the literature, five key issues pose a challenge to inclusive management strategies:

Challenge (1)	Differences in culture complicate how we define and approach natural resource issues [146].
Challenge (2)	Current policy processes often fail to recognize Indigenous knowledge as a legitimate source of information to be used in partnership with Western science to inform decision making [23, 147, 149–150].
Challenge (3)	The distribution of environmental impacts and the burden of conservation challenge perceptions of ethics [23].
Challenge (4)	Institutional racism constructs a power imbalance within policy processes such that decision making is stacked against the marginalized [149].
Challenge (5)	There is further room to foster mutual trust and respect of each other's perspectives and interests [23, 149].

The next section aims to provide a foundation on which to understand these five challenges and unpack how a difference in perspectives must be addressed within the policy process. A greater focus will be placed on *Challenge (1)* and *Challenge (2)*.

Drivers of wildlife management and policy

There are always conflicts of interest in natural resource use and management. Before discussing the challenges to collaboration, it is important to determine who has a stake in the game and how decisions are made. Fig. 6.1 illustrates how wildlife resource users and wildlife resource managers drive Arctic food security. The diagram focuses on how Indigenous drivers and conservation drivers play a central role in the management process. These drivers may be groups of people, methods of analysis, or philosophies.

In considering resource use, wildlife provide food for Alaska Native communities and industry while tourism and conservation value wildlife alive [146]. The first challenge then becomes allocating some distribution of resources across users. This allocation is complicated by not only the spatial and temporal realities of wildlife population dynamics but also by natural and anthropogenic pressures, such as natural disaster and poor management strategies.

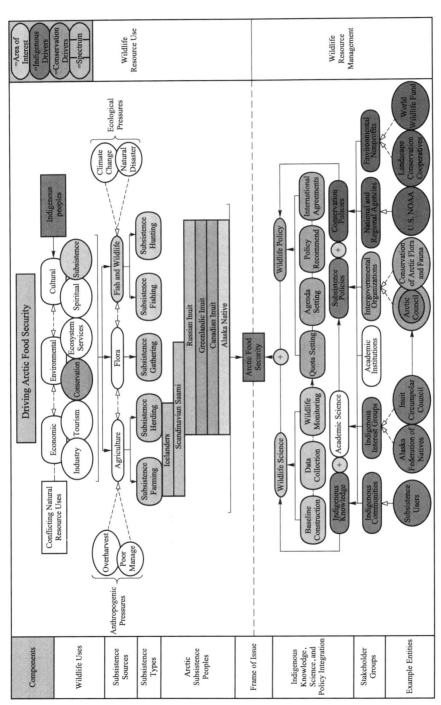

Fig. 6.1 Components of Wildlife Resource Use and Management Related to Arctic Food Security.

The author developed this figure based on information obtained from numerous sources, notably those developed by the Arctic Council and its several working groups [23, 144, 151].

In considering Western natural resource management, decisions are made through *wildlife science* and existing policy. Here I use wildlife science to encompass both Western science and Indigenous knowledge. Many Alaska Natives recognize a lack of respect for Indigenous knowledge within wildlife science [149], as do some researchers [147, 150]. In short, wildlife science uses baseline data, data collection, and wildlife monitoring to inform wildlife policies that take the forms of subsistence quotas, institutional agendas, policy recommendations, and international agreements. Many interests are represented through different groups contributing to the accumulation of information and the decision-making such as the Inuit Circumpolar Council, US federal agencies and state departments, and environmental organizations such as the World Wildlife Fund. This story will consider the integration of Indigenous knowledge, academic science, and policy from a wildlife perspective.

How Alaska Natives define natural resource issues

As *Challenge (1)* states, the first key issue to address within a collaborative framework is differences in the cultural definitions and approaches to public policy and decision-making. Studies on water management policies across countries show that our cultures and society determine how we define and approach natural resource issues [152]. How Alaska Natives and Western society practice environmentalism and natural resource management is summarized in Table 6.1. Obvious differences include perceptions of the human role in the environment and coexistence as opposed to stewardship. Acknowledging these cultural differences in perspectives is essential for developing culturally applicable management strategies.

Alaska Natives define natural resource issues differently than most environmentalists. Problems are generally framed from cultural or economic perspectives rather than from conservation perspectives, despite the overall desired environmental outcomes being healthy wildlife population dynamics. For instance, some Alaska Natives will use terms like cultural value rather than intrinsic value and cite the economic burden of a decrease in subsistence sources [23]. Thus, some people may argue that Alaska Native communities fail to partake in environmental agendas because they tend to frame problems under different terms. This is surely not the case; rather the region has a complex history of conflict between Indigenous people and early environmentalists that have shifted many Natives away from using similar vernacular and problem definitions. Many environmental organizations such as Greenpeace and Sea Shepherd have been, or still are, major opponents to the cultural harvest of important food species such as whales and other marine mammals. Other environmental groups have opposed rural economic development in a region in which Western society is responsible

Table 6.1 Differing perspectives on environmentalism and resource management

Characteristics of Environmentalism	Alaska Native Practice	Western Practice
Human Role in Environment	Part of ecosystem	Removed from ecosystem
Human Role in Management	Coexistence	Stewardship
Perceptions of Traditional Conservation Philosophies	Burden on communities	Necessary for resource protection
Alternative Conservation Philosophies	Conservation through use	Conservation science
Historic Approach to Resource Use	Use many resource simultaneously and sustainably	Move to a new resource after exploiting a previous one
Approach to Scarcity	Management decisions adept at dealing with a scarcity of certainty and information	Management decisions adept at dealing with a scarcity of resources
Approach to Management	Management decisions adept at dealing with boom-bust cycles	Management decisions adept at dealing with overexploitation
Resource Ownership	Do not usually own the resource but may have legal rights to its use	Usually owns the resource and has responsibility to uphold legal use of it

Note: The author developed this table using participant observation of Alaska Native perspectives, Indigenous knowledge from sources such as the Inuit Circumpolar Council Alaska, and academic literature on natural resource management.

for bringing economic development and also fails to supply viable alternatives. Others are intolerant of natural resource use in a region that outdated environmentalists feel should be "preserved" rather than conserved [146].

In fact, even how Alaska Natives and environmentalists define *conservation* differs in nuanced ways. For the Western world, conservation in its simplest terms is the protection and restoration of the natural environment and its resources, made necessary through a concern over exploitation. For Alaska Natives, conservation is more a practice of *conservation through use* [153]. Traditional Indigenous economies in Alaska relied heavily on this sustainable use of resources such as wildlife species in a way that avoided exploitation [144]. Conservation through use works in part through cultural norms as well as through the time constraints inherent to a seasonal climate with limited availability to store and process food. Families generally accepted the cycles of feast and famine and stored only what could keep in the months between plenty. Alaska Natives are adaptive and can make boots-on-the-ground decisions when wildlife populations unexpectedly fluctuate. This fluid movement between resources as ecological conditions change how Alaska Natives have limited their impact on the environment to levels sustainable over the millennia. In contrast, Western societies have a history of using a resource until it is depleted and then moving on to another. This is the scenario that gave birth to conservation practices in the early 20th century.

Alaska Natives obviously care greatly for the environment even if we don't frame natural resource issues the same way as Western societies. Environmental issues have been seen more frequently amongst Indigenous interest groups' priorities. In 2016, the Alaska Federation of Natives issued their annual priorities report which included the support of important policies such as funding and investment for climate change mitigation in Alaska Native communities, protections from toxic algal blooms, implementation of federal subsistence review, funding for the Kuskokwim and Yukon Inter-Tribal Fish Commission, and disaster funds for subsistence fishery and wildlife population collapse [154].

Despite differences in framing natural resource issues, Alaska Natives and environmentalists are beginning to work together. Indigenous people are both resilient and adaptive and our presence at the table is more important now than ever. In recent years and under the threat of climate change, Alaska Natives and conservationists have begun reconciling and aiding one another through their collective resources and information [147]. Collaboration on wildlife research and planning has been of particular interest to these groups [147, 150].

Western and Alaska Native approaches to management and decision-making

Historically, Alaska Natives have practiced a form of adaptive ecosystem management guided through cultural practice, Indigenous knowledge, and a reciprocal relationship with natural environmental cycles. Many Indigenous people consider themselves to be a part of the ecosystem in which they live [146] and the two have evolved together in the few thousand years that Native communities have occupied the landscape. During this time, we've formulated ideas about environmental systems and how best to make natural resource decisions.

> *"As an Elder explains, the Arctic environment is like a puzzle, with all pieces having a place and all pieces necessary to make up the entire picture. These pieces include Inuit languages, retention of Indigenous knowledge, animal health, oceans and rivers, etc. This description of the environment helps explain how the Arctic ecosystem is made up of multiple parts. Scientists may also understand this explanation in terms of systems. Each puzzle piece can be envisioned as a system that together makes up the entire ecosystem. The Inuit culture is a system within this larger ecosystem, just as the hydraulic system is part of the*

same ecosystem. And just as the Arctic ice system is interlinked within that system, so is the Inuit culture interconnected with all aspects of the larger ecosystem." [23]

As stated before, one of the greatest cultural differences that affect how we conduct policy processes is how we approach problems and make decisions. The practice of conservation through use rather than a conservation of resources is one such example. Perhaps the cultural construction of conservation through use does not resemble a Western management strategy in its formality, but it has been effective at supporting healthy ecosystem dynamics for generations. As Table 6.2 illustrates, the Alaska Native approach to decision-making is different than the typical Western approach. For instance, ecosystem dynamics have relied heavily not only on our adaptive ecosystem management, but also our bottom-up decision-making approach that is made on the ground by hunters, their families, and their communities. Under rural and remote circumstances, individuals must make decisions at the margin. They use their better judgment to determine

Table 6.2 Differing perspectives on environmental decision-making

Characteristic of Natural Resource Decision-making	Alaska Native Practice	Western Practice
Primary Decision-making Entity	Local Community	Bureaucratic Agency
Trusted Authority	Elders and Indigenous knowledge holders	Environmental technocrats
Ways of Knowing	Indigenous knowledge	Western science
Common Direction of Decision Making	Bottom-up	Top-down
Responsiveness of Decision-making	Adaptive	Reactive
Information Needed	Best available information and knowledge	Scientific evidence, peer reviewed preferred
Primary Consideration in Analyzing Policy Alternatives	Impacts to sovereignty and culture	Economic impacts
Risk Aversion	Comparatively more risk averse	Comparatively less risk averse
Risk Mitigation	Convene a long time to reduce the risk of making a bad decision	Recognize risks and set aside mitigation funds

Note: The author developed this table using participant observation of Alaska Native community structure and decision-making processes and consultation of academic literature on bureaucracy [155].

whether the weather is good to dry fish without it molding or whether the time is right during the migration to hunt and prepare caribou. These communities' resource extraction patterns shifted season-to-season and year-to-year depending on availability and the health of the animal and habitat, allowing for robust and biodiverse ecosystem dynamics through time.

Making decisions at the margin allows Alaska Natives to practice adaptive place-based management strategies that consider both spatial and temporal realities. How Alaska Natives and Western society build policy for management purposes is briefly described in Table 6.3. Rural communities in the Arctic are adept at using place-based models while also navigating landscape-scale environmental impacts. Western policies, especially those drafted by bureaucracies, rely heavily on generalized models. These generalized models are used when information is scarce and are constructed by considering aggregate variables. The goal is to optimize the model for combatting similar problems across many landscapes. These generalized models are likely not as effective as place-based models that are built on a greater understanding of the immediate environment.

Table 6.3 Differing approaches to the policy process

Characteristic of Policy Approach	Alaska Native Perspective	Western Society Perspective
Policy Process Entity	Local community	Legislative branch
Model for Policy Decisions	Place-based model	Generalized model
Regulatory Mechanism	Social regulation	Formal regulation

Note: The author developed this table using participant observation and drawing conclusions from case study analysis [156].

Some mechanism must control adherence to decisions once they have been made. For Alaska Natives, natural resource use is traditionally culturally-constructed as social policy rather than dictated by formal public policy. This social mechanism has roots in the well-being of a wider community. We only take as much as we need and are sure to share with our elders and others that may be unable to provide food for themselves — there must be no waste and enough for all. In contrast, Western society relies on rigid rules and laws born in part from the overexploitation of resources. These rigid policies make place-based management very difficult. Some of the resistance to stepping away from rigid policies is the Western policy process' lack of recognition for the role of Indigenous knowledge in natural resource management.

Indigenous knowledge in environmental decision-making processes

As *Challenge (2)* recognizes, any conversation about collaborative governance between organizations and Indigenous communities must consider and accept the place Indigenous knowledge has in the decision-making process. While science is validated through the scientific process and peer review, Indigenous knowledge is validated through our long-standing cultural practice and the intergenerational exchange of oral tradition [23, 150]. While Western science has been the dominant driver of science-based policy, Indigenous knowledge is slowly becoming accepted as an accredited source of natural resource knowledge [147, 149–150]. It also appears more frequently alongside Western science in research findings and policy briefs across disciplines [144, 151, 156].

The inclusion of Indigenous knowledge in policy processes has been driven by several limitations of Western science. Researchers, scientists, and policy makers have limited time and money to dedicate to the many environmental problems across the world. In the case of the Arctic, resources are often poured into expensive and time-consuming scientific research and monitoring programs, such as those managed by the Arctic Council in collaboration with the eight Arctic nations. These projects have immense value to natural resource use and conservation efforts but are limited by the Western scientific process that stresses replication and the need for a robust number of observations. This makes time a limiting factor in the accumulation of acceptable data. The Arctic landscape is remote and seemingly inhospitable to those who have not lived there which further limits data collection efforts. For Western conservation, time is of the essence, and as it is a crisis-based science, it relies on the use of the best available information to make reactive decisions. There is not always time to gather more robust data and thus the use of Western science alone struggles to support holistic conservation initiatives.

As many Indigenous people have occupied the same landscapes since time immemorial, they have incredible reservoirs of place-based natural history knowledge that is passed on both by oral tradition and cultural practice. Archaeological records show that Alaska Natives have occupied some hunting grounds and settlements for millennia. A research project, through the Barrow Arctic Science Consortium, has shown continuous use of a peninsula called Nuvuk as a settlement, graveyard, and important hunting ground for many wildlife species for over a thousand years [157]. Alaska Natives rely heavily on wildlife as a primary source of food and thus their knowledge base has inextricable valuable for wildlife conservation [147]. The Arctic's Indigenous people recognize that research should include as many perspectives as possible [149]. The question becomes how researchers and decision makers can bring Western science and Indigenous knowledge together to collaboratively protect and manage these natural resources (*see* Chapter IV).

A self-proclaimed successful case, the Northwest Arctic Borough Subsistence Mapping Project entitled "Documenting Our Way of Life through Maps", was a 5-year project that brought together Indigenous knowledge holders in an effort to improve subsistence information for land use decision-making [156]. Mayor Richards of the Northwest Arctic Borough wrote in the forward of the volume:

> "The purpose of the project was to use traditional knowledge to identify places people depend on for their traditional food resources and the current status and locations of important subsistence species. In some sections of this atlas, researchers incorporated data from other sources (for example, satellite whale tracking studies), but, even in those cases, the study results were reviewed by local experts who verified and augmented the findings. This project therefore breaks new ground in building a strong body of scientifically defensible evidence on a broad, solid foundation of local and traditional knowledge."

The conclusive evidence provided by Western science in the Arctic is lacking as many scientific papers will claim in their concluding remarks. There is a limited body of information on species' distributions and migration patterns because the landscape is remote and challenging for conducting research. In the case of the mapping project, Indigenous knowledge supported the community's ability to tell more intricate stories about species' natural histories and dynamics than were available through academic literature and reports. Subsistence users in the region were able to map important geospatial realities for subsistence species such as high-use feeding, spawning, birthing, rearing, and molting grounds as well as high-density migration routes and other important dynamics that should be considered in landscape-scale conservation planning and land use decision-making. This knowledge has been built over the generations of interactions between Alaska Natives and the plants and animals of the Arctic. Indigenous knowledge holders play an undervalued role in partnering with scientists in the production of mutual knowledge [149–150].

The mapping project has immense conservation value and to reach its depth would have taken years of expensive scientific methods to replicate. This combination of Indigenous knowledge and Western science was compiled into a landscape-scale ecosystem analysis of land and water important to subsistence species across seasons. While conservation efforts are not directly mentioned in the report, discussions around land use and wildlife protections may inherently lead to formalized conservation practices. Researching and monitoring ecosystems at this depth is not feasible through western scientific methods but can be greatly informed through the inclusion of Indigenous people and their knowledge of the landscape.

There are several ethical questions that conservationists should ask when attempting to benefit from Indigenous knowledge. One consideration is whether or not conservationists are exploiting this knowledge by using it to provide protections for species that further marginalize those Indigenous knowledge holders [147]. If researchers learn that a species of seal is appearing less and less at an important breeding ground, is it ethical to forbid subsistence hunters to continue hunting in that spot? It would be hypocritical if Western society did not first attempt to remedy the problem on their side by examining the practices and the impacts they have on Arctic wildlife. If a species is under threat of local extinction, is it an agency's place to forbid a father from teaching his daughter how to have a meaningful cultural and spiritual relationship with that species through hunting? What happens when we sever the connections that Indigenous people have to the land and the living? What happens when Indigenous knowledge and respect for the myriad dynamics of our ecosystems are suppressed and eliminated? We can only guess at the future impacts unfavorable laws would have on Alaska Natives and our culture, but our communities are certainly feeling the strain [146].

A major barrier to the inclusion of Indigenous knowledge in decision making is the Western idea that data must be supported by a standardized scientific method and peer review. Many Alaska Natives do not think we should have to legitimize Indigenous knowledge because the teachings and practices that are built on this way of knowing have been in use for much longer than modern science [149]. In fact, many subsistence users think Alaska Natives should validate the findings of Western science for accuracy. However, communicating the value of Indigenous knowledge to Western scientists will undeniably require some standard of analysis until the western world is ready to accept Indigenous knowledge as a valid way of knowing [149].

The question becomes how decision makers can integrate Indigenous knowledge into the Western policy-making process. This is something that the Inuit Circumpolar Council has an invested interest in pushing into the environmental discourse through international forums such as the United Nations and the Arctic Council. Most Indigenous people are in agreement that they need to be at the table more often and earlier in the process [149]. The collective gathering of information may be the optimal stage for beginning collaborative efforts because everyone will enter the decision-making process with the same information.

Baselines, data collection, and ecological monitoring

Some academics and intergovernmental organizations have pushed for greater Alaska Native collaboration in research, particularly for baseline construction, data collection, and ecological monitoring [151]. Constructing environmental baselines entails recreating historic environmental realities which allow scientists to quantitatively measure environmental changes over time. Data collection entails the production and accumulation of information while ecological monitoring entails the observing, estimating, and forecasting of environmental trends. Each of these three activities is incredibly valuable to the creation of science-based wildlife policy, and each way of knowing brings its own strengths and capabilities.

How the different ways of knowing uniquely contribute to wildlife science and policy are illustrated in Fig. 6.2. Both Western science and Indigenous knowledge are considered ways of knowing. Each has its own information tools and contributions that can be used in a mutual knowledge exchange to inform policy. Through a mutual knowledge exchange, scientists and communities can share their information ethically. Lacking mutual consent, Alaska Natives may consider some uses of Indigenous knowledge as an exploitation of information for scientific gain that may ultimately hurt the communities that provided it [147].

As for its contributions, Indigenous knowledge is unique in its ability to provide information on historic ecological baselines with which to compare anthropogenic impacts [150]. Constructing baselines through modern science requires advanced archaeological work and there are few scientists dedicated to studying historic wildlife population dynamics. While neither archaeology nor Indigenous knowledge can provide the complete picture, the intergenerational nature of Indigenous knowledge allows scientists to track changes over the last few decades [150]. If scientists invested more energy in engaging Alaska Native communities in constructing baselines, decision-makers would have better estimates of the rate of change and could advocate for the appropriately stringent policies.

Engagement in subsistence practices support Alaska Native inclusion in data collection. Alaska Natives can help scientists' better track the spatial and temporal patterns of key species by recording observations and other variables as they subsist. The previous example of the Northwest Arctic Borough Subsistence Mapping Project shows evidence of the complementary role of subsistence with other technological information tools such as geographic information systems, modeling, and tracking [156]. A combination of these tools is also important for ecological monitoring. Monitoring relies on information on spatial and temporal patterns, climate change projections, ecosystem services, and environmental impacts assessments, to determine the scale and rate of environmental change. These more technical tools and methods of assessing information are the strong suits of Western science. Management strategies may be more successful in the long term by continuing to incorporate new information and knowledge about the environment into decision making.

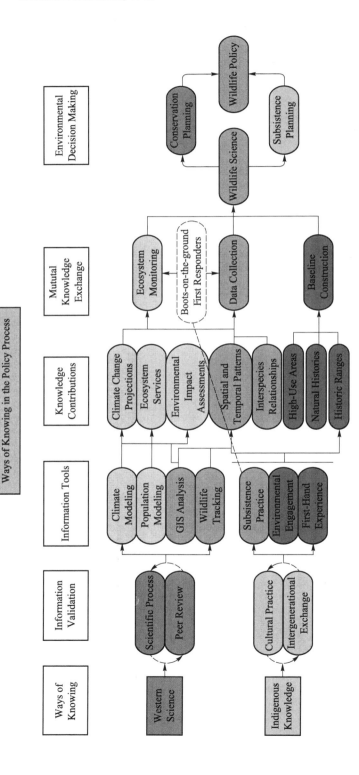

Fig. 6.2 Integrating ways of knowing in wildlife policy.
The author developed this figure based on information obtained from several reports and case studies [149, 151, 156].

6.3.2 Partnership — A Role for Nonprofits and Agencies in Conservation of Native Lands in Alaska*

John Wros.
Source: John Wros.

The confusion over Alaska Native lands is analogous to that in the "lower 48" states (contiguous continental U.S. states excluding Alaska and Hawaii). The map of Alaska reflects cumulative land actions dating back to the turn of the 20th century. This division is ever changing, and to the present day those that benefit from owning a slice of The Great Land commit themselves to solving a legally complex and often nebulous puzzle of split estate rights, easements, pre-statehood actions, and Federal and State acts unique to it. The two contemporary focal points of this division are the *Alaska National Interest Lands Conservation Act*, universally referred to as ANILCA, and the *Alaska Native Claims Settlement Act*, referred to as ANCSA. ANILCA was passed in 1980, while ANCSA was passed in 1971, but I intentionally refer to them out of chronological order; although ANCSA predates ANILCA, conservationists in Alaska predominantly work in an ANILCA-centric reality.

That said, it's often the Alaska Native lands established under ANCSA that conservationists prize, representing the best of Alaska's wildlands. The critical disconnect is the distinction between what Alaska Natives call their traditional lands and those lands that are patented to Alaska Native groups. The overwhelming majority of all lands patented to Alaska Natives are patented to Alaska Native Corporations, which will be described later. Overlap between traditional lands and patented lands is common, but as I have come to understand it, only marginally comprehensive.

---

* By John Wros. John Wros is the Alaska real estate associate for the national nonprofit, The Conservation Fund, with more than a decade of experience in the field working with land trusts and conservancies. His current work is focused on conservation easements and acquisitions with Alaska Native groups, private land owners, and state and Federal agencies. John is a member of the Forest Ecology Interest Group at the University of Washington School of Environment and Forest Sciences, studying the evolution of public-private partnerships in conservation.

That disconnect is necessarily described by the members of the Alaska Native community who have been disconnected and has been captured most recently in environmental and social activist movements surrounding the Arctic National Wildlife Refuge. For those interested in Alaska Native land rights issues, I recommend investigating the ongoing work of Karlin Itchoak, Chief Administrative and Legal Officer for the Ukpeagvik Iñupiat Corporation. The following pages exemplify the niche role that conventional Western conservation has played over the last decade or so in the evolution of Alaska Native land status since ANILCA, specifically on the collaborative advantage gained in partnership.

Alaska Native lands, traditional lands, ancestral lands, and historic lands are all terms used interchangeably. In this section I'll refer to traditional, historic, and ancestral lands as *traditional lands*, those lands patented to individual Alaska Natives as *native allotments*, and lands patented and managed by Alaska Native Corporations as *native corporation lands*.

Alaska Native traditional lands

By the broadest definition, traditional lands are the historic and current cultural and subsistence use lands of all Alaska Native peoples. I do not know if anyone would make a case that every square mile of Alaska should be considered traditional lands. In fact, my own home lies in the historic no-man's land between the Prince William Sound Eskimos and the Dena'ina people. That said, early house pits, cemetery sites, and fishing and hunting camps have been verified from the Arctic Coast to the most remote islands in the Gulf of Alaska. Many of these areas are near existing villages, but not all. An example of this extreme distribution are the Gwich'in — Athabascan Indians who call themselves "the caribou people" — and nomadic Inupiaq Eskimos, who between them followed ancient animal migrations that spanned the from the Canadian border to Norton Sound. Many of these places are still used by native peoples today. Yet it is the exception that traditional lands are owned by Alaska Natives. Of Alaska's approximately 375 million eligible traditional use acres, Alaska Native individuals and villages own less than 1%. This figure does not include Native Corporation lands, for reasons explained in the next section.

To weigh the benefit of partnership it is necessary to understand the essential incompatibility of traditional values and real estate. Native allotments, village sites, and Native Corporation lands are rigid real estate designations constructed on a foundation of Alaska Native traditional land-values that have changed over the centuries with the course of rivers, movement of caribou herds, and the state of Tribal relationships. For example, the complicated land selection criteria of ANCSA was often necessarily boiled down to identifying areas encompassing "home". As you might imagine, that concept is not grounded in township, range, and section. The arc of human habitation in Alaska spans more than 10,000 years, and over the millennia the geographic scope of traditional use is beyond knowledge. The current map tells the story of how Alaska was appropriated. Practically, these static real-estate designations result in situations where Tribes do not have the ability to grow beyond the extent of existing patented lands or move away from village sites threated by climate change and diminishing land values of native allotments that succumb to erosion.

It is easy to understand that the natural habitat and aesthetics that drew Alaska Natives to an area are the same qualities targeted by conservationists. **The conservation benefit in partnering with Alaska Natives lies in the fundamental principle that these groups have sustainably lived on and managed these lands for thousands of years. They will continue to do so with management practices grounded in core values more conducive to a healthy environment than non-Native parties.** This is substantiated by the growing acknowledgement by academia and government of traditional knowledge. The fields of landscape ecology, land conservation, climate change mitigation, and wildlife management have observed that the traditional knowledge and land practices of Natives often utilize sustainable harvests, seasonality, and a greater capacity for flexibility than parcel and management unit-based systems. A second conservation benefit lies in the availability of Alaska Natives in remote areas of the state, where no other suitable conservation land manager might exist. While the logic of returning lands to native management is broadly appreciated, the social, political, and economic reality of such actions is difficult to overcome, and the cases of significant transfers of land back to Alaska Natives for conservation and management has been so limited as to be considered negligible. Why would a public or private landowner relinquish ownership of lands? Are traditional management practices sufficient if adjacent to an area or connected to a system not managed in the same way? Are Alaska Native villages equipped to enforce management for the challenges of increasingly non-Native use of lands? Do they State and Federal legal systems in place support this type of ownership and management?

As explained, the benefit to Alaska Natives cannot come from buying traditional lands (Non-Native Corporation) from them, limited by available acreage and ethics. By any mechanism, conservation must aim to restore ownership and control of traditional lands to Alaska Natives. Let us initially assume that the above questions can be satisfied and consider the following hypothetical example of the benefit of partnership with Alaska Natives: Of the traditional lands we have considered, all can be impacted by eminent domain to the detriment of those resources. Tribes in the United States have the condition of being sovereign. Functionally, the only way one sovereign takes the land or rights of another sovereign is to go to war over them. In this scenario, a conservation organization would be incentivized to purchase a parcel and place a conservation easement on it, then transfer title of the land to the sovereign Alaska Native Tribe. For conservation, this partnership achieves assured conservation by way of a conservation easement, with the threat of eminent domain negated. For Alaska Natives, this partnership restores ownership and management control for the use of traditional lands — barring an act of war. To my knowledge this method has only been employed once in Alaska, but as we will discuss, the mechanisms used to achieve it are well established.

A discussion of traditional land and conservation must include native allotments. The *Alaska Native Allotment Act* of 1906 allowed Alaska Natives to acquire up to 160 acres of unallocated land from the United State Government, including all surface and subsurface rights (Fig. 6.3). These lands have inherent protection, as they cannot be sold or transferred without the review and authorization of the US Bureau of Indian Affairs. These parcels are generally in species and habitat rich areas, due to their likely proximity to subsistence harvest. While these sites are already owned and managed by individual Alaska Natives, they can certainly be sold to non-Native individuals. Frequently this occurs when passed to the next generation. It is often the case that heirs to an allotment do not have interest in continuing to own the land, and in some cases, are unaware they have inherited these traditional lands. Native allotments are a logical place to employ conservation.

Alaska Native corporation lands

The majority of Alaska Native lands are owned and managed by Alaska Native Corporations, Alaska's alternative to the Native American reservation system used in the lower 48. A Native Corporation was established for each verified Alaska Native group, supported by an allocation of land and money in settlement from the United States Government. These corporations are broadly comparable to other private sector corporations, with the distinction of having their village members as the shareholders. Lands, being the central component of the settlement, were selected by the representative corporation based on the geographic proximity of a Native village and the historic use of that Native group. It is the mission of these corporations to develop their selected lands in such a way as to develop revenue to distribute to their constituent shareholders. In example, members of the Native Village of Eklutna receive dividends from the revenue of Eklutna Inc. earned by developing the lands selected under ANCSA — natural gas, oil, mining, timber, lease, etc. I use this example because Eklutna Inc. is a uniquely conservation minded Native Corporation. The obvious flaw is that in order for Native village members to receive financial benefit the resources of their traditional lands must be developed.

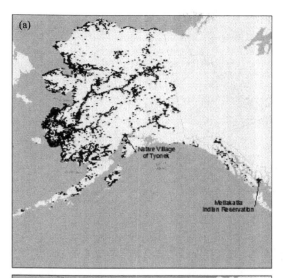

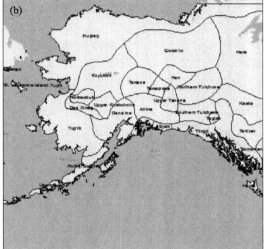

Fig. 6.3 (a) Alaska Native allotment points (1–160 acres) and sovereign lands of Metlakatla and Tyonek. (b) distribution of unique Alaska Native peoples.

Map data from the Alaska Department of Natural Resources, April 2006. Source: John Wros.

Not coincidentally, these lands again represent much of the best habitat and aesthetic value in Alaska. This is exemplified by the actions following the 1989 Exxon Valdez Oil Spill. The resulting financial settlement created the single largest pot of money for habitat restoration and conservation in the history of Alaska. To date, the vast majority of the funds spent to purchase "large parcel" conservation easements or land title was done so on Alaska Native Corporation lands. The primary benefit to Alaska Natives in the case of an easement or fee deal is monetary, which may be positive to groups that are land rich and cash poor. Such easements or land purchases may not be positive where finite subsistence and traditional use resources and values outbalance monetary value. Now nearly fifty years after ANCSA, it is not necessarily the case that the Native village leaders are the controlling members of the corporations.

Conflict arises when a Native Corporation develops traditional use and sacred lands contrary to the wishes of their village shareholders. It is further complicated by split estates, where an Alaska Native Village Corporation may own and control the surface rights, but a separate Alaska Native Regional Corporation owns the subsurface and mineral rights (Fig. 6.4). Subsurface rights may include access through the surface, the surface is exposed to ecological disruption and potential monetary degradation. In combination, these factors have created rifts between Native villages, village corporations, and regional corporations. This is one more example of the disconnect of Alaska Native traditional values and the contemporary real estate system.

Conservation partnership has proven to be an effective way to unify villages, village corporations, and regional corporations. Consider the scenario of a third-party conservation easement: A land trust or other entity may purchase and hold a conservation easement to the development rights on a given tract of native corporation lands, incentivized by the resulting protection and stewardship and natural resources in perpetuity. The benefit to the Native village members is now two-fold, including the same monetary gain as would be achieved through development activities as well as assured management and access to those lands consistent with traditional uses. The benefit to the village corporation is successful fulfillment of its mission to profit from their lands. In the event that a duplicate conservation easement is also purchased on the subsurface rights, the same benefit of fiscal mission fulfillment is achieved by the regional corporation, and the conflict off subsurface access through the surface is negated. In this way, cooperative conservation can and has relieve the burdens of ANCSA and repair the relationships of the Alaska Native system. Importantly, these actions have returned management control of traditional lands to Alaska Natives while preserving some of the richest and most pristine natural systems on the planet.

Nonprofit and agency partners

It is worth noting that all of the actions so far discussed can be completed without significant non-Native partnership. In the case of villages purchasing land and gaining sovereign rights, no partner is required. In the case of a conservation easement, a Native village or nonprofit branch of a Native Corporation can hold conservation easements. Other options for traditional and Native Corporation lands in Alaska include fee-to-trust transfers, carbon easements under

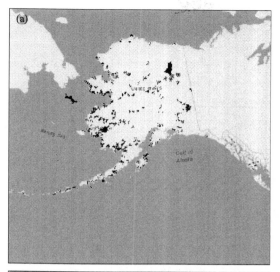

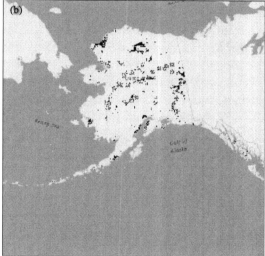

Fig. 6.4 (a) Alaska Native Village Corporation Lands. (b) Alaska Native Regional Corporation Lands.

Map data from the Alaska Department of Natural Resources, April 2006. Source: John Wros.

the California Carbon Offset Program, sustainable forest harvest planning, wetland mitigation, land exchanges, and restoration. The role of partnerships is in creating a *collaborative advantage* for both the conservation partner and Alaska Natives. The premise of this concept is that something is achieved that neither group could achieve on their own.

Returning to the first question in evaluating the merits of partnership, why would a public or private landowner relinquish ownership of lands? As with any real estate action, the likely answer is sale for monetary gain. Even if we consider the known economic benefits of ecosystem services, purchasing land for no tangible return is a nebulous concept for an individual or private organization. While philanthropic giving is one means to do so, historically it has been achieved using government funds set aside for wildlife habitat, sport fishing and hunting, or recreation access programs. To name a few in Alaska: The Land and Water Conservation Fund, Forest Legacy Program, Cooperative Endangered Species grants, North American Wetlands Conservation Grants, the U.S. Coastal Program. Nearly every Federal and State agency has programs and funding available for such projects.

Both nonprofits and agencies have the potential to provide mutual conservation benefit in partnership with Alaska Natives. In terms of land, the collaborative advantage gained from Alaska Native groups is traditional knowledge and opportunity, while the collaborative advantage gained from agencies and nonprofit partners is funding and the specialized skills necessary to put it into real estate transactions. Nonprofits have proven to be particularly effective amidst the underlying strained relationship between Alaska Natives and the state or federal governments. In my experience, this is due to several factors in the following order: Flexibility offered outside of government programs that is more compatible with traditional values and personalities, approachability, credibility, specialization, and distrust of government. Agencies have acknowledged that that the solution lies with the increased involvement of nonprofits and non-governmental organizations (NGOs) as intermediaries to Alaska Natives.

In the terms of dollars and cents, mission-based nonprofits and other non-governmental public service providers are the leanest example of effective partnership in conservation, leveraging federal and state resources, but acting independent of direct government oversight to deliver goods and services efficiently and effectively. In contrast to collaborate advantage is the concept of *collaborative inertia*, referring to collaboration where output is either negligible or extremely slow. Viewed objectively, the aforementioned complications result in Alaska's agencies requiring more public resources to achieve a particular conservation outcome. This is due to limitations in capacity and a strict procedural process. For this reason, nonprofit-agency partnerships have been prevalent in conservation since in 1980s and are now built into most government programs.

But do not discount the value of agencies as a partner. In answer to our remaining evaluating questions, agencies are key to enabling both Native Alaskans and nonprofits. Where adjacent land use is beyond the control of an Alaska Native landowner, we must rely on ecoregional species management, environmental laws, and real estate law. In this case a specialized third-party easement holder, such as a nonprofit land trust, may again prove to be a bridge between traditional management and agency enforcement of violations. This is also the case regarding enforcement of violations such as trespass and poaching by non-Natives on Alaska Native lands. The collaborative advantage gained by Alaska Natives is assistance in managing their lands in a contemporary landscape that includes non-traditional uses. The collaborative advantage to agencies is the opportunity to include these lands and habitats as part of a larger conservation plan without the burden and cost of owning them. The state and federal policies that currently exist to support this type of land action are relatively limited, but robust. The primary tool in Alaska is a conservation easement, supported by the *Federal Universal Conservation Easement Act*. In addition, any conservation project done using Federal grant money comes with monetary assurances, such that a major misuse of land and habitat can result in a requirement by the responsible party to repay those grants plus interest. And as previously mentioned, the Bureau of Indian Affairs offers protection and assistance to sovereign land owners and Native allottees. Perhaps the most powerful conservation tool available is not legal at all — the optics of undermining or extorting Alaska Natives beyond what has already occurred are so ghastly that I have heard it called political suicide.

The future of partnership

Estimates of the global conservation estate necessary to preserve biodiversity and ecosystem services have incrementally risen 500% in the last three decades. E.O. Wilson echoes scientists and policy-makers when he says, "*Half for the world of humanity, half for the rest of life, to make a planet both self-sustaining and pleasant.*" Half of the Earth's lands and waters is a hedge against the difficulties of managing landscape-scale ecological processes across a mosaic of landowners and jurisdictional boundaries. Growing concern for species decline, rapidly shifting climate, crop diversity, working lands, and the vanishing cultural resources and iconic landscapes may well mean 50% is a low goal.

We must consider two factors: (1) Where are the available conservation opportunities to meet this goal, and (2) Who is going to do it? Framed this way the collaborative advantage for conservation achieved by partnering with Alaska Native groups is clear; we cannot reasonably ask our governments to be the single owner and manager of such a magnitude of land, wildlife, and natural resources, so must look to partners with compatible values and a long-term connection to the landscape.

Partnering with Alaska Natives as we have discussed supports a true ecosystem-based management approach, recognizing all interactions with an ecosystem, including humans. After decades, the experience of this philosophy suggests that opportunity for conservation is best achieved when social nuances are considered over a broad time scale, calling for different points of implementation at different points in the project life cycle. Less demanding social outcomes, such as communication, preceded more complex ones, such as trust, and procedural and social improvements often precede ecological change. Nonprofits and NGOs are particularly effective in non-threatening outreach and communication, in many cases offering the very same government sanctioned programs and incentives and an agency, but thorough personal relationships and trust can create unique opportunities to implement them.

These individual relationships are then the grounds for successful collaborative monitoring and stewardship in perpetuity. Conservation groups and agencies at all levels face the challenge of programmatic and fiscal protection in perpetuity. Limited funding likewise limits state and federal government from incorporating many ecologically connected lands into management units. Scientists and land managers now recognize that social systems operating around land reserves play and important role in the ability of individual units to achieve their goals. Ecosystem-based management requires collaborations that build strength through trust across jurisdictional boundaries. In Alaska, the unifying group best suited to do this are Alaska Natives. Not only will this reduce the costs to government of monitoring and managing land, but also increased the effectiveness of preservation and the durability of conservation projects by increasing the region's community members and stakeholders.

6.4 Federal Agency and Tribes: Continuing Challenges to Tribal Rights

Federally-recognized Tribes have a unique legal and political relationship with the United States. Tribal Sovereignty is rooted in the formation of the United States. Tribal governments are fundamental to this relationship. This relationship is affirmed under the U.S. Constitution, as well as by treaties and agreements, congressional acts, executive orders and judicial rulings. Tribal governments have a government-to-government relationship with the United States federal government.

The U.S. government has a trust responsibility to American Indians, which was a promise made as treaties and agreements were made with Tribes. The U.S. government acquired almost all of its land through negotiation of treaties, executive orders and agreements with Tribes. The U.S. government's trust responsibility to Tribes was affirmed in the 1830s by the Supreme Court. When the government entered into treaties and the government also agreed to protect and enhance Tribes. The treaty-making era ended in 1871, primarily because the government had acquired all of the land that it wanted and the treaty making process created conflicts. Thereafter, Tribes were moved onto reservation by executive order and agreements. Many Tribes continued to seek federal recognition up to present times.

> *"Perhaps the most basic principle of all Indian law, supported by a host of decisions... is the principle that those powers lawfully vested in an Indian tribe are not, in general, delegated powers granted by express acts of Congress, but rather inherent powers of a limited sovereignty which has never been extinguished. Each tribe begins its relationship with the federal government as a sovereign power, recognized as such in treaty and legislation." (Felix S. Cohen, 1942, p. 31)*

The Power of Congress over Indian affairs is supreme. Federal Indian treaties and statutes do not need state approval before becoming the supreme law of the land (Dick v. U.S. 1908). Even if a state did not exist when the treaty law went into effect, the state still must obey it. For example, any hunting, fishing, or water rights given by the federal government to Indians before a territory became a state, must be honored by the state (Puyallup Tribe v. Department of Game, 1968). In recent years, the United States has said that every federal agency has an obligation to ensure the protection of Tribal governments, even though the trust relationship is administered primarily through the Bureau of Indian Affairs. For example, the Department of Energy has a responsibility to ensure that the energy needs of Tribes are being met, and the Environmental Protection Agency has a responsibility to ensure that the environments within Indian reservations are protected.

There is a legacy of mistrust between U.S. government agencies and Native American Tribes that goes back over 500 years when European colonialists first arrived on the North American continent. The basis of this mistrust can be traced back to many broken treaties that were signed to guarantee Indian rights to hunt, fish and collect customary resources for the lands they gave to the U.S. federal government. These lands were appropriated for the Euro-American settlers. Consistently, Tribes have to go to the courts to maintain their rights [13]. Many battles were fought by Euro-American settlers to eliminate the culture and way of life of Native Americans which still impact Tribes today [13]. This legacy of mistrust, especially with federal agencies, continues to be an issue. This creates problems for federal agencies who need to consult with Native people when conducting resource and environmental assessments during the permitting process.

EPA (Environmental Protection Agency) is the organization that works with Tribes on the regulatory side. They need to address environmental justice issues found on reservations. The EPA is called a regulatory agency for a reason because they are authorized by the US Congress to write and enforce environmental laws and regulations passed by Congress. To protect human health and the environment typically focuses on land management of solid and hazardous wastes and any of a number of toxic chemicals (e.g., asbestos, lead, PCB's, etc.) as well as the quality of drinking water, ground water, watershed restoration[1]. These are important factors to manage but are less able to measure other non-point sources of pollution that decrease the health of the land. They also focus on specific hazards that may not be particularly relevant when exploring the impact of climate change on land quality or health. Further, an organization such as EPA has a mandate and they are very effective at responding to issues focusing on environmental and human health. They should not be expected to address land quality issues unless they are the result of toxic chemical contamination.

The EPA is also required by law to consult with Native American Tribes/Native Alaskans so that any action to protect the environment and public health does not infringe on treaty rights. Ultimately EPA has to ensure that federal environmental laws are followed on Tribal lands (Accessed on July 11, 2016, Region 10 Tribal Priorities. May 23, 2013, Final. EPA). As part of this, the **EPA has to work with Tribes to address environmental justice issues on reservation resulting from their actions. The key words are that EPA has to consult with Tribes which does not mean that they have to listen to or do what the Tribes want. They are supposed to modify their plans if they are incompatible with Tribal treaty rights or impacts Tribal sovereignty.**

In 2011 the EPA staff developed a policy on how to work on environmental justice issues with federally recognized Tribes and Indigenous people[2]. In February 2016 EPA published its policy to guide discussions on Tribal treaty rights,

> "...initial step in EPA's efforts to improve the methods and processes in place to meet the commitment to honor and respect Tribal treaty rights and resources protected by treaties. The Guidance is the first of its kind for any federal agency."[3]

The latter sentence is most interesting since it is the first federal agency that has developed a policy to respect Tribal treaty rights and resources (as well as access to resources) and the date this document was published was February 2016. As summarized in this policy, the EPA needs to ensure that their actions do not conflict with treaty rights since:

---

1 e.g., accessed on July 27, 2017 at www.epa.gov/r10earth/tribalwater.htm
2 Accessed on July 6, 2016 at https://www.epa.gov/environmentaljustice/environmental-justice-Tribes-and-indigenous-peoples
3 Accessed on July 6, 2016 at https://www.epa.gov/tribal/tribal-treaty-rights

> "The U.S. Constitution defines treaties as part of the supreme law of the land, with the same legal force as federal statutes. Treaties are to be interpreted in accordance with the federal Indian canons of construction, a set of long-standing principles developed by courts to guide the interpretation of treaties between the U.S. government and Indian Tribes. As the Supreme Court has explained, treaties should be construed liberally in favor of, giving effect to the treaty terms as Tribes would have understood them, with ambiguous provisions interpreted for their benefit. Only Congress may abrogate Indian treaty rights, and courts will not find that abrogation has occurred absent clear evidence of congressional intent."

These EPA policies are recent so determining whether it will make a difference for Tribal interaction with this agency is premature to discuss. It should guide the process by which EPA factors in protecting treaty rights or communicating better with Tribes when a proposed EPA action may impact environment and/or public health. What is clear is that the EPA recognizes that they have a commitment to ensure treaty rights are protected and they need to consult with Tribes on any action that would impact Tribal sovereignty. A common concern raised by Native Alaskans when the EPA was determining whether a mine should be built in Bristol Bay Alaska was the lack of consultation during the early stages of the permitting process and during critical stages of the evaluation process (K Cook, personal communication).

The continuing challenges to Tribes, their sovereignty and having input on decisions made on public lands is playing out today. The *High Country News* (March 19, 2018) describes how having a White House friendly to industries is opening up several areas of Alaska for drilling for oil in the Arctic National Wildlife Refuge (ANWR), cutting trees in an old growth forest in the Tongass National Forest, and building a road through the Izembek National Wildlife Refuge [153]. Not allowing these activities to occur has been a priority for some Native groups, conservation and environmentalist groups because they are "home to polar bears, migratory birds and caribou calving ground" [158]. As summarized by Langlois [158]:

> "Opening ANWR to development is part of a broader effort to ramp up America's Arctic oil production, which has decreased from 2 million barrels a day in 1987 to around 500,000 today. That and the low price of oil have plunged Alaska into a fiscal crisis and cost the state close to 10,000 jobs. With Murkowski's blessing — and possible urging — Interior Secretary Ryan Zinke has also made every eligible acre of another massive Arctic refuge, the National Petroleum Reserve-Alaska, available for leasing. And the Trump administration has proposed open-

ing nearly all of Alaska's oceans to offshore drilling, including Arctic hunting grounds important to Iñupiat whaling communities."

The most important point here is how evidence-based science is not being used to determine whether to pursue these activities. **These are politically motivated decisions. It also means that it depends on who is in office on how the lands will be respected. Such a decision process is not sustainable or ensuring the viability of these landscapes. These political decisions are suggested to be attempts to revive industries that are no longer viable.** Langlois [158] wrote:

"1960s. At the time, logging drove the economy of Southeast Alaska, but today, fewer than 1 percent of the region's jobs are connected to timber. Instead, fishing, tourism and other industries are booming."

Tribal interactions with federal agencies is a government to government interaction. The U.S. federal government has a trustee responsibility for Tribes because of the numerous treaties that they signed with different Tribes. The federal government is not meeting their fiduciary responsibilities towards Tribes (*see* Section 5.3.1) despite all the signed treaties. For the Washington State level context *see* Section 6.2.

In general, federal agencies are not including and consulting Tribes even when their decisions will impact Tribal lands/water and resource viability as well as Tribal members health. Tribes can provide comments on federal actions that might impact their lands or water, but they are mostly excluded from the process (*see* Section 6.4.2). An excellent example demonstrating the lack of a voice of Native people in the decision process is whether a mine should be developed in Bristol Bay Alaska. EPA is the agency responsible for conducting an assessment whether a proposed project or infrastructure should be approved. EPA does consult with Native Alaskans but typically occurs at the end of the process when a decision has already been mostly made. When the company initially wanted to build a mine in Bristol Bay, Native Alaska communities were able to stop the mine from being built because of its impact on Sockeye salmon runs and their Indigenous subsistence lifestyles. Data show that almost 50% of the global sockeye consumed around the world comes from this one Bay which makes this a significant revenue generating business for many non-Native fisher people.

The consulting process is now being overwhelmed by politics and where the Indigenous people have no voice in the process on building a mine in Bristol Bay. The decision has vacillated on whether the U.S. EPA would approve the mine — mostly driven by political decisions. The current U.S. federal President has reversed the earlier decision. Now the revenue and employment opportunities at Bristol Bay has made the mine a viable option again. The loss of the Indigenous subsistence lifestyles doesn't seem to matter anymore or the fact that this Bay provides the habitat for so much of the salmon consumed around the world. The lack of consultation of Indigenous communities means they do not have a voice in land, water and resources issues on their lands. Further, if the earlier decision was evidence based, the decision should not change because new legislators are voted into office.

Decisions made at the federal level also are not place-based but top-down decisions that are little influenced by processes occurring at a regional or local level. Similar issues resulted from the IFMAT assessments where the federal government is not meeting its fiduciary responsibilities for Tribes (*see* Section 5.3.1). This continuing trend is unfortunate since Tribes are managing their lands and resources sustainably because they retain a balance in their management decisions where protection for wildlife, cultural resources and sovereignty have priority over the generation of revenue (*see* Section 5.3.2). This contrasts lands adjacent to Tribal lands where economics dominate the decision-making process. This does not mean that revenue generation is not important for Tribes, but the environment has a higher priority. The scarcity of knowledge related to climate change impacts suggests that management agencies at both the state and federal should explore the framework used by Tribes related to lands, water and resources.

The next story on the role of Indigenous people in the US national forest planning (*see* Section 6.4.1) is revealing in how federal governments review their process for obtaining advice from Tribes. It also highlights that many of the topics are re-occurring points that continue to be issues that are not being addressed.

6.4.1 Indigenous People's Role in National Forest Planning[*]

In April 2012, then Secretary of Agriculture Tom Vilsack announced a new planning rule that governed the management of 193 million acres of national forest system lands. The new rule was submitted to an intense period of comment before it was finalized however relatively few of the nearly 300,000 official comments submitted were from Indigenous people. The USDA Forest Service must do a better job of gaining input from Indigenous people.

Of those that commented, there were general themes that can be summarized as:

- Indigenous people are allowed to submit comments but some feel their comments and the consultation process do not really affect decision-making for projects on the ground.

---

[*] By Melody Starya Mobley.

Melody Mobley.
Source: Melody Mobley.

- There is concern that Alaska Native Corporations are given the same influence as federally-recognized Tribes.
- Some Indigenous people feel that their youth are the "eyes, ears, and voice" of the people, especially Tribal leaders, yet they are not included in the planning process.
- The USDA Forest Service culture of moving personnel from locations rather quickly generates a concern among Indigenous people that they must provide their input and knowledge of place time and again without fair compensation.
- There is concern that Indigenous people are not adequately involved in consultation and coordination, nor have adequate government-to-government relationships been established or maintained.
- Some Indigenous people believe the Forest Service planning rule must adequately protect any culturally and spiritually sensitive information, such as location of the resources.
- There is a general concern about how Forest Service planning affects or addresses treaty rights and obligations.

Most decisions that affect or coordinate specific projects with Indigenous People are made at the local, or district, level rather than at a national level as part of a planning rule. While there is merit in this approach, it does not ensure consistent treatment of Indigenous people, federally-recognized Tribes, nor Alaska Native Corporations nationwide.

Forest Service planning rule tribal comments from official forest service website[1]

The USDA Forest Service responses to Indigenous people's comments on the proposed rule and draft environmental impact statement are summarized next. These are related to the new planning rule for the national forest lands as reflected in the final programmatic environmental impact statement. The responses to the comments were made by the Forest Service.

INTRODUCTION: Multiple Tribes and Alaska Native Corporations (ANCs) submitted comments during the public comment period for the proposed rule. To provide Tribes and ANCs with information on how their concerns are addressed in the preferred alternative [Modified Alternative A of the final *Programmatic Environmental Impact Statement* (PEIS)], this document summarizes those comments and provides a response for how they were considered during development of the preferred alternative.

(1) Status of Tribes and Alaska Native Corporations Comment: Some respondents commented that Alaska Native Corporations should not be given the same status as federally recognized Indian Tribes.

> **Response:** *The statutory provisions of 25 U.S.C. 450 note require that Federal agencies consult with ANCs on the same basis as Indian Tribes under Executive Order 13175. While the preferred alternative would require consultation and participation opportunities for ANCs, the Department engages in a government-to-government relationship only with federally recognized Indian Tribes, consistent with Executive Order 13175.*

(2) Youth Engagement Comment: Some respondents commented that it is very important to engage Native American people and youth, and that youth are the ears, eyes, and voices of Tribal leaders.

> **Response:** *The Department recognizes the need to engage a full range of interests and individuals, including Native Americans and youth, in the planning process. Section 219.4 of the preferred alternative includes requirements for encouraging participation from youth, low income, and minority populations and includes requirements for providing participation and consultation opportunities for federally recognized Indian Tribes and ANCs.*

---

1 https://www.fs.usda.gov/Internet/FSE_DOCUMENTS/stelprdb5349464.pdf, edited

Comment: One respondent requested to meet with the Forest Service to develop a collaborative effort to engage Tribal youth.

> **Response:** *Please contact your local Forest Service office to inquire about existing and potential future opportunities for engaging Tribal youth in your local area.*

(3) Cultural Resources and Traditional Knowledge Comment: One respondent suggested that a provision for sharing conservation knowledge be included in the final rule. One respondent stated that it should not be the responsibility of the Tribes to provide information on Tribal knowledge, Indigenous knowledge, or land ethics information to a Forest Supervisor or District Ranger seeking such information for consideration of a project or planning activity. These individuals are frequently are reassigned to another area and therefore a loss the institutional knowledge locally. It is burdensome for a Tribe to repeatedly provide this information for consideration.

> **Response:** *The preferred alternative includes a requirement for the responsible official to request information regarding native knowledge, Indigenous ecological knowledge, and land ethics as part of the participation and consultation provisions provided in Section 219.4. The Department realizes that it may be difficult for Tribes to work with the Forest Service when their points of contact and leadership representatives for their national forest frequently change. This is an issue that the Agency is aware of and is reviewing through performance standards, in collaboration with additional tools such as handover memos that describe important issues and relationships to the incoming manager.*

(4) Coordination, Consultation, and Government-to-Government Relationship Comment: Some respondents commented that the responsible official should actively engage in coordination with Tribal land management programs for the purposes of development or revision regarding any consultation or collaboration endeavor with their specific Tribe. Some respondents requested specific coordination with Tribal programs. Other respondents commented that the rule weakens the requirements to coordinate planning with Indian Tribes. One respondent requested that the Tribal coordination provisions from the *Federal Land Policy and Management Act* of 1976 [43 USC 1712(b)] be included in the final rule, while another suggested including the complete text of 43 USC 1712 in the rule.

Response: *The preferred alternative would require the responsible official to provide participation, consultation, and coordination opportunities for Tribes during the land management planning process, under Section 219.4. This section of the preferred alternative also states that the responsible official would coordinate land management planning with the equivalent and related planning efforts of federally recognized Indian Tribes and ANCs. A citation to 43 USC 1712(b) has been added to the preferred alternative at Section 219.4(b)(2). Participation in a collaborative process would be voluntary and would supplement, not replace, consultation. The preferred alternative explicitly requires that plans must comply with all applicable laws and regulations.*

Comment: One respondent requested that further Tribal consultation occur to discuss and assist with drafting the type of agreement the responsible official would need to implement to support the language of protecting confidential information. This respondent also requested that the agreement not be an umbrella document that all regions must implement. It should be flexible to each region and local unit because the issues and types of protection may vary from region to region.

Response: *Development of agreements is outside the scope of a national planning rule. To request consultation on this specific issue or to discuss an agreement, please contact your local Forest Service unit. Agreements between Tribes and a local unit are currently entered into at the local level and we expect agreements to continue to be entered into at the local level to provide the flexibility for an agreement to be developed that best meets the needs of the local parties involved.*

Comment: One respondent felt that the proposed rule does not go far enough in identifying the unique government-to-government relationship between Tribes and the Forest Service. One respondent requested that the final rule recognize and provide for direct consultation regarding forest plan amendments and revisions with affected ANCs and Tribal organizations, in addition to federally recognized Indian Tribes.

Response: *The Department recognizes the unique government-to-government relationship that the federal government has with Tribes, and has engaged Tribes throughout the rulemaking process. The preferred alternative contains modified wording regarding trust responsibilities to ensure accurate recognition of the relationship between the Federal Government and federally recognized Tribes. This modified language at Section 219.4 says "The Department recognizes the Federal*

Government has certain trust responsibilities and a unique legal relationship with federally recognized Indian Tribes." In addition to providing opportunities for engagement to both Tribes and ANCs, under Section 219.4 of the preferred alternative, the responsible official would provide both federally recognized Indian Tribes and ANCs the opportunity to undertake consultation consistent with Executive Order 13175 and 25 USC 450 note. Section 219.4 of the preferred alternative also states that the responsible official would coordinate land management planning with the equivalent and related planning efforts of federally recognized Indian Tribes and ANCs. There are no requirements for the Department to provide direct consultation with Tribal organizations. Federal agencies are required to consult with ANCs and federally recognized Indian Tribes per Executive Order 13175 and the statutory provisions of 25 USC 450 note. While there are no specific requirements to provide consultation to Tribal organizations, under the preferred alternative, these organizations would be able to participate in the land management planning process through the public participation requirements of Section 219.4.

(5) Comment: A few respondents requested that the final rule specifically encourage Tribal co-management with the Forest Service.

Response: *The Agency may not delegate its decision-making authority to other entities. The preferred alternative does not address the concept of co-management. However, the preferred alternative would provide opportunities for Tribes to participate during the planning process and to discuss opportunities to meet shared objectives. The Department acknowledges the importance of Tribal participation in the land management planning process. The preferred alternative would provide opportunities for consultation and participation early and throughout the land management planning process at Section 219.4.*

(6) Treaty Rights and Obligations Comment: Several respondents commented that the final rule must recognize reserved rights and treaty rights. One respondent suggested that the rule should outline specifically how treaty obligations or other legally binding obligations to Indian Tribes will be addressed and assured in the planning rule and subsequent development of forest management plans. Some respondents disagreed with the language in the rule that the rule would not "affect" treaty rights.

Response: *The preferred alternative recognizes and does not interfere with prior existing Tribal rights, including those involving hunt-*

ing, fishing, gathering, and protecting cultural and spiritual sites. The preferred alternative would require the Agency to work with federally-recognized Indian Tribes, government-to-government, as provided in treaties and laws and consistent with Executive Orders, when developing, amending, or revising plans. The Department does recognize that Tribes often have reserved rights and Tribes can discuss reserved rights during consultation and collaboration opportunities afforded by the rule. If reserved rights are an issue for Tribes, they would be able to discuss them during the participation and consultation opportunities provided by Section 219.4 of the preferred alternative.

Section 219.1 of the preferred alternative states that it would not affect treaty rights or valid existing rights and that plans must comply with all applicable laws and regulations. The language in Section 219.1 of the preferred alternative that "This part does not affect treaty rights or valid existing rights established by statute or legal instruments" does not imply that the Agency believes land management planning does not affect Tribes. Rather, this language means that the preferred alternative would not interfere with treaty rights or valid existing rights and that the land management planning process would comply with these rights.

(7) Cooperating Agency Status Comment: Some respondents requested clarification for why federally recognized Indian Tribes would be encouraged to seek cooperating agency status and what the benefit would be to Tribes. They emphasized that the federal government already has trust responsibilities to Tribal Nations and therefore they do not understand the purpose of requesting cooperating agency status. One respondent also requested a definition for cooperating agency status.

Response: *The preferred alternative retains the provisions for encouraging federally recognized Indian Tribes to seek cooperating agency status in the NEPA process for development, amendment, or revisions of a plan, where appropriate. The opportunity for federally recognized Tribes to seek cooperating agency statues provides an additional opportunity for Tribes to be engaged in the planning process and provides further avenues for Tribes to provide input during the planning process. Tribes would have the opportunity to seek cooperating agency status in addition to the other opportunities that would be provided to Tribes for participation and consultation during the planning process. Cooperating agency status would not replace nor supersede the trust responsibilities already in place; rather, it would provide another opportunity for Tribes to be engaged in the planning process.*

> The Council on Environmental Quality regulations for implementing the National Environmental Policy Act *define a cooperating agency as "Any Federal agency other than a lead agency which has jurisdiction by law or special expertise with respect to any environmental impact involved in a proposal (or a reasonable alternative) for legislation or other major Federal action significantly affecting the quality of the human environment. The selection and responsibilities of a cooperating agency are described in [40 CFR] Section 1501.6. A state or local agency of similar qualifications or, when the effects are on a reservation, an Indian Tribe, may by agreement with the lead agency become a cooperating agency." (40 CFR 1508.5)*

(8) Cultural/Spiritual Resources, Sacred Areas, and Confidentiality

Comment: One respondent requested that the final rule ensure protection of cultural resources. One respondent requested that the rule include acknowledgement of Tribally-valued resources, access, and spiritual and cultural practices and locations and that these should be acknowledged during the development of land management plans, without disclosing any culturally and spiritually sensitive information.

> **Response:** *Under the preferred alternative at Section 219.10, plan components for a new plan or plan provision would provide for protection of cultural and historic resources and management of areas of Tribal importance. When developing plan components for integrated resource management, the responsible official would consider cultural and heritage resources and ecosystem services (Section 219.10). By definition, ecosystem services include cultural services such as educational, aesthetic, spiritual and cultural heritage values, recreational experiences and tourism opportunities. Under Section 219.4 of the preferred alternative, the responsible official would request information about native knowledge, land ethics, cultural issues, and sacred and culturally significant sites during the planning process. Section 219.1(e) of the preferred alternative would require that, during the planning process, the responsible official comply with Section 8106 of the* Food, Conservation, and Energy Act *of 2008 (25 USC 3056), Executive Order 13007 of May 24, 1996, Executive Order 13175 of November 6, 2000, and other laws and requirements with respect to disclosing or withholding under the* Freedom of Information Act *(5 USC 552) certain information regarding reburial sites or other information that is culturally sensitive to an Indian Tribe or Tribes.*

Chapter VI Tribes, State and Federal Agencies: Leadership and... — 331

Comment: One respondent requested that the rule include a requirement to build an effective strategy for the management and sustainability of culturally-important resources and places, developed in cooperation with affected treaty Tribes.

> **Response:** *Strategies for the management and sustainability of resources and places on National Forest System lands will primarily be developed at the local level. During the land management planning process under the preferred alternative, Tribes would be provided opportunities for participation, consultation, and coordination. To address concerns regarding cultural sustainability, Section 219.8 of the preferred alternative would require plan components to consider social, cultural, and economic conditions relevant to the area influenced by the plan, as well as cultural and historic resources and uses. At the national level, a sacred sites policy review is currently ongoing, and Tribes have been provided with information on opportunities to participate in and consult on this initiative.*

Comment: One respondent requested that the final rule outline a strategy for the protection and privacy, when needed, of sacred areas that the Tribes depend upon, and as outlined in treaties and federal laws.

> **Response:** *Section 219.1 (e) would require that, during the planning process, the responsible official would comply with "Section 8106 of the* Food, Conservation, and Energy Act *of 2008 (25 USC 3056), Executive Order 13007 of May 24, 1996, Executive Order 13175 of November 6, 2000, laws, and other requirements with respect to disclosing or withholding under the* Freedom of Information Act *(5 USC 552) certain information regarding reburial sites or other information that is culturally sensitive to an Indian Tribe or Tribes." Regarding sacred areas, policies and procedures relating to Forest Service management that may affect sacred sites are currently being reviewed in a separate USDA Office of Tribal Relations/Forest Service initiative. Tribes have been provided with information on opportunities to participate in and consult on this initiative.*

(9) Other Comments: Some respondents requested that funding be identified to conduct outreach processes with Tribes and develop deeper relationships with Tribes. One respondent requested that the Forest Service honor the respondent's requests for resources and assistance to be able to participate in Forest Service projects and respond to Agency requests for coordination or other participation.

Response: *A national planning rule does not allocate funding for specific uses and resources. Procedures for implementing the participation and consultation requirements of any final rule will be developed as part of the Forest Service Directives System, and there will be additional opportunities for Tribes to provide input on the Directives.*

Comment: One respondent requested that the rule include a requirement to consider roads and their development, maintenance, and decommissioning at a forest-wide level that addresses the needs for treaty rights access and fish and wildlife protection, along with other considerations.

Response: *Decisions regarding the development, maintenance, and decommissioning of specific roads occur at the forest level during project planning. Under the preferred alternative, responsible officials would consider appropriate placement and sustainable management of infrastructure, such as recreational facilities and transportation and utility corridors.*

Comment: One respondent requested that the rule consider integration of the evolving Special Forest Products Policy and take into consideration any treaty-protected Native American gathering rights when developing land management plans applicable to forest products and determining commercial harvest levels by non-treaty harvesters.

Response: *Further clarification regarding gathering non-timber forest products is being developed in a new special forest products rule that is outside the scope of the planning rule. Under Section 219.10 of the preferred alternative, the responsible official would consider, in collaboration with federally recognized Tribes; ANCs; other Federal agencies; and State and local governments, habitat conditions for wildlife, fish, and plants commonly enjoyed and used by the public for hunting, fishing, trapping, gathering, observing, subsistence, and other activities. The preferred alternative also includes requirements for coordination with the land management planning efforts of federally recognized Tribes and ANCs.*

Comment: One respondent asked whether a specific site permit for activities on or uses of a national forest or grassland would be available to Native Americans.

Response: *A planning rule and forest plans do not address permitting, as permitting requirements are addressed in other regulations. Issues regarding use and access on a specific forest or grassland will be considered at the local level. For information regarding permits on*

Chapter VI Tribes, State and Federal Agencies: Leadership and... — 333

National Forest System lands in your area, please contact your local Forest Service office.

Comment: One respondent requested that the following standards be included in the final rule: (1) Forest management shall not threaten or diminish, either directly or indirectly, the resources or tenure rights of Indigenous people; (2) Sites of special cultural, ecological, economic or religious significance to Indigenous and other peoples shall be clearly identified in cooperation with such peoples, and recognized and protected by forest managers; (3) If specific traditional knowledge is requested from Indigenous people in the course of forest planning, those peoples shall be compensated for the application of their traditional knowledge regarding the use of forest species or management systems in forest operations. This respondent also stated that traditional practices associated with sites of special cultural, ecological, economic, or religious significance must be allowed.

Response: *The preferred alternative would not affect treaty rights or valid existing rights. The Department of Agriculture acknowledges the importance of the land to Indigenous people. A policy regarding sacred sites is being developed through a separate initiative from the planning rule, and Tribes have been provided the opportunity to consult in this initiative.*

Section 219.1 of the preferred alternative recognizes that National Forest System lands provide spiritual and cultural benefits. Under Section 219.10 of the preferred alternative, plan components would provide for the management of areas of Tribal importance and protection of cultural and historic resources. The Department does not have the authority through a national planning rule to implement a policy for compensating the exchange of information. Tribes and the public are not required to participate in the planning process or to share information; however, they would be encouraged to do so and would be provided opportunities to do so under Section 219.4 of the preferred alternative.

CONCLUSION

When I (Melody Mobley) first joined the USDA Forest Service in 1979, I heard these same concerns about rule-making and planning from Indigenous people. Indigenous people do not believe that their input is affecting Forest Service decisions nationally or at the local level. Traditional methods of gaining official comments from these people are not working. People of color in general are not involved in these decision-making processes for the management of public lands.

6.4.2 USDA Forest Service Use of Culture in Land and Resource Management Planning Decisions*

The Forest Service is responsible for managing the lands and resources of the National Forest System, which includes 192 million acres of land in states, the Virgin Islands, and Puerto Rico, to improve and protect federal forests. The system is composed of national forests, national grasslands, and various other lands under the jurisdiction of the Secretary of Agriculture. Sustainability of these lands and resources was supposed to be the essence of Forest Service land and natural resource management from the very beginnings of the National Forest System.

Over the years, Congress enacted several laws directed toward protecting or improving the natural environment, conserving natural resources to meet the needs of the American people in perpetuity, and providing for greater public involvement in agency decision-making.

Since the 1960s, the number of federal, state and local agencies, Tribes, members of the public, and interested groups wanting to be involved in planning decisions and share stewardship responsibilities has skyrocketed. In some cases, Forest Service personnel have been able to learn significant information, create new understanding, build trust, obtain new resources for implementation and monitoring, and diffuse potential conflicts by engaging these parties more effectively in the planning process through collaboration. While collaborative approaches do not end conflict or necessarily result in consensus, by engaging people and identifying key issues early in the process, they potentially enable the Forest Service to make better decisions and to manage conflict more effectively.

Melody Mobley.
Source: Melody Mobley.

---

* By Melody Starya Mobley.

Similarly, independent scientific and public review can greatly enhance the credibility of planning and validate the soundness of stewardship decisions and the reality of achievements. Effective collaboration can enable the Forest Service to identify key scientific and public issues and to target its limited resources on trying to resolve those issues at the most appropriate time and geographic scale. The Forest Service can improve its planning and decision-making by effectively collaborating with a broad array of citizens, other public servants, and governmental and private entities.

A final planning rule was developed to facilitate greater public collaboration in all phases of the planning process. The rule expanded on existing requirements for collaboration to increase management choices, create new understanding, build trust, obtain new resources for implementation and monitoring, manage conflict more productively, and more fully inform decision-making to ensure the long-term sustainability of the multiple resources of national forests and grasslands. The rule encouraged land managers to more actively engage the American people, other federal, state and local agencies, Tribes, and interested groups in the planning and management of the national forests and grasslands. In collaborative settings that provide opportunities for early, open, and frequent public involvement, responsible Forest Service officials would serve as process conveners, facilitators, leaders, participants, and decision-makers, as appropriate. The final planning rule created opportunities for people, communities, and organizations to work together in the identification of key issues, discussions of opportunities for contributing to sustainability, and development and promotion of landscape goals.

However, public trust in and the credibility of the Forest Service has waned over the years. Various groups believe they do not have an equal voice, that their issues and concerns are not given the same weight as other federal and commercial interests. In general, and in some locations, outreach conducted to elicit place-based or culture-derived concerns is completely ineffective. A finite group of individuals and groups is used time and again to represent diverse input during the planning process. There is still a boiler-plate mentally used as forest plans are developed.

Requirements for science-based input, especially silviculture, outweighs culture-based concerns. Top-down direction overwhelms input at the local level. Commercial interests prevail over traditional values from Indigenous groups. Lack of adequate funding, inadequate training on soliciting input from diverse cultures, and assigning the solicitation of place-based concerns a lower priority than others severely limits outreach and inclusion of individuals to help identify and solve issues before they mature into intractable problems, pitting people against each other, and encouraging litigation rather than generating mutually beneficial results.

The agency feels there are no significant negative civil rights impacts in the way it solicits input during the forest planning process, but culture-based concerns are not consistently or adequately addressed, and some Indigenous people and underserved communities are completely left out. Within the agency, staff members who share these concerns and express them are routinely ridiculed by their peers with the support of management. There is a culture of giving lip-service to non-traditional values that does not appear to be changing. The agency itself is beleaguered by civil rights complaints from its own employees and a systemic hostile work environment for people of color, including Indigenous people. In light of these issues, the agency lacks credibility even when it does seek input from Indigenous people and others who have traditionally been left out of the planning process.

6.4.3 Working as an Individual within a Federal Corporate Culture*

Federal organizations are typically large, bureaucratic organizations where loyalty to the organization is everything and production is highly valued over individuality and interpersonal skills. As a person of color, my natural resource management skills were always suspect. Not only were my skills tested time and again, I had to be three times as good as the traditional employees—White males. Lip service is given to building diverse teams where everyone has an equal voice when in fact people of color and, to a lesser extent, women on these teams is typically window dressing. They are often expected to be seen but not heard, and when they do speak up their ideas are often denigrated but a White male sharing the same innovative idea is praised.

So, how does someone different from the majority population in a large bureaucracy retain their individuality while succeeding professionally? The first step is to clearly understand the mission and vision of the organization and make sure that your unique contributions are consistent with them. It is no good having brilliant ideas that don't contribute to the existing organization mission and vision.

Second, choose your battles. It is difficult to be silent when you feel you have a wonderful idea, but you must if it has no chance of being implemented within the current organization, or it does not have any political support or fit within the mission of the organization. If you are going to really push for your idea to be implemented, make sure it is clearly linked to the mission or vision of the organization.

Third, determine whether any staff within the organization is more team-oriented and determine if there is a place for you there based on your own culture and skills. Large, bureaucratic organizations have many parts or staff

---

* By Melody Starya Mobley.

groups. Some are clearly focused on production of goods while others are more focused on service. To be of highest value to the organization, look for ways to integrate your unique ideas, talents, and cultural strengths into the organization by sharing ideas that introduce better service to production-focused staffs and production into service-oriented staffs.

Fourth, look for ways to build allies or mentors who are in leadership positions at your level of the organization. It is often difficult to find people of color in leadership positions who are not modelling White male behaviors and parroting traditional agency positions so that they can be accepted and to be successful. Do not limit yourself to mentors who share your race and culture. Look for mentors and allies who demonstrate that they support cultural diversity within the organization.

Fifth, develop a thick skin. If you are the only one working on a project or team who is not a White male, it is not uncommon for culturally insensitive statements to be made or inappropriate joking to take place. Colleagues and leaders may make very offensive statements about your views and ideas, or your racial or cultural identity, that you may or may not be in a position to retort or correct without severe retaliation. It is unfortunate that in this era one is subject to overt prejudice or blatant ignorance, but people of color commonly are still today. Not only must one have a thick skin but, in this case, also choose your battles.

Sixth, of last resort, do not be afraid to file an equal employment opportunity (EEO) complaint if, and only if, one is willing to suffer the consequences. This is the most serious step because in a federal organization one can achieve a satisfactory settlement agreement but lose all chances to further pursue a career or progress in the agency. One essentially "wins the battle but loses the war". Retaliation for filing EEO complaints is rampant throughout the federal government. Anyone who denies this in advising an employee to file an EEO complaint is uninformed or being dishonest.

The most successful federal agencies and organizations fully take advantage of the innovative ideas that culturally diverse people bring to an agency, team or project, when everyone is allowed to fully participate and contribute.

6.5 Inter-Tribal Collaborations: Increase Tribal Role in Natural Resource Planning

It might come as a surprise to some that Native Americans did not disappear after being colonized by European settlers — even though many attempts were made to eliminate them and their cultures [13]. Today Tribes in the Pacific

Northwest of the US are leaders in restoring habitat and mitigating land-use issues impacting the health of customary lands and waters. Tribes are forging inter-Tribal collaborations to manage and support habitat restoration projects for fish or wildlife in the Pacific Northwest U.S. This is not the action of a few individuals but multiple Tribes as well as they are collaborating with their non-Tribal neighbors. Without Tribes, it is not clear if any attempts would be made to develop solutions for declining salmon runs. Tribes have

> "the legal authority, reaffirmed by the U.S. Supreme Court, to defend natural resources for the benefit not only of Tribes, but all citizens of the Pacific Northwest".
>
> Tribes are committed "to cooperation in managing the region's natural resources. Cooperative management—a coordinated approach of governments, agencies, industry and the public—has led to improved integration of management responsibilities and more efficient use of limited funds and staff."[1]

Tribes learned early on that they needed to cooperate to address reduced salmon runs, and the impacts of climate change and development. Salmon have no boundaries so multiple stakeholders need to be involved in the search for solutions. Tribes have become the leaders in these efforts through their sovereignty, treaty rights and Indigenous knowledge.

A few positive examples of the resiliency of Tribes and how they are adapting to protecting the resources that are important to them are provided next:

- The Tulalip Tribe has successfully entered into an MOU with the Forest Service which has allowed them to enhance huckleberry areas using traditional methods such as pruning and planned controlled burns of vegetation. The Tribes have also involved Tribal youth as workers on several projects to get younger generations back onto ancestral lands. They have partnered to bring needed Tribal and federal grant funding to the forest for collaborative research and restoration efforts. In this way the Tribe is sharing their resources for road repair, forest prescriptions, and managing cultural and natural resources.
- Recently, the Nisqually Tribe welcomed back the saltwater to a 100-acre expanse of pasture land that hadn't seen the tides flow in and out since it was diked for agricultural use more than 100 years ago. This large-scale dike removal project was designed to restore some of the richest biological reserves in Puget Sound, places where the rivers meet the sea and hundreds of species of aquatic plants, invertebrates, fish and sea birds congregate to feed, seek refuge and energize the Puget Sound ecosystem.

---

1 Accessed on June 13, 2018 at http://www.nwifc.org/w/wp-content/uploads/tribal_obama_recommend.pdf

- In 2005 The Northwest Indian College began a training program about traditional foods and medicines. The Traditional Plants and Foods Program offered monthly meetings with a host Tribe and invited other Tribes to come and learn from Elders and other practitioners about traditional medicine and food plants.

Participants learned not only about the cultural importance of the foods, but also how and when to gather them. Cultural specialists, health care workers, educators, and Tribal elders from over a dozen Tribes came to the event. Theresa Parker (Makah) helped organize the very first gathering.

> "It was such a beautiful day. After everyone toured our museum and cultural center, we led them on a hike on the ethno-botanical trail that was created by our youth and elders. For lunch, we had halibut stew, herring row on eel grass, mussels, goose-neck barnacles, Ozette potatoes, berries, and horsetail fertile shoots! The elders just lit up while they shared stories about the foods that were such an important part of their childhood (Northwest Indian College 2013)."

- The Umatilla Tribe in Northeastern Oregon State began a program called "First Foods First". First-food refers to the traditional foods used by the Umatilla, Walla Walla and Cayuse Tribes all who reside on this reservation. At one time, Native people drank clean water from streams, harvested roots, fish and deer which provided healthy nutrients, proteins, and natural sugars. These can be considered cultural resources or treaty right resources. The Umatilla Tribe is encouraging their Tribal members to gather healthy natural foods and to not eat so much processed foods or fast foods that can cause diabetes.

    They have also begun a program that places their traditional foods first, when developing management plans such as grazing, timber harvesting, agriculture, etc. They are also working with federal and state land management agencies to protect first-foods.

    PENDLETON — Traditional foods are so important to the Umatilla, Walla Walla and Cayuse Tribes that they've centered their entire land-management strategy on protecting them.

    The Confederated Tribes of the Umatilla Indian Reservation have adopted what they say is a one-of-a-kind plan for their rugged 178,000 acres that aims to save the "first foods" — salmon, wild game, roots, berries and clear, pure water (Cockle 2009).

The Northwest Indian Fisheries Commission is a clear success story where Tribes are taking leadership roles in habitat restoration for salmon. These 20 Tribes have and are collaborating to restore or mitigate fisheries management on lands and water that are not healthy for salmon. Information on NWIFC can be found at https://nwifc.org/ and a few facts are summarized below:

> "NWIFC members: Lummi, Nooksack, Swinomish, Upper Skagit, Sauk-Suiattle, Stillaguamish, Tulalip, Muckleshoot, Puyallup, Nisqually, Squaxin Island, Skokomish, Suquamish, Port Gamble S'Klallam, Jamestown S'Klallam, Lower Elwha Klallam, Makah, Quileute, Quinault, and Hoh."

> "NWIFC was created following the 1974 U.S. v. Washington ruling (Boldt Decision) that re-affirmed the Tribes' treaty-reserved fishing rights. The ruling established them as natural resources co-managers with the State of Washington with an equal share of the harvestable number of salmon returning annually."

> "The Northwest Indian Fisheries Commission (NWIFC) is a natural resources management support service organization for 20 treaty Indian Tribes in Western Washington. Headquartered in Olympia, the NWIFC employs approximately 65 people with satellite offices in Burlington and Forks."

Another inter-Tribal collaboration is the Columbia River Inter-Tribal Fish Commission (CRITFC) where four Tribes partnered on the restoration of salmon runs: Yakama, Umatilla, Warm Springs and Nez Perce.[1] As described in the CRITFC website,

> "Salmon are one of the traditional "First Foods" that are honored at Tribal ceremonies. The other First Foods are wild game, roots, berries, and pure water." Further, "The songs sung and ceremonies held in the longhouses today are the same ones that have been performed in honor of the sacred First Foods for thousands of years."

Also, Tribes collaborate with state and federal agencies to restore salmon runs and have oversight of over the large ecological restoration projects being managed by non-Tribal groups, e.g., NOAA (National Oceanic and Atmospheric Administration) Fisheries, PNNL (Pacific Northwest National Laboratory), Bonneville Power, USDA Forest Service to name a few. These projects range from culvert replacement, building fish hatcheries and restocking streams with salmon, riparian habitat restoration, addition of large woody debris to create habitat, dam removal and salmon reintroductions.

Tribes in Washington State also support Tribes in other states when they face serious challenges because of land- or resource-use decisions that have the

---

[1] Accessed on June 13, 2018 at http://www.critfc.org/salmon-culture/tribal-salmon-culture/

potential to contaminate the land or water. They fight for nature. The inter-Tribal help in protecting resources like water is shown in what just happened at Standing Rock Sioux in North Dakota. Tribes in the Pacific Northwest supported Tribes fighting to stop the building of the Dakota Access Pipeline as "water protectors" (Fig. 6.5). Many non-Tribal people also supported the Standing Rock Sioux against the building of the pipeline, e.g., U.S. veterans and even as far away as Palestinians.

Fig. 6.5 Opponents of the Dakota Access Pipeline in October 2016. Photo: Rob Wilson.[1]

Another approach to have the Native voice heard is written by John McCoy (*see* Section 6.6). It shows how an individual is making a difference and moving Tribal goals forward (also *see* Section 6.2.1). The common thread among these stories are that Tribes are fighting for the environment and their culture and sovereignty. Tribes are not going away.

6.5.1 The Water Protectors: Protest at Standing Rock*

Native American protestors who called themselves "Water Protectors" had the largest gathering of Native American protestors ever at the unlikely spot of Standing Rock, North Dakota in 2016. The 1,200 mile long Dakota Access Pipeline (DAPL) was planning to cross the Missouri River at a point near the Standing Rock Indian Reservation. The Texas-based company announced that it had permits to complete the project which would deliver Bakken oil fields crude

---

1 Accessed on June 12, 2018 at http://www.indianz.com/News/2016/12/07/tiffany-midge-dont-shame-the-standing-ro.asp
* By Dr. Mike Marchand.

oil in the amount of 450,000 barrels per day and would be complete by the end of the year. The Water Protectors were concerned about pollution of water from the river crossing and also were concerned about the practice of fracking used to develop the oil wells on the other end of the pipeline.

This started a series of lawsuits and protests and mass arrests. Standing Rock Chairman Archambault was arrested along with others that summer for marching on the highway. Native Americans started gathering at Standing Rock on a large scale in the summer of 2016. A large number of Tribal leaders joined in the protests in August of 2016. Myself and a couple of other Colville Tribal leaders, my brother Edwin Marchand and Willie Womer went to Standing Rock to join the massive protests along with many other Tribal members (Fig. 6.6). Tribal leaders met and conferred with the Standing Rock leaders and spiritual leaders. Tribal leaders were given a microphone and gave their concerns and prayers. As Tribe's entered they carried their Tribal flags. Eventually the flags were placed on an avenue of Tribal flags.

The camp rapidly expanded. There were cook tents, first aid tents, legal aid tents, and countless tents and shelters for the Water Protectors who stayed there for the next several months. Donations to support the camp poured in from all over the country. My Tribe donated logging truck loads of firewood for the camp. We also donated smoked salmon. Other Tribes donated what they could. Prayers were said. Songs were sung. Ceremonies were conducted. This was to be a spiritual gathering. Violence was not condoned by the leaders.

The numbers swelled to over 10,000 Native Americans by Labor Day weekend in September. Dale American Horse, Sicangu Lakota, chained himself to a

Fig. 6.6 Photo of Dr. Mike (center), Edwin Marchand (right) and willie Womer (left)—all Colville Tribal members supporting and protesting against the Dakota Access Pipeline crossing the Standing Rock Indian Reservation.
Photo Source: Dr. Mike.

pipeline backhoe in protest and was arrested, during peaceful protests. Many more arrests were to follow. The main road to Standing Rock was shut down by the State of North Dakota, cutting off access to the nearest city of Bismarck. The Tribe's casino, a main source of Tribal income, had major revenue losses.

Law enforcement and contract security personnel used an assortment of violent methods over the next several months, ranging from shooting potentially lethal rubber bullets at the heads of protestors to spraying water cannons at peaceful protestors in the cold sub-freezing North Dakota temperatures to sending attack dogs against the Protectors. Some of the attack dogs switched side and attacked the law enforcement people. There was a lot of chaos. Native Americans would send drones overhead to shoot video to feed into the Internet, since there was often no media coverage. Law enforcement would shoot the drones down with guns. Local law enforcement facilities were swamped. Many arrestees were detained in makeshift holding facilities including dog kennels.

The legal skirmishes have continued and the tide has turned against the Tribes. President Obama did take some steps in support of the Native Americans, but the new Administration under President Trump has since reversed those decisions and the pipeline is now in business.

6.6 Intra- and Inter-Governmental Affairs and Public Policy Process*

Dedication: Dedicated to Wayne Williams and Stanley G. Jones, Sr. for having faith in me and for their service to The Tulalip Tribes.

——*John McCoy, August 3, 2017*

John McCoy.
Source: John McCoy.

---

* By John McCoy.

6.6.1 Preface

The approach I took in my work for the Tulalip Tribes and the approach I'm using to illustrate intra- and inter-governmental relationships is based in the culmination of my life experiences tied together by Stanley Jones, Chairman and Wayne Williams, Business Manager, both Tribal Elders of The Tulalip Tribes. When I came back to work for Tulalip, Wayne explained the vision of the elders. He used his ability to tell stories to make it relevant to the process, to myself and others. Wayne Williams set the goal, and I used experiences from my life to make it happen, always with Cultural Resources and Natural Resources by my side.

If you plan right and you network, both intra- and inter-governmentally, if you are flexible to changes as needed, the direction will come together and you will succeed. This is a story of how we developed a plan, networked and succeeded in implementing the vision of our Tribe.

I think it is important to underscore that word: Implementation. My mom had a strong work ethic. Growing up my mom would always encourage me to do things, to think outside the box, to get things done.

> This councilman from Hoh graduated from Haskell University. He said that he wanted to take the Master's class in public administration that I was teaching at Evergreen. I said, *"Great. Happy to have you. Look forward to teaching you."*
>
> I asked him, *"Why do you think you need to have this course if you graduated from Haskell?"* I assumed they would have come at it from an Indian point of view, since I knew a couple of the instructors there. He said, *"Well they taught me all the academics, but they didn't tell me how to use the information."*
>
> And it just so happened that a week later, I ran into one of the professors. I said, *"Dan. I just talked to a graduate of Haskell University. He said that you taught him all the academics, but you didn't tell him how to use it."*
>
> Dan looked at me, thought for a second and said, *"He's right."*
>
> I was crushed.

I am really bothered by that because it is difficult to get into college, do the work, and pass your classes. Yet that isn't enough; you still need to know how to tie it all together and make it work. If you don't get that in your education, you are doomed to make all the mistakes the rest of us make. Why do that? We should give our students a head start.

This section contains information. More importantly it contains the stories of how to use that information and succeed.

6.6.2 Intra- and Inter-Governmental Affairs and Public Policy Process

The vision

You must understand the mission of the organization and the direction they want to go. Wayne Williams did a good job with me in understanding when I came back to Tulalip[1]. He gave me the vision and told me about how far it went back — to the 1910s.

Wayne gave me the vision for what the Tribal leaders of the 1900s through 1920s and 1930s wanted. First and foremost, they required self-governance and self-determination. Our Tribal leaders knew we were going to get there through economic development. In their infinite wisdom, our elders knew way back then that there was going to be a big trading post in the North-East corner of the reservation.

The framework

Everyone needs a sounding board to bounce ideas off to see how valid they are. I used Mike Taylor as a sounding board. I asked some real mundane questions of Mike Taylor and he always had a good response for me, sometimes with a smile and sometimes with a look of "are you kidding me?" We worked well together.

Municipalities or cities have a fiscal tax base. I asked Mike one day, "*What is the structure of what we are doing? What would it look like if we asked the State to step back from taxes?*" Mike went off, working with students, and came back with a very simple plan. The answer would be: The Tribe had to own the land; the Tribe had to own the business, and it had to be managed by an Indian.

We then created a municipality recognized by the federal government.

The plan

It takes time to develop a plan, and you must give the plan the time to develop. You can't rush it. You also need a solid goal to focus the plan. Wayne Williams, Business Manager, gave us that.

I recommend starting with intra-governmental affairs. You need to know who in your Tribe or your organization you should pay attention to. You need to know who your allies are going to be and who your detractors. This takes networking.

---

[1] In my early years, I was a Tribal commercial fisherman. I left the Tribes to join the US Air Force and served my country for 20 years as a computer operator and a programmer. After my service, I worked for Sperry UNIVAC/UNISYS as a systems integrator for another 12 years.

Networking is understanding and building relationships; there is a lot of networking to be done, both intra- and inter-governmental. You'll need to continually work those relationships if you are to get anything done. Like President Trump is finding out: If you don't work with the Congress, you don't get anything done.

That takes time; it takes effort. It also means you'll need to be able to give and take. If you are developing a plan for something, you should have a fall back plan because not all plans are perfect. As you move along through the process, no matter whose process it is, you are going to have to adjust. You keep your goal in mind, but you make you adjustments on how you get to the end goal. At the end of the day, you'll compromise to make it move forward. We were given a simple mission: Maintain self-determination and self-governance. The problem was getting there.

We created the framework: Quil Ceda Village. Now we had to weave the fabric. We had to understand what we had and what we needed and how we were going to go about this.

One of the first things we did was have the Human Resources and Planning Departments do a physical and human asset inventory. What did we have? We needed to know what we had physically. That is, what were the buildings, trees, streets, water supply — the whole nine yards? And, for the human assets, what kind of skills did we have? Learning about the skills available is when we realized we needed to do a lot of educating.

The fabric: Weaving intra-governmental and inter-governmental relationships

In this story, the intra-governmental effort was with the Tulalip Tribes, the inter-governmental relationship building was with everyone else: The city, the county, the State, etc. My story will interweave these elements, but they are both there and they are both necessary. My listener/my reader will notice this as we explore the path taken to implement the vision of our elders. You'll see this in the stories of the Salish Networks and the Sewage Plant.

Salish Networks

Developing our plan took up the first couple years. If we were to be successful in economic development, I thoroughly realized through my life experience we would have to include technical improvements; we needed to have the internet on the reservation. The Internet is a necessary utility just like electricity and water.

Right now, the Internet is treated as a profit center. Well on the reservation, you don't have enough people to make it a profit center; therefore, it is a utility. You need that utility to do education, government work, communications and economic development. Because the outside world, the business community, is all about just-in-time inventory, they don't want a huge inventory on the shelf. Yet they want to be able to get the product and have it shipped in. That reduces their cost and saves on valuable shelf space. They need the Internet, so as they sell something, its replacement is on the way. That is what business needs.

The telecommunications companies wouldn't come onto the reservation, so we had to develop the communications systems ourselves. To bring the Internet to the reservation, we started off with Everett Community College (EVCC) and asked them to help us. They in turn went and chatted with University of Washington Bothell and brought them in. We used U.W. and EVCC students to deliver the network. It was a three-year project, but good things take time.

I use students because when I was in the outside world, in the Air Force and UNISYS[1], whenever I needed studies done and didn't have the internal resources, I would have to contract with expensive firms, extremely expensive firms. Very often, because I worked for UNISYS and all my projects were supporting the Federal Government, a lot of those dollars were coming from the government. Still, I couldn't believe the amount of money I was paying these people — $200 to $400 an hour.

In those days, if I would hire a firm, I would issue a Request For Proposal. I would send the RFP to UNISYS, Hewlett Packard, IBM and the integrator houses, like Boeing. But I knew what their rates were. I knew they were $200, $300 or $400 per hour depending upon the type of person you were asking for to do the study. I could be spending hundreds of millions of dollars just to bring the Internet.

The hardware and software are predictable costs; those are static costs. We knew what they were, and we knew we could live with them. But it was the personnel costs that would kill us. I knew from my days at UNISYS that I was going to have people shuffling in and out. That is how I had to manage my contracts when I was at UNISYS. I was shuffling people in and out and their rate of pay was different. To stay within the dollar amount of my contract, I had to worry about how much I was paying someone within my budget.

Knowing the unpredictability of personnel costs and knowing that we had to educate Tribal members, we thought why don't the personnel go to school in the morning and work with us in the afternoon? Why can't they take what they are learning and put it to work? We would pay them, but we didn't have to pay them as much as people already finished with their training.

A total of 1,300 students went through this program and all 1,300 got a job upon graduation because they had a project they could point to that was successful and completed. The program began as Technology Leap. It is now Salish

---

[1] UNISYS is a global information technology company that specializes in providing industry-focused solutions integrated with leading-edge security to clients in the government, financial services and commercial markets. They developed the first commercial computer produced in the U.S.

Networks. We employed not only the students, but also Tribal members in various aspects of the deployment of the Internet on to the reservation.

We had to bring in all the skills of intra- and inter-governmental relationships. We had to keep the Tribal members happy. We kept them included so that they knew we weren't off doing something unrelated to the Tribe, that we were doing something for them. Tribal members wanted to know that a non-Tribal company wasn't making money or that individuals weren't making money on top of their wages. Tribal members were supportive of almost all we did, but we got push back about putting towers up, so there are very few towers on the reservation.

The Sewage Plant

Inter-governmental relationship building might have a lot of titles, but it comes down to the same thing: Networking, working with folks that can deliver. In our inter-governmental networking with the city of Marysville and trying to keep development costs down, we attempted to negotiate sewer services with the city of Marysville. We already had a water wheeling[1] agreement; getting water was not a problem. Getting rid of the waste was a problem.

> Peter Mills and I negotiated for two or three years, trying to get sewer services. Marysville just fought us and fought us. Finally, one night after some painful negotiations which didn't go anywhere, Peter and I were out in the parking lot. We were getting ready to leave, and their chief negotiator came out and was walking across the parking lot.
>
> I stopped her, and I said, "*All right, damn it. We need to come to something here where I can say yes or no. What is the bottom line, what is it that you need to make this deal successful?*"
>
> I was shocked at what she said, "*You have to pay for 75% of the upgrade. We will not guarantee capacity.*"
>
> And I said, "*Why in the hell would I do that?*"
>
> "That's the deal."
>
> And I said, "*Well, you drove me to no. We're done. Those other meetings we set up, forget them. No deal.*"
>
> That was late at night, so first thing in the morning I called Daryl Williams, Natural Resources, and asked him to come up. I talked to him and told him that negotiations totally tanked. I told him what they were asking for. I said, "*I can't subject the Tribe to that. Go out and find the most environmentally, technically correct sewer system. We'll do it ourselves.*"
>
> And he did. About six months later he came back, and he had two systems. There was a Xenon that was in Denver. We went and saw

---

[1] Water wheeling is transporting water. In this case Marysville was delivering our water "right", our treaty right, by piping water from Marysville to the businesses located on the Eastern border of the reservation.

that one and looked it over. But before we went, I asked, "*Where's the other one? Where do we go to see it?*"

He said, "*There's none in the United States.*"

"*What?*"

"*There's none in the United States.*"

"*Why?*"

"*Well, they haven't licensed it here yet. The technology, it hasn't been licensed.*"

We found out where they had the functioning facilities around the world. It was developed in Japan by Kubota. Then there was a system in England that just happened to be 50 miles from where Peter Mills' mom lived. I talked to the Board and said, "*Let's go to England. Peter can see his mom, and we have an on-board translator.*" The Board approved the trip, and we set it up.

Herman Williams Jr. was chair at the time and he came with us. We also had Pettit Engineering with us. We ended up visiting three different plants in England. It was amazing. We were there in England for seven days touring these three plants.

When we were in Heathrow Airport, getting ready to leave, Herman said, "*Well have you made a decision on which system we are going to have one John?*"

"*Yep.*"

He said, "*Which one?*"

"*Kubota.*"

He said, "*I thought that was probably the one you'd to choose. Okay.*" Then the Pettit engineer said, "*How long do I have to get this done?*" And I looked at him and said, "*18 months.*"

He said, "*No. It's going to take us at least two years.*"

I said, "*If you want the contract, you got 18 months. In two weeks, I want your initial plans on how you are going to make this happen.*"

A couple weeks later, he came back. He brought the initial plan and was telling me about all kinds of problems they are having trying to meet 18 months. He was telling me about the obstacles they had to overcome.

I told him, "*That's too bad. A second casino opens in 18 months, and the waste has to go someplace!*"

The engineer and I were talking about the size of the plant we needed. I said, "*That's taking an awful lot of ground. Do they do it double decker instead of spread out?*"

So Pettit went back to our contact in Japan and asked their engineering department. They came back and said, "*We've never done it,* but there is no reason why it wouldn't work. So we would say, if that is what you want to do, do it."

Decision made. *"Double decker. Make it happen. And you still have that original date, so get to it."*

It all came together on the appointed date and time. And that engineer, the one I told at Heathrow Airport that he had 18 months to complete, all he does is install Kubota systems all over the US. He's traveling all the time, making nice money. He told me, *"John. You gave me one hell of a job! I'm making nice money. I'm not working as hard as I used to, but I'm staying busy."*

We wanted to bring in the very first system to be licensed in the United States, so we included the federal government and the state government as a part of our inter-governmental networking. We brought all the Feds: The EPA, the Army Corps of Engineers, the Department of the Interior, the State Department of Ecology, the Department of Commerce and Snohomish County into a meeting in Seattle. We gave them our sewage treatment proposal. We showed them what we were going to do, what we wanted to build, and how we were going to control it. We let them know that the system came with a Supervisory Control and Data Acquisition system, a SCADA[1] system.

A SCADA system has a lot of sensors or probes in the sewer system that provide data. It even tells you if a generator is not functioning properly. All this data would be pumped out over the Internet where the Japanese had a computer system that would use it to determine how the system was functioning.

We told the Feds, the State and Snohomish County that they would have access to this SCADA system and the software to analyze how well the system was functioning. We asked for a temporary license so that we could prove the system. We suggested that systems like these could be deployed all over the United States.

The Feds gave us a provisional license. After a year of operation in Tulalip, the Kubota system was licensed all over the U.S.

The Tulalip Tribes invested 40 million dollars. They invested on something not proven in the U.S. and not licensed and not guaranteed. They did this because we gave them a plan that made sense. They had wanted to know what we were going to do with the water. In the plan, we told them we would inject the water back into the aquafer. This meant we would replenish our water supplies. We would irrigate with it. And we would put it in a wetland that fed a stream.

Sometimes you get more than you expect too. Now the stream is salmon bearing.

---

[1] A system of software and hardware elements that allows industrial organizations to control industrial processes locally or at remote locations and to monitor, gather, and process real-time data.

Flexibility

Through all this work Hank Gobin, Cultural Resources, and Daryl Williams, Natural Resources, were sitting on our right and left sides. We asked them to not let us go down a path where they were going to make us stop. We asked them to tell us about any issues in the planning stage so that their changes wouldn't impact the progress.

Every project must be flexible because you always run into things you didn't account for. They are always there: Murphy's Law. Murphy is always on the job; he comes to work early and stays late. You must be flexible since no plan is perfect. There are holes in everyone's knowledge. Consequently, where there are holes, things slip through in the planning. You try to talk to as many people as you can, and then you stay flexible.

All through our projects, we had to do a lot of chatting with Tribal citizens to understand what they were looking for. We also talked to the surrounding community. We really didn't want something that was out of character with the community, either on or off the reservation.

We asked staff and others to bring their concerns up as soon as they encountered a situation where we might need to make an adjustment. We told them that the sooner they brought it up, the less expensive the adjustment would be. And they listened: We were able to keep 99% of adjustments minor so they didn't cost too much.

I felt we were successful in being flexible and adjusting. I felt we met the requests of our community physically, culturally and spiritually.

> The way we were laying out Quil Ceda Village created a cultural concern. In our culture, we have a path that runs on the west side of I-5. This path is the path that the elders on the other side take. They traverse that path quite often. Our little ones see them all the time. They let us know when they are there. They let us know which direction they are going.
>
> The way we were initially laying out Quil Ceda Village blocked that path. We had to adjust it a little bit to allow this path to remain unobstructed where the elders on the other side could traverse on through the village without being unimpeded. This we did, and our elders are free to walk as they desire.

Walking in multiple worlds

To recap, the vision came from the elders. We had to buy into that vision. We had to have flexibility and the sensibility to let others work. We had to walk in two, perhaps more, worlds. There are many worlds, depending upon career field, government, or whatever culture an organization has created. To be effective, you should learn those cultures and be able to move back and forth between them. You need to understand the terminology because a word from one cultural environment could mean something totally different in another.

> I learned that when I was working with individuals who used to work for Burroughs. In 1986, Burroughs and Sperry-UNIVAC merged. Although we were two computer companies, you would think we were totally different empires because of the culture they had built and the terminology they used. One of my tasks of the merger was understanding Burroughs' people and translating to the Sperry-UNIVAC people.
>
> I was successful in making that happen at the technology level, so management asked me to help them in some areas. I was basically living/working in two or more worlds. I needed to understand the other cultures to be able to operate effectively in all.

The Tulalip culture is very distinct as are all Tribal cultures. Even among the Tribes in the Puget Sound where we are all coastal Salish, we are all a little bit different. Working in multiple worlds is understanding those differences and being able to work with all of them.

This is not only in language and understanding each other; it is also in other areas like dress. If you dress a certain way in a particular culture you could offend someone by it and vice versa. It is understanding those things and being able to move back and forth so as not to affront but to appreciate.

> In my job in the legislature, I'm on several external committees mandated by the legislature. On the K-12 Education Committee, I've noticed that education is devoid of culture, and it is deliberate. You can tell it is deliberate because if you get rid of a person's culture, you can totally change them. Unfortunately, kids see this, and that is what they are rebelling against — not all — but most of them.
>
> I and a few others have begun the process of having deliberate conversations around relevant culture. I'm telling them that every child that walks through the door has a different culture. It is up to the school district to find out what that is and meet it if they want to be successful with that child. Whether it is a native child or a Muslim child, it doesn't make any difference. There are differences and you must adjust to those. I'm quite blunt with the educators. *"This assimilation has to stop."*

Dealing with children or with businesses is the same; you are walking in many worlds. To navigate multiple cultures with success, it is crucial to do the following:

(1) Listen: You must learn to LISTEN. You aren't going to understand the other culture unless you listen. It is the most difficult and the easiest.

(2) Be respectful: You must listen and take the information in. But when you use the information, you should use it respectfully. If you don't respect it, you may seem demeaning or superior.

(3) Ask questions: If you listen and you don't understand, you ask questions. You ask respectfully. Be prepared to broaden your "box". You'll learn something new every day.

The importance of relationships never ends

During the project, we'd answer questions from Tribal members and take them on tours. It's important to keep the Tribal population informed. It is an important aspect of intra-governmental relationships and underscores how you should continually listen and nurture those you have built.

> When we got back from England during one of the first development meetings we had, the Chairman, Herman Williams Jr. asked, *"Well we've decided where the treatment plant is going to be. Does your wife approve?"* The reason he asked that question was because my family and I lived across the street from where all the development was going to be. *"What about the odor?"*
>
> Jeannie, my wife, had been to England with us. The English have a law that their buildings are required to blend in with the landscape. This facility we were touring was out in a pasture, so they had to make it look like a barn. The sewer plant was inside the barn. We had a tour and they did a good job of explaining what everything was. We were outside the building and just then the maintenance crewed showed up to do their work. My wife stopped one of them and asked one of them a question. He proceeded to take her on another tour of the plant. I followed along behind. I just wanted to know what she was asking if the subject came up again, I'd have an answer for it.
>
> Jeannie was asking relevant questions, technical, about the functionality of the plant, what was keeping the odors down, etc. I was impressed with the questions she was asking. When she was done with that second tour, she rejoined the group and we moved on.
>
> Later in the day, I needed something and had to go home. I told Jeannie, *"The Chairman asked if you approved of the sewage plant*

being across the street from the house." She looked at me. She looked out the window, and I could tell she was thinking about England. And she said, *"I'm fine with it. I'm okay."* And we moved forward with the project.

Sustainable infrastructure

Quil Ceda Village is a continuing project. We had to establish a sustainable infrastructure so the Tribe could move forward with new projects without having the expense of building new basic infrastructure. We had the original plan, but we knew more would follow and we prepared for the future.

So how did our plan fare? Damn good.

Today there is a very big "trading post". Quil Ceda Village was created as the economic arm for The Tulalip Tribes. In the Northeast corner of the reservation are the resort casino, the convention center, the Premium Outlet Mall with 135 stores, the strip mall, gas stations, Walmart, Home Depot, Cabela's, restaurants, and banks. The hotel and one of the restaurants are the only 5-star hospitality businesses in Snohomish County. It continually grows, adding new stores, restaurants and casinos.

Most importantly, we manifested our elders' vision of self-determination and self-governance through Quil Ceda Village our 21st Century trading post, a federally-recognized city.

Chapter VII
Native People's Knowledge-Forming Approaches Needed for Nature Literacy to Emerge among Citizens

"Every society needs educated people, but the primary responsibility of educated people is to bring wisdom back into the community and make it available to others so that the lives they are leading make sense."

~Vine Deloria, *'Red Earth, White Lies: Native Americans and the Myth of Scientific Fact'* (1997)~

"Imagination is more important than knowledge. For knowledge is limited to all we now know and understand, while imagination embraces the entire world, and all there ever will be to know and understand."

~Albert Einstein~

7.1 Why We Need New Education Tool for Nature Literacy for the Masses

Students, citizens and politicians lack science literacy skills needed in today's information economies [159]. Science literacy is not just being able to remember facts but an ability to critically evaluate, contextualize and determine what science data are needed to address a problem generated in our environmental information world. The 2010 National Research Council report summarized science literacy in education as teaching *"adaptability, complex communication/social skills, nonroutine problem-solving skills, self-management/self-development, and systems thinking"* to meet the challenges imposed by our 21st century information economies. Most educational programs do not teach these skills [160] so science knowledge becomes a "hidden half" of the information economy. Even when decision-makers recognize they need to include science and consult with scientists, an environmental problem is so narrowly defined so its ecological, social or cultural scale and context are not made transparent. It is not an option to avoid including the sciences in today's information-based economies since "Science is part of the wonderful tapestry of human culture, intertwined with things like art, music, theater, film and even religion." [161] Science literacy is not just for scientists but are essential skills needed by students, citizens and politicians.

The 2013 special issue of the journal *Science* identified several grand challenges in science education. They wrote that to address these challenges a more diverse group of students will need to be attracted to major in the sciences and to engage them in learning how to form knowledge during the educational process. Also, programs are needed to ensure that they do not drop out of the sciences once they are attracted into these programs. Science education should be youth creating knowledge instead of being "spoon fed" knowledge educators think they need to learn. Further, **not only is there a need to train more scientists with skills to address science-based problems but non-science employment sectors need employees to be grounded in Science, Technology, Engineering and Mathematics (STEM) education knowledge. Private sector businesses want to hire employees capable of applying** *"abstract, conceptual thinking to complex real-world problems—including problems that involve the use of scientific and technical knowledge—that are nonstandard, full of ambiguities, and have more than one right answer"* **[162]**.

There is wide-spread interest by educators in K-12 schools or college programs to create science education programs that strives to remove barriers to forming interdisciplinary knowledge, e.g., STEM (links science, technology, engineering, mathematic in science education). Despite this interest, STEM programs are finding it challenging to attract students. According to the US Department of Education, only 16% of high school students are interested in STEM-related careers or majoring in STEM programs when they matriculate at institutions of higher education [163]. Many demographic groups (e.g., Hispanics, African Americans, American Indians, and Alaska Natives) may account for a quarter of the U.S. population but only 11% of the STEM degree holders [164]. This means that these demographic groups are less able to pursue careers with high career opportunities compared to all other occupations [163]. The number of students in all demographic groups expressing interest in pursuing science STEM occupations decreased from 23% in 2012 to 22% in 2016 while the interest in computer sciences and mathematics increased from 9% to 15% in 2016 [172].

Not only are STEM programs finding it challenging to attract students into these programs, they need to attract a broader segment of the student body to take science classes. Students currently focusing on obtaining degrees in non-science fields need to also take STEM classes. The STEM programs need to expand and diversify the student population interested in acquiring some level of science knowledge competency. Within the last 10 years, STEM has evolved to include the arts or writing as part of science education. These programs have evolved into STEAM (adding arts to STEM teaching to stimulate student creative thinking) and now STREAM is being explored (adding writing to STEAM) [173]. These changes are still not resulting in increasing student matriculating into science programs or into programs preparing students for employment opportunities that require some STEM knowledge.

If the trend of fewer high school students being motivated to matriculate into STEM programs at universities [163] continues, private sector businesses will not be able to find home-grown science literate employees. STEM in its current form is not attractive to certain demographic groups (e.g., Hispanics, African Americans, American Indians, and Alaska Natives) — data shows that they are a quarter of the US population but only 11% of the STEM degree holders [164]. Those who do enter STEM programs often drop out. Classes in the science are perceived by university students to be too difficult because of the need to learn the language of each discipline. They also seem to be disconnected and not relevant for their educational interests. Few STEM programs are contextualized to include holistic forms of knowledge, e.g., spiritual as well as art, plants, animals, geology as well as physics, or tools to critically think about knowledge across disciplines. Also, students do not link employment opportunities as needing science literacy in today's information economies.

Today's youth do not have the tools and skill set to understand and communicate the complex and transdisciplinary knowledge that is needed to address climate change and land-use change impacts on the environment. Today is a time where technology has made it possible to produce massive amounts of data but not the tools for how to deal with it. This results in science knowledge being held by a small group of scientists who have the science language skills to critically assess these data. This is the worst situation you can have since scientists are not the ones making environmental or natural resource policy decisions. This means that we need the masses to hold some level of nature literacy. This goal will not be reached if the approaches used by Western science teachers continues today. Western science is too linear and mostly disciplinary focused which means the training of future scientists or amateur scientists follows the same traditional approaches used in the past by science educational institutions. Western societies need to learn how to form knowledge of nature that explicitly addresses the massive amounts of data that is available today and how to critically assess data under a "scarcity of knowledge" umbrella. We need to learn how to form knowledge just like Tribal people who have dealt with complex and transdisciplinary knowledge to make decisions for several thousand years. Further, we also need to have youth realize that STEM science knowledge is something they can grasp and communicate using today's technology.

> *...Science knowledge being held by a small group of scientists who have pursued graduate educational opportunities [but]... scientists are not the ones making environmental and natural resource policy decisions... the masses [need] to hold some level of nature literacy. Since Tribal people have dealt with complex and transdisciplinary knowledge to make decisions for several thousand years... we need to look at what they are doing that is allowing this to occur.*

7.2 Massive Amounts of Fragmented Data in STEM Sciences

To sustain resilient environments and societies while making environmental, societal and resource trade-offs require credible knowledge and a process that links historical and current science with cultural and economic data. The challenge today is the massive amounts of scientific data that is instantaneously available on our mobile devices but where educators, scientists and decision-makers do not have robust tools or frameworks to assess the quality of the data or to holistically and mechanistically link data generated by different disciplines. Further, few high schools or college students are learning the required tools and knowledge-forming processes to comprehend and utilize massive amounts of disciplinary-based science knowledge in a holistic manner to make decisions. As fewer high school students are motivated to matriculate into STEM programs in universities, the lack of science knowledge reverberates throughout society as the lay public is inundated with science news that they cannot critically assess [165]. The explosion in social media and instant communication of knowledge confounds the problem of "fake news" being used to support environmental trade-offs [166]. Economic indicators therefore become the primary tool to decide how to develop our economies and to identify indicators of human well-being and environmental impacts. The challenge for economic tools is that they are not designed to include volatility and unexpected events [42] which are normal events in coupled socio-ecological and economic systems.

Western science also has to be part of the solution to address today's data overload problem. Despite the lack of place-based knowledge, Western science has produced a growing body of knowledge on the impacts of extreme climatic events on species and ecosystems. This knowledge is needed by Indigenous people whose locally-based context need to be recalibrated for today's new weather and ecosystem conditions. Indigenous people are the most impacted by and need to adapt to climate change because of climate induced loss of species on the lands their lands [167]. Indigenous people are "place-based" and continue to gather cultural food as an important part of their customary practices. They depend on collecting natural resources for their livelihood and to maintain their cultural practices. Tribes have "legally protected rights to economic, cultural and spiritual uses of natural resources" but which extreme weather changes cause "range shifts in plants, animals and aquatic species" from reservations and customary lands [168]. Tribes are not decoupled from their land and resources compared to many Western world citizens who do not know where their basic resource needs come from, e.g., toilet paper. Even though Indigenous people are very adaptable and retain knowledge passed down as stories related to natural boom-and-bust cycles, they need to form new knowledge to address cultural resources located on altered lands and/or waterways.

Linking art/culture to science [24] has received much attention as an approach to teach youth to make better or more creative decisions in our personal lives while developing our economies. Such an approach still does not provide a tool or framework that functionally links different types of knowledge. The current trend has been to link art to science knowledge, but the art is still an inanimate object that represents science. So art activities may be fun for the general public but it does not facilitate them to make more environmental and culturally-based decisions using the massive amount of existing data. It is not enough to read a book similar to *How to Make Friends and Influence People* written by Carnegie in 1936 and become friends with a diversity of people from different cultures. The continuing challenge is to explicitly link the different disciplines in mechanistic ways and where the students learn to "create" knowledge and use evidence-based approaches so "fake" knowledge is not accepted. Students need to develop their critical thinking skills in the sciences. Using Native people's ways of forming holistic knowledge can teach youth critical thinking skills.

7.3 Critical Analysis Lacking in Environmental Education

John Gordon (former Dean at Yale) wrote how environmental problems persist for decades and are increasingly costly to society while providing few benefits. Thus, environmental managers and decision-makers will want to acquire storytelling approaches to more quickly resolve a problem or build consensus on an issue. Educational institutions should also be a consumer of such a tool since it can be used to support deeper learning in US schools, especially for students wanting to matriculate into college. Students need to learn to apply knowledge they learn to address problems emerging from our rapidly changing economies. Tools like what is proposed in Chapter VIII would facilitate students and adults to use technology, or a process, to explore their understanding of complex scientific, societal and cultural problems. Too many students are not exposed to this critical thinking process.

When combining knowledge from multiple disciplines, students and non-scientists need to learn how to "create" knowledge for themselves instead of "receiving" knowledge that is delivered in a pre-packaged format and summarized according to someone else's perspective. The Harvard Graduate School of Education study wrote that there is urgency for high school students to *"create knowledge, rather than receive knowledge"* if youth will be able to vote, get a job or deal with ethical issues [183]. But, instruction on deep learning are mostly found in private schools or schools in affluent neighborhoods. According to Harvard GSE [183] 80% of high schools have no programs to expose students to this thinking process. Deep thinking or critical analysis is seldom part of the K-12 curricula especially for youth living in less affluent neighborhoods.

Even at the university level, curricula frequently do not emphasize critical thinking while teaching disciplinary topics well. Universities use undergraduate capstone projects as a mechanism for students to make the connections between the disparate and varying knowledge types that they have learned during their first three years of College. The capstone projects only occur in the last year of a student's program, so students are less immersed in critical thinking or problem solving approaches until the end of their academic program. By this stage, students have already been indoctrinated into disciplinary forms of thinking that teaches them to focus problem-solving within a narrower framework. Some programs (e.g., iSchool, U.W.) teach critical thinking in their capstone projects since students learn to design and build computer applications. In these programs, students learn the Design Process (NASA Design Squad) [184] to come up with solutions for a problem. This process follows a succession of steps to define the "real" problem which then is used to identify the context of the solution.

The scientific method is taught to university students as a way to form knowledge but it is not designed to explore problems holistically. Even scientists and educators struggle with understanding or learning about knowledge formed in other disciplines. Therefore they focus on their disciplinary strengths and explore environmental problems in a disciplinary vacuum. When they communicate their science to the public, it further confuses citizens trying to make decisions because too much information is communicated that is decontextualized from the problem. Even disciplinary scientists are challenged by the immense amount of information that is available today. According to Biswas and Kirchherr [185], *"Up to 1.5 million peer-reviewed articles are published annually. However, many are ignored even within scientific communities — 82 percent of articles published* in humanities are not even cited once. No one ever refers to 32 percent of the peer-reviewed articles in the social and 27 percent in the natural sciences." This suggests that much of this science is not used in making decisions and in forming knowledge on managing nature.

Since critical thinking skills are essential tools for anyone entering the job market today, we need decision-makers to adapt holistically to form knowledge to respond to our dynamic social, economic and natural environments. It is not enough to memorize facts but there is a need to be able to critically think through information. It would be an impossible task to memorize all the massive data that exists today. This doesn't mean some people are really good at memorizing facts, so they are a "walking dictionary or encyclopedia" of bits of information. We have all met people who are repositories of information and can provide a factual response to any specific question. These people would do well in television game shows. Unfortunately, these people frequently do not understand how these facts interconnect. So they don't solve problems, especially complex environmental problems. They are too involved in the detail of the problem so they do not see the context nor are they able to critically think through the knowledge they have. As an old expression states "they can't see the forest for the trees." This results in the "symptom" of a problem being addressed instead of the "disease".

People can learn to critically think about environmental problems if they learn practical skills and learn from their elders or grandparents. It is not their fault that they are not exposed to holistic approaches to explore complex problems. We know they can learn new skills since non-scientists adapt throughout their life to new situations. This is evident in how most people learn new employment skills as they shift to new jobs. Making career changes is also not uncommon. The average person makes a career change at least 5 to 7 times in their lifetime and will hold about 10 jobs by the time they are 42 years old.[1] Thomas Friedman wrote in 2013 [185] that the youth graduating today do not need to just "find" a job in the marketplace but will have to "invent" a job. Friedman quoted Tony Wagner — a Harvard education specialist who wrote:

> "Today," he said via e-mail, "because **knowledge is available on every Internet-connected device, what you know matters far less than what you can do with what you know. The capacity to innovate — the ability to solve problems creatively or bring new possibilities to life — and skills like critical thinking, communication and collaboration are far more important than academic knowledge. As one executive told me, 'We can teach new hires the content, and we will have to because it continues to change, but we can't teach them how to think — to ask the right questions — and to take initiative.'"**

---

[1] http://www.careers-advice-online.com/career-change-statistics.html

7.4 Native People's Storytelling Practices to Communicate Holistic Science

Today, science knowledge is poorly communicated to non-scientists and decision-makers. Science is communicated as a series of facts and decision-makers do not have the tools to critically think through a problem. The complexity of environmental problems makes science knowledge appear contradictory to non-scientists, especially since they hear about so many exceptions to the rule or the same facts are used to defend polarized views of a problem. Despite these issues, science knowledge needs to be an essential part of decision-making by society. Society depends on natural resource for their survival and economic development and health. To continue to collect natural resources requires a resilient nature. This is where we need to educate youth to critically think about complex problems and not wait until they attend a university. But it is not just learning how think holistic but there is a need to communicate holistic knowledge to decision-makers. The complexity of the sciences can be communicated by teaching storytelling skills to youth as practiced by Native Americans. The stories need to be based on the holistic ways that Native people form knowledge. Youth can then use technology to digitize these stories using several different multimedia products. This is a wonderful way to communicate the complexity of an environmental and social problem as well as the challenges facing solutions.

Linking Indigenous ways of forming knowledge to that used by the Western science approach is possible when Indigenous people are equal partners in forming new knowledge [174]. Many successful collaborations have occurred between Indigenous and Western scientists especially in conservation management. Conniff [175] provided many global examples of community managed areas providing greater conservation success than establishing the traditional park models developed by Western world scientists. Leonard et al. [176] reported how effectively the Miriwoong people in Australia used Indigenous knowledge to adapt to climate change before the impacts of climate change were noticeable in ecosystems. In Australia, commercial fire management programs were combined with aboriginal approaches that reduced Greenhouse Gas (GHGs) emissions by 37.7% when fire frequency and intensity were managed [177]. Rick et al. [178] determined that Indigenous people who lived around the Chesapeake developed practices to harvest oysters sustainably in spite of climate change, sea level rises and increases in the number of oysters harvested; these historical practices have since been supported by biological data to reconstruct a new approach to harvesting oysters. These examples need to be better communicated so a broader segment of the population is aware of these success stories. Youth need to learn the process for how knowledge was formed during these collaborations, so they fine tune their own critical thinking skills.

The non-scientists need to know these stories to make environmental policy robust and to make ethical trade-offs needs evidence-based knowledge and the participation of a broad segment of data collectors, policy makers and stakeholders [179]. Stories can help communicate the uncertainties in the scientific knowledge while making science education culturally relevant to non-scientifically inclines decision-makers and Indigenous students [180]. This lack of relevancy helps to explain why parts of our society are chronically underrepresented in STEM-related fields [181]. Adapting science educational approaches to attracting Indigenous students as well as other demographic groups into the sciences is essential if we are to promote a conversation between Indigenous and Western science. These dialogues support education, capacity building, and community-based natural resource management in a culturally relevant environment.

Forming Indigenous knowledge requires us to learn how to tell stories. Stories can teach, in an intertaining manner, youth about the land and its resources across multiple spatial and temporal scales [13], as well as to coordinate behavior related to the collection of limited resources [188]. Stories need to be holistic where our environmental and human health decisions are viewed through a lens of our relationships with nature. A holistic approach acknowledges that ecosystems experience unpredictable recurring cycles and people/environment/ecology are all interconnected. Indigenous people provide us a successful model to teach science literacy to non-scientists. Further, **Indigenous storytelling approaches are ideally suited to be communicated using the emerging new tools developed by computer scientists and media experts. For science literacy to thrive and attract more students, educational programs need to embed science in popular cultures [161].**

There have been several very successful scientists who have described or communicated science for the general public. Johnson [161] wrote a graphic novel entitled *The Dialogues. Conversations about the Nature of the Universe* that has been very popular reading even though he may be describing physics phenomenon such as string theory. He is uniquely qualified to counter that science should not be separate from "art, intuition and mystery" and the *"special talents: Required in order to engage with and even contribute to science are present in all of us."* [161]. He further describes *"Science is one thread of culture — and entertainment, including graphic books, can reflect that."*

Tribal approaches to storytelling and communicating their knowledge can be linked to building games or producing other multi-media products. This is the future of communicating science and attracting a diversity of youth to want to matriculate into the sciences. We can't continue to use the past approaches central to Western education models to address complex, multi-layered but decontextualized approaches to include culture in the knowledge-forming processes.

Building games to communicate science knowledge should be able to expand the number of youth, especially from the demographic groups that are less represented in science programs today, to be attracted to science learning. The potential can be seen with the success of Duolingo — a language application. Duolingo has *"over 60 million registered users, it teaches languages to more people than the entire U.S. public school system"* [169–170]. If the games being developed are able to attract or expand the number of players, it should significantly contribute to attracting and motivating youth into the STEM programs. There is also a need to diversify the demographic groups represented in the sciences [171]. It is especially important to attract Indigenous youth into STEM programs since our decision frameworks need to integrate massive science and societal data while retaining place-based knowledge passed down through multiple generations. Youth are ideal for developing new frameworks because of their facility with computer technology and playing games. Also, by linking Native people's storytelling with Western science knowledge of altered environments, games can be designed that entertain while motivating youth to explore complex problems facing decision-makers in coupled socio-ecological systems.

Chapter VIII
Learning Indigenous People's Way to Tell Circular Stories

If Indigenous knowledge is viewed as "... *a collection of data about the environment that could complement and be integrated within the existing data sets used by state management systems*" [182], this knowledge forming process would be useless to Western science. The current Western science methodology cannot be adapted to build tools needed by future environmental decision-makers (*see* Chapter IV for a discussion of the differences between the Western science and Tribal people knowledge forming process). A Western science assessment tool doesn't just need to include more data held by Native people. It has to be built to allow decision-makers to critically and holistically think through a problem. It also needs to be compatible with Tribal cultural values. This means that elements of each knowledge forming process cannot be lumped together to build a better tool. To change how knowledge is formed and communicated by Western sciences, Indigenous people will have to be equal partners in building a new framework that is fundamentally different from any currently used in any of the Western science disciplines. The Western science groups cannot dominate and determine how a new tool is built. The standardized tools commonly used in the Western sciences are not designed to "... *acknowledge the value system and cosmological context within which this traditional knowledge was generated*" [182] and therefore are unable to integrate Indigenous knowledge except as a narrative in research.

We need to teach how to be critical thinkers. Morris et al. [159] wrote people are not born with scientific thinking skills and learn these skills as they are "scaffolded via educational and cultural tools". Therefore, tools are needed to easily and critically contextualize science knowledge and to develop a platform that teaches these thinking processes to citizens without science degrees. We do not need to build a new framework or tool since Indigenous people have taught these skills to their youth for hundreds of years [186]. As Reyes-Garcia et al. [187] wrote,

> "*'Traditional ecological knowledge' emphasizes the historical continuity of such bodies of knowledge, not only their local embeddedness, a characteristic that seems to contribute to the long-term resilience of social-ecological systems by providing a pool of information and practices that improves societies' adaptive capacity to cope with recurrent environmental or social disturbances.*"

Today technology is an essential tool to learn how to learn critical thinking skills. By learning to tell circular stories like Native people, youth would learn critical thinking skills. When we say stories, we are not talking about learning how to tell European fairy tales where the ending is always the same and very predictable. Section 8.4 and 8.5 present a discussion of how to digitize Native people stories without pickling culture and how Navajo tell stories.

8.1 Technology to Digitize Stories Part of Popular Culture

Our solution is to learn from Indigenous people who already create knowledge "*... across biological, physical, cultural and spiritual systems*" and "*includes insights based on evidence acquired through direct and long-term experiences and extensive and multigenerational observations, lessons and skills*". [23]. Indigenous knowledge takes the principles of ecosystem management ideas developed in the mid-1990s [22] and makes them "adaptive" as well as including people as part of natural ecosystems. What makes our solution novel is that keeping track of science and Indigenous knowledge would be overwhelming to any person and it is impossible to test some relationships without damaging or reducing the health of our lands. But today's advances in computer technology and application development has made it possible to take large meta-files of quantitative and qualitative information and link different types of data or knowledge and deliver to a user friendly tool such as a cell phone. Technology is allowing people to create knowledge and to change the manner in which we learn and integrate knowledge.

The European Union and some programs in the United States have reversed declines in enrollment in their STEM programs by using gamification to attract and retain a higher number and a more diverse group of students to expand the size of their programs [189-190]. These programs report that students are very engaged and motivated to learn to critically think about complex and interconnected forms of knowledge from both the natural and social sciences, as well as to learn to work in groups to design solutions for complex problems [189-190].

Morris et al. [159] suggested three mechanisms by which video games support science thinking and increased science literacy: First, games are designed to have

> "a number of motivational scaffolds, such as feedback, rewards, and flow states that engage students relative to traditional cultural learning tools. Second, there are a number of cognitive scaffolds, such as simulations and embedded reasoning skills that compensate for the limitations of the individual cognitive system. Third, fully developed scientific thinking requires metacognition, and video games provide metacognitive scaffolding ("examine assumptions or beliefs of how the system works") in the form of constrained learning and identity adoption."

The challenge in game design is to balance the delivery of knowledge-creating tools while motivating and engaging youth to continue to play a game. Games have been successfully developed to teach computer science knowledge in technical universities [190], as virtual laboratories in biology, microbiology and electrical engineering [189, 191], and, also to engage and motivate non-scientists in ecology [192]. Games are very effective at teaching knowledge where the basic principles are well understood, e.g., how electricity is produced, but the dynamics of the environment, why people make the decisions that they do, is more challenging. Current games are based on topics and relationships that are better understood and where choices do not have repercussions for your survival [193].

According to the Green Games website, their games are designed especially for kids to "explore green living, sustainability, ecology, environmental care, and social equity" [194]. These games are designed to motivate children to play while also creating knowledge of complex natural and social environments. Green Games will probably not capture a customer base as large as Pokémon Go where estimates suggest **9.55 million total daily U.S. users play this game** [195]. It is unrealistic to expect green games to generate interest among the general public since they focus on specific externalities of our growing economies, e.g., better manage garbage or wastes, and reduce food wastes or calculate your carbon footprint, calculate your energy consumption and make a choice of how you want to travel to work. These games are already being played by people who already value green environmental behavior which is estimated to be less than 20% of the US population [196]. To motivate and expand youth interest in STEM programs requires expanding beyond this niche market to attract the general public.

Until 2014, games were not designed to link the land/environment and decision made by people solving a problem. A game may use ecological principles to change how the game protagonists behave (e.g. Darwin's Demons where you are fighting aliens [197]), but they do not represent place-based reality, knowledge of the land or the culture of the people who live on the land that is passed down inter-generationally. Thus, most of the games do not develop your critical thinking skills so you make better or more ethical decisions based on playing such a game. The first video game that beautifully introduced a place-based and culturally-based game, characterized as a puzzle-platformer game was "Never Alone".[1]

Western science is still working on developing tools and frameworks capable of mechanistically linking biological, physical, cultural and spiritual systems, e.g., [198-202], therefore, it is understandable that communicating a holistic approach to creating knowledge is not part of STEM, STEAM or STREAM programs. In contrast, Indigenous knowledge already "includes insights based on evidence acquired through direct and long-term experiences and extensive and multigenerational observations, lessons and skills" [23]. These frameworks are based on a group of people's understandings of a place that are based on long-term knowledge retained by a community.

Indigenous knowledge-forming frameworks are not siloes of disciplinary knowledge and includes a process for critical thinking about complex and coupled socio-ecological systems. These are the same people that are disproportionately vulnerable to the impacts of climate change [167] so they have a vested interest in "getting the science right". This is why including Native American ways to form knowledge with Western science is crucial to introduce into our educations systems so future leaders develop their critical thinking skills.

Further, gaming science can contribute to scientific literacy [159] by using tools familiar to most youth. Therefore, Indigenous knowledge is an excellent vehicle for youth to create place-based knowledge while driving on multiple "roads" built by Western scientists who have a solid understanding of the relationships between land-use changes, loss of productive capacity of crops and trees due to degraded soils and alterations of the climatic patterns. Youth building knowledge frameworks and playing games will instill critical thinking and link social, cultural and natural science knowledge into STEM programs.

8.2 Digital Technologies Part of Popular Culture

The application of digital technology to facilitate learning Indigenous languages, stories or environments or ecological keys are flourishing around the world. These apps are teaching tools to learn about their environment or location and were mostly developed by non-Indigenous people for Tribes. Examples of digital technologies are:

---

[1] http://neveralonegame.com/

Chapter VIII   Learning Indigenous People's Way to Tell Circular Stories ▬ 369

- Google Earth has databases and games that use geotagged digital images that can be downloaded and explored by the public.[1]
- Mobile Learning Incubator group, University of Wisconsin has developed plant and bird apps which requires the user to identify a plant or bird.[2] They also developed a "Sustainability U" app where an "app directs players to find signs posted in campus buildings and scan a code with their mobile device to pull up information, videos, or games related to waste, energy, water or transportation".[3]
- National Geographic's Project Noah—allowing citizens to explore and document plants and wildlife around them that using their cell phones. This is a form of citizen science who geotag, identify species that are collected in a database.[4]
- Learning language games developed for at least 24 different Indigenous languages in Central and South America.[5]
- SAI Cuentos Mágicos de los Ancestros embeds Indigenous stories in an interactive space that kids and their parents can play and SAI Actividades Infantiles is a game that introduces kids to the territories inhabited by different Colombian native ethnic groups.[6]
- "Vamos a aprender Purépecha" is a game that teaches Purépecha which is a native language in Mexico.[7]

A few years back, games were not designed to realistically represent Indigenous people and their cultures. For example, "The Spirit of Spring" is an example of a game where the main character or hero is portrayed as being Native but the story is addressing bullying. The latter game is described as the first empathy game *"an adventure empathy game with puzzle elements"* built to teach prosocial skills.[8] Any demographic group could have played the main character since Native practices or cultures are not relevant for the storyline.

The scarcity of games designed and developed by Tribal people reflects the low number of game developers who are Tribal members. In 2011, when the International Game Developer's Association surveyed to see how many Indigenous game developers were part of their association, only 41 out of the 6,438 responders (∼0.6%) were indigenous (Starkey 2014) [198]. Until Indigenous game developer's lens to represent Tribal people are used to build game, games will

---

[1] Accessed on February 6, 2017 at http://www.makeuseof.com/tag/6-google-earth-games-fun-frolic-globe/
[2] https://mobile.wisc.edu/teaching-and-learning/
[3] Accessed on February 6, 2017 at https://www.chem.wisc.edu/content/mobile-sustainability-game-helps-students-take-environmental-action
[4] Accessed on February 6, 2017 at http://www.projectnoah.org/
[5] Accessed on February 6, 2017 at https://rising.globalvoices.org/lenguas/
[6] Accessed on February 6, 2017 at http://www.colombiagames.com/wordpress/en/index.html
[7] Accessed February 6, 2017 at http://ccemx.org/vamos-a-aprender-purepecha/
[8] http://www.weareminority.com/about-minority/

represent Tribal people in an "obnoxious" or false manner and "games are inherently about conquest or progression. My games focus on building communities and establishing lasting settlements." [203]. If the current Tribal game developers were developing games representative of Indigenous cultures, perhaps players would be exhibiting less aggressive behavior [127].

Many Indigenous communities have prioritized revitalizing their languages and youth learning their Native languages [13]. Despite Tribal people being forbidden to speak their language in the late 19th century, 196 living Indigenous languages still exist in the US [204]. American Tribes have taken ownership and are leading the efforts to revitalize and record their language knowledge. For example, Thornton Media Inc (TMI) was started in 1995 by a Cherokee Nation citizen as a language learning company. TMI it has helped over 170 American Indian Tribes to build language games to revitalize and teach their languages. In 2013, TMI released "a new 3-D video game ... named "Talking Games" gives players the chance to learn Cherokee while playing a video game" [205]. In 2013, Shoshone youth enrolled in the University of Utah's Shoshone Language Program built the first language video game that used Shoshone stories to teach their language (Indian Country Media Network 2013). (NOTE: The link to TheEneeGame.com is no longer live even though the website noted that the video is still being tested and developed.)

The Tulalip Tribe also teach the Lushootseed language that includes teaching songs and stories spoken in their language. All of this information is readily available on the Tribal website [206]. Some of the video games built by Tulalip Tribes include: Names of plants where you see the word written and also spoken; hearing the sounds made by a fox, brown bear, porcupine, raccoon, moose and mole; learning to sing songs and tell stories through traditional story tellers or historical recordings made of Tribal elders in the Lushootseed language [206]. In Canada, the Aboriginal Media Lab (AML) has produced an alternative reality game that focuses on players healing themselves using knowledge of plants, medicines and healthy food as well as healing the Indigenous territories around Vancouver, British Columbia [207]. One of AML's objectives is to have aboriginal and non-aboriginal people interact on what are healthy practices.

Several game developer companies are collaborating with Native people to build games while also educating and building capacity in the Native youth to build games. This is occurring because Native communities still lack the programming and game development skills needed to build their own games. For example, Big hART[1] is an arts and social justice company in Western Australia that is teaching aboriginal youth to build digital storybook or games to maintain their culture. Other companies show Tribal art or translate other games into native languages[2]. Pinnguaq is also developing curriculum to teach Native

---

[1] http://yijalayala.bighart.org/
[2] http://pinnguaq.com/localization/

youth how to make games as well as working on an interactive fiction story that is available in the Inuktitut and English languages.

The 2014 release of the video game entitled "Never Alone" changed the role of Indigenous people in using games to teach their culture and languages. This game was built from a collaboration between the Cook Inlet Tribal Council and E-Line Media (non-native game developers) that converted the Iñupiaq story into a game-based cultural storytelling [203]. This is an Indigenous educational game representing the Inupiaq culture and is a beautiful scripting of their stories to portray their culture and the environmental challenges they face. The Alaska Native Community was involved in every aspect of the game being developed. Their storytellers, the script translated for the game and artists created the storyline and its imagery and sound. Thus the game avoided becoming based on western world stereotypes of an imaginary Native people. "Never Alone" is called a "puzzle-platformer" and "a two-dimensional game in which the player crosses gaps and avoids enemies while crossing from one side of the level to the other side" [208].

In April 2016, another video game entitled Huni Jui was built to educate outsiders on Kaxinawá or Huni Kuin people culture and myths; the Huni Kuin people live in Western Brazil and Eastern Peru [209]. This game is similar to "Never Alone" in that it was a collaboration between an anthropologist at the University of São Paulo and Huni Kuin people from 32 villages. The villagers designed the story that was narrated by shamans and created all the sound effects for the prototype that was developed [209]. The game is available in the Kaxinawá language *hatxã kuin* and Portuguese with Spanish and English subtitles. This game was designed to be educational to teach about the culture of an Indigenous group and not a fun game. It has a role to educate but is not game-based cultural storytelling that would motivate players to repeat play the game.

8.3 Challenges in Communicating and Telling Circular Stories

Indigenous knowledge Indigenous people have used storytelling to teach their youth about the land and its resources across multiple spatio-temporal scales [13] as well as to coordinate behavior related to the collection of limiting resources [188]. These stories are holistic where our environmental and human health decisions are viewed through a lens of our relationships with nature. A holistic approach acknowledges that ecosystems experience unpredictable recurring cycles and people/environment/ecology are all interconnected.

Indigenous people provide us a successful model to teach science literacy to non-scientists. Further, Indigenous storytelling approaches are ideally suited to be communicated using the emerging new tools developed by computer scientists and media experts. For science literacy to thrive and attract more students, educational programs need to embed science in popular cultures [161].

8.3.1 Science Literacy Needs to be Circular and Not Linear

Teaching science literacy today can be problematic because there is a need to teach "the interconnected nature of fundamental concepts for earth, space, life and environmental sciences" [11]. This is a **holistic view of the world that "emphasizes the historical continuity of such bodies of knowledge, not only their local embeddedness, a characteristic that seems to contribute to the long-term resilience of social-ecological systems by providing a pool of information and practices that improves societies' adaptive capacity to cope with recurrent environmental or social disturbances" [187].** However, science is taught using a narrow disciplinary lens by STEM programs which de-contextualize information that do not teach science literacy skills. As Reyes-García et al. [187] summarized: *"Scientific knowledge... goal of being universal, transferable, mobile, and not tied to a singular place."* This is also a very linear thinking approach to forming science knowledge and where some disciplines dominate the discourse. This mental construct of science today can be viewed as a seesaw slide that moves you up and down at the same spot and leads a decision-maker to adopt a pre-determined end-point or solution.

The challenge is that stories told by Indigenous people are "circular" and include multiple interconnections between "earth, space, life and environmental sciences" [11]. Circular thinking is difficult to integrate with the linear approach to science that dominates the Western approach to science today. Rose [10] describes these differences:

> *"With linear thinking, we rely on logic, institutions, and others to try to protect ourselves. Circular thinking originates from the earth, the universe, and the Creator; we are all connected and all safe."*

This makes teaching a Native people's knowledge of nature and its circular story telling approach a challenge for someone who is not an Indigenous person.

It is inadequate to tell a class populated by non-Native youth to start writing a story about nature and how to live and respect the land. These youth do not have sufficient knowledge to write a nature story since they have not learned to take complex scientific knowledge held by Native people and deliver an ethical thinking process. So any program teaching youth nature knowledge needs to have youth digest large quantities of transgenerational knowledge and be able to justify the circular path their knowledge might take. The storytelling process needs to provide a tool to youth to address their relationship to nature that is based on knowledge on how to live on the land.

8.3.2 Science Literacy Is an Information Problem

Today, the fragmented use of science to teach science literacy does not allow science knowledge to be used effectively to respond to environmental, economic and social boom-and-bust cycles [13, 42]. The standardized tools commonly used in the sciences are not designed to "...acknowledge the value system and cosmological context within which this traditional knowledge was generated" [182] and therefore unable to integrate Indigenous knowledge except as a narrative in research.

From a Western world perspective, it probably seems strange to think that one can write stories about science and translate all of its complexities. Western European stories were written as fantasies or make-believe worlds where the "good" people overcome "bad" people. In these stories, class structure doesn't matter as a person born on a farm can become the leader of a country, e.g., win the attention of a princess by being clever enough and saving a kingdom from bad people. These stories were not written to tell a morale tale or teach ethical behavior in nature. They were written to entertain the reader and to escape the trials and tribulations found in our world [210]. This contrasts Native American Tribes where stories entertain but also deliver a moral held by the community. Elaine Emerson, a Colville elder and fluent speaker, recounted to University of Washington students and professors that each story has an ethical and moral basis to it.

Digitizing native stories is not simply a process to take stories, writing storyboards and then putting them in an application. The framework and tools to teach circular storytelling are not readily available, and the tools are emerging to transfer this knowledge into modern digital formats or popular media products. Circular stories to address science literacy are practiced within Tribes and held by Tribal elders gifted in the culture and linguistic heritage of each tribe. This form of storytelling must become part of the popular media and culture if we are going to be able to translate science to the next generation of decision-makers. The issue of knowledge transfer is unique, especially when the context is cultural, and the target demographic are young people whose knowledge framework culturally differs from Tribes. The three issues facing the

digitizing Indigenous stories are:
- No roadmap or curriculum exists today that delivers a program designed to teach youth how to write circular stories which are inherently culturally-based and have a transdisciplinary science context.
- Circular storytelling is an information problem where a story has to include knowledge of the interconnected nature held by Tribal members and ecosystem scientists in academic settings.
- Technological and cultural media tools are just emerging to translate circular stories and where the target demographic group are young people whose knowledge framework culturally differs from Tribes.

8.3.3 Indigenous Stories Are Not Linear but Cultural and Transdisciplinary Science Knowledge

On the surface, Native people's stories transmit knowledge of nature and where nature is not told as a linear story that moves from a challenge to a predictable end point. Native people's stories are similar to a mystery novel where pieces of evidence suggest who are the "bad" players while revealing bits of information that ultimately reveal knowledge to recognize who are the guilty parties. These are circular stories and not linear. In Native American stories, animals and humans can transform into other shapes [211] and the conclusions are not predictable and repeatable. The coyote — a trickster — is a common animal in these stories as described by Jay Miller [211]:

> "He embodies all the human traits: Laziness and patient industry or frantic exertion; foolishness and skillful planning; selfishness and concern for others.... He is the incomplete and the imperfect.... Coyote has fur, and it is the incomplete and it is fun to hear about him and his exploits. What he lacks in dignity he makes up in sheer exuberance."

Native Americans have long oral traditions that transmit knowledge that is transgenerational and may include descriptions of natural phenomenon that are important to know to live in balance with the natural world. Native Americans have continued to tell their stories of knowledge held by the tribe even after they were forcibly removed from their lands but have continued. This has kept their customs, values and languages alive within the Tribal community. Frequently these learning moments occurred where a story became part of the transgenerational transmission of Tribal knowledge. It was also a way to tell a story that could have several different endings so they became learning moments for Tribal youth to learn to respect nature, how to live on the land that includes its cycles of change and how to read the land to reduce its vulnerability to land-uses. It is a learning tool where elders teach youth with their stories.

Despite Native American traditions of telling stories of how to live on the land and of nature, few Western European fairy tales are designed to teach the science of nature to youth. In most cases, the surroundings are not important to the story nor is there a discussion of impacts on nature from some human activity. Western fairy tales are human centric stories. Few Western European stories have endings that are not clear and predictable, i.e., the stories are linear and predictable even when insufficient knowledge to write a predictable story. Western world stories are fantasies where a person can escape into another world and where there are no repercussions of the decisions that a person makes.

8.4 Digitizing Native Stories without Pickling Culture: Interview of JD Tovey*

JD Tovey (left) and Dr. Phil Fawcett (right).
Photo Sources: JD Tovey, Phil Fawcett.

Kristiina: Why don't we get Phil to introduce himself? Can you introduce yourself and describe your background? Why are we even talking to you?

> **Phil:** *I am Phil Fawcett. I have had over 35 years' computer experience. Most of that time was with Microsoft. I live on 25 acres east of Seattle. I help defend rural land rights or at least keep it rural and from development. With all the tech industries growing in the area, there is a lot of pressure to urbanize or to create these opportunities for people to economically build houses and to take huge acres and to convert it*

---

* By Alexa Schreier, Dr. Phil Fawcett and Dr. Kristiina Vogt.

to suburbia. We are trying to figure out how to preserve that and keep the feel of the rural area.

There is a lot of prejudice about if you live in a rural area that you are not as sophisticated as someone who lives in the urban side of things. We found the opposite to be true. Working at Microsoft, I had a hard time with the joke that I lived in Western Idaho. That I lived way out there and there was not anything there. There was no bandwidth.

The last several few years I have focused on getting my PhD. I spend half my career in Microsoft research transferring ideas from researchers worldwide. Really smart people with several PhDs from Stanford didn't know what to do with their lives because they were so brilliant. My role was to take those folks ideas and getting them into some form that could be monetized for products. This required me to figure how to do that across cultures. I traveled to China, to Europe and other places and looked at different academic cultures and met with different scientists to make that happen.

I have also done a lot of volunteer work. One of my personal values is to give 10% of my wealth and resources back. That is my goal. So, I have done 25 trips to different countries like Mexico and Russia volunteering my time helping in orphanages, to build homes for people who don't have homes that are living in plastic houses and all sorts of crazy things.

Now I am working in the capstone program at U.W. and I really like it. I finished my PhD in 2017 at U.W. I like those surfaces where you are not one or the other, but you are trying to bring both of those worlds together. I teach across two institutions in computer sciences.

I like to look at gaps between x and y, whether it is economics or land-use or ecological. One of the projects that I been a sponsor on and have been involved with is a capstone project that worked with the American Tribes to help preserve their culture through games and design. We have done some proof of concept of taking Native American stories and putting them in a gaming format. A game that really helped us thinking about this was "Never Alone" started by the Cook Inlet Native Alaska community as a way to preserve their culture. They were able to get a huge commercial investment of about $15 million dollars by a gaming studio and investors. They maintained the core idea of the animals in the game.

Chapter VIII Learning Indigenous People's Way to Tell Circular Stories

> *I am trying to explore how we can bridge the gap here between the elders and the youth and different cultures and technical differences between the two, possibly through gaming.*

[Phil to JD]: I want to ask you about digital preservation and bridging the gap. Can we preserve the story and the culture and the language in a digital way to bridge the gap between elders and the youth? When we bring these stories into a digital format, we now have a lot more flexibility to blend stories, to change the role so it might have been a coyote and now it could be a crow or all sorts of things. So now we can build a new set of stories that could be passed on to the next generation. My hypotheses is that if we develop the technical capability of youth and then connect them with the elders to create the storyline for the games, could that be used for the long term for the preservation of the culture and the language. Or are we going to get too much of a distortion of the original idea or is that happening anyway as it is passed on from elder to elder. How are stories passed on in Tribal cultures and is there a need to better connect elders with youth? Is there a digital mechanism to better to do that so we have digital preservation?

JD: *I don't know. What is that game?*

Phil: "Never Alone."

> **JD:** *Some of my Board members at a retreat I was at were talking about language. One of our board priorities at the last session was the preservation of language. Language preservation and teaching language. I brought it up with a couple of the board members and they loved it. But they then began to tell a story that was not the game's actual story. I remember thinking at that time, "Is it important to have the story generally right or is it also important to protect the culture by not telling the story?"*

> *So even on our reservation we have this conflict to not tell anyone about where the camas are because all we have left is that knowledge. They can't take that away from us. They can take trees away or other things away but they can't take our knowledge away. I knew the background of the game was generally a story that was a boy and now it was changed to a girl.*

Phil: The other issue with this story is that when they switched from just the Cook Inlet driving it to the commercial game studio, it changed 180 degrees. It morphed into something that could win an Academy Award but not necessarily preserve the culture.

[**Kristiina to JD**]: I just want you to respond to something you said at the Colville meeting during our summer 2017 when you said "**Sharing knowledge is an important role that you see for yourself and for your generation.**" So how do you factor that in?

> **JD:** *In our ceded territory, 6.5 million acres, there were probably close to over a million Indians living there pre-1500s. The Cayuse have a language isolate. The only way that linguist's say you can have a language isolate for 15,000 years or more is if you have a population size of 100,000 to 300,000 people at a minimum. Now there are 250,000 of us and few of us are full blood. My grandmother is probably the last of the dozen full blood Cayuse. She is listed as a full blood Cayuse. So there is all this land. This landscape was chuck full of Indians before **the Great Dying** which happened about 1520 to 1600 C.E. All the diseases came here over hundreds of years before the White people arrived with Lewis and Clark. We had the Russians, the Spaniards, the French and the Scots. People were coming from all directions. There is a term that describes this — **the Great Dying**. Lewis and Clark arrived some 200 to 300 years after the Great Dying. There was a massive cultural roll over at that time.*
>
> *So, all these people who were care takers of the lands died. There was the **Great Promise**. All of the plants and animals could talk at this time. A messenger came to the people and said we have this new creatures coming. They are people, and they are going to be like babies barely able to walk and they don't understand anything. But they were going to be really special. All the animals lined up in order and the salmon said I will sacrifice my body in exchange for being cared for. The deer then stood up. So, we have a serving order of the food that sacrificed themselves in order to take care of us. But it is an exchange. They will sacrifice their bodies as long as we take care of them. This was the **Great Promise**.*
>
> *But after the Great Dying we physically could not take care of everything. The situation now is where we have culturally linked our souls to this **Great Promise** that we are supposed to take care of everything. You still see that now in our DNR project where it is more about taking care of our ecosystems but we can't physically do it. We are in this weird space where the only way we can care for now with other organizations. All these other organizations, e.g., BLM, USFS, are not doing it. They are slowly coming around where they are starting to include human behavior in those land management organizations. They excluded human behavior for 200 years because it was weird and there-*

fore was not a driver of natural resource management. Now that we are able to start including human behavior back into natural resource management, it is closer to how the Indians managed lands for thousands of years.

But it is a slow process and the Indians who have this **Great Promise** *can't physically take care of all these lands they hold dear. The only way they can do it is transfer that knowledge to new caretakers. That requires two things. One is to give up some knowledge even if it is the philosophical part of the* **Great Promise***, even if it is not the exact location of where the camas roots are. There are two different knowledges there — one is the philosophy and the other is hard data. The other thing is that they have to admit that they can't do it by themselves. That is the hardest thing to admit that we physically can't do this. Does that admit that we are dying? Does that admit that we are going away? These are all loaded questions. The only way we can take care of this land is through the new caretakers. Does it admit something that no one wants to admit?*

Phil: Explicitly or implicitly or both ways.

JD: *The question then is "How to do that?" Then it comes back to the game.* Is the feeling of the game, the story or the lesson of the game morality more important than the accuracy of the game? *What you need out of it is the morality. It doesn't matter if it is a boy or girl.*

Kristiina: Don't the characters in the stories morph anyway?

JD: *I have seen the same story but with different characters. But if you ask an elder, the story has to be always right. My grandmother tells me some of the coyote stories and she says they have to be right. She will correct a story and say no it didn't happen like that. It has to be like this.*

Phil: So there is a pull and a push between those two things.

JD: *It is like knowledge. It is something we have and you can't take it away from us. It must be right. There is more than getting the words in order. There is a coyote story where he throws his eyeballs up into the air. He runs around and catches them in his own head. But they land in his head and "ah" which means they fit just right. But it doesn't make sense in English. But an eagle comes and steals his eyes and he is chasing the eagle around to get his eyes back. But it has to be right.*

Phil: It gets out of order.

JD: *If it gets out of order, it is not the same story anymore.*

Phil: So what you are saying is that principles of the story can be abstracted out and passed on and continued. But the absolute storyline needs to be exact and if it is exact it becomes a passing on of the knowledge which in some cases is sacred but in some cases not. You made the statement that they can take away everything but they can't take away knowledge. So would a digital preservation, even though it preserves the language from dying out as the elders continue to pass away and do not pass it to the youth, is that an ok thing to happen? Is that a natural thing to happen or are not? Can you make a decision what elements should be preserved or what should not or is there some middle ground?

JD: *I don't know. This is something I wrestle with. It was the vice chairman who was playing the game (Never Alone) and he was retelling the story. But knowing the history of it, I knew that it was not the actual story.*

Phil: It was gamified which were also some of the constraints.

JD: *Another person was playing the game and was telling the story from the game but I knew it was modified because it had been gamified. The story is not the same but he was telling the story like it was.*

Kristiina: When we were at an NCAI (National Congress of American Indians) meeting, a program director mentioned about how they were getting Indian youth in the west coast U.S. to write new stories. We live in a highly altered environment so the story may be different functionally today, even if certain basic facts are the same. When I think of Native Americans, I think of them managing landscapes. Mike talked about how they would do something on a specific piece of land but would not come back to it for two years. They were managing the one piece but at a landscape scale. Of course this was also when the European settlers arrived and they thought this was wilderness. No one was managing the land so they could take it. But Tribes had a longer-term temporal and spatial knowledge.

The moral of landscape management would be that you don't do certain things on the land. Remember when we had the language expert from the Colville and she said that each story has a moral. So in some cases you can write a story that tell that. When you say it is important to share knowledge maybe it is not specific information but it is the relationships or interconnectors. We have been looking a lot at stories and the stories are circular (*see* Section 3.2). You don't have this fixed answer so if you try to put that into a story it won't work.

Rodney has talked about stories where an animal dies and falls into a fire pit. It then morphs into something else. He said it was a beautiful morphing of the other animal into a white bird.

Phil: But the idea during the gamification of the stories is that we took three different stories and the Colville purposely didn't give us the original story. So there already was an abstraction. The principles were there about working with the ecology. That was there. So I thought that was a good morphing of it. So the idea is to preserve some level of knowledge while core knowledge will always be passed on in traditional stories. But in my mind the idea that there is another way of looking at the world and doing it through a digital format. Some of the limitations that we ran into, for example, "*When you try to digitize a piece of native art it is pretty stagnant. It is a jpg file. Then you have to use these other tools to add elements before you can use it in a gaming environment. So that morphing technically means it won't have the same properties as if you were using your imagination to tell the story.*"

> **JD:** *See I love that stuff. On one hand I am all for it because one thing* **I am very critical off is the pickling of culture. When you pickle a cucumber, it is not a cucumber anymore. It is a pickle. So we have a cucumber that we are going to pickle to preserve it. But it is not what it was anymore. Now it is something else. But now we have it saved it in a jar on a shelf somewhere. We have a whole shelf full of pickles.**

Phil: So it is almost an exchange idea.

Kristiina: Most people when they see an Indian story they really don't understand it. I don't think our group working on trying to digitize the Colville stories really understood what the stories were all about.

Phil: They were a bunch of information scientists.

Kristiina: They converted it into a quest. But if I think about the coyote story and that he wanted a good name and not what was left over, they are different stories. I remember reading about the coyote where he used toothpicks to hold his eye lids open. He was going to stay awake so he could be at the front of the line when names were being given out. He fell asleep. His wife didn't wake him up.

> **JD:** *He had multiple wives.*

Kristiina: His wife, who was not even a coyote, but more like a mole. She didn't want to wake him up because he was really tired. When he woke up all the names had already been given out. But I was thinking about what you just said. If you are a western European, you would this is a little odd story. But maybe the story is that the coyote was using the wrong approach, e.g., using toothpicks to hold your eyes open.

Phil: Or you don't stay up beyond the point where you can't sleep.

JD: *We have kids growing up now where they take language classes, Nez Perce, Umatilla, or Walla Walla. These are immersion classes, so this is all they learn from Head Start to kindergarten. So they can speak those languages. When they go home, they continue speaking these languages. They tried teaching adults but they just weren't learning it. So now all these kids learn these languages and the parents are going to be obligated to learn them to communicate with the kid. So the kids doing their homework are going to have to learn the language. So this is a more effective way.*

What is interesting is that we have this weird Native Pidgeon English that has developed. A lot of this language has words that are shorter than Nez Perce. For concepts that are shorter in our language they prefer using this Pidgeon English. The youth were always made to be strict on how they spoke but now they are creating their own language and using it. It is not just something from a book.

On another note, we do a lot of recordings of elders but they are all locked away in a safe. For my dissertation I have to access whole parts on villages. I have to go to a room in a building. The doors are locked and they bring me the books. I have said, I need this type of stuff and they will catalogue it and bring me the books. It is just knowledge that is in these books. I just have to sit there and read it. I can't make copies of the book. I can't leave with it and have to do all my work right there. Who sees these? This is our culture.

Kristiina: This is like the French Academy of Languages that didn't want any none French words to permeate into the French language in the 1960s and 1970s. They wanted French people to only use and speak in the original French language. But they don't think about the fact that there were no computers when the language first developed. They need to use four or five French words to describe a computer. This is a dead language when it doesn't change. It is like Latin. If you don't change, no one will speak it. When I learned English the biggest thing I had to deal with were all the different meanings of one word. I

Chapter VIII   Learning Indigenous People's Way to Tell Circular Stories —— 383

couldn't figure out why people were laughing when I first started talking English because I was not making a joke. But the common use was something other than what I had learned. It makes me wonder what I was saying at that time.

Phil: I have a question on your last comment on needing to be in the room to look at important documents. Is that knowledge that the tribe considers should remain in that room?

> **JD:** *Yeah.*

Phil: Is part of that knowledge something that should be shared? Can it influence the rest of society by making them look at something using a different world view? Can we get a different view by gamifying part of the ecology?

> **JD:** *I would like to think so. So much of it is so strange because it challenges our own beliefs. For example, we have an urban development where they found ancestors buried beneath it that were more than 8,000 years old. All of our burials typically face east. These burials were so old they were not facing east. They were looking in a different direction and they were buried very deep.*

Phil: More than 12 feet deep?

> **JD:** *Well we are at the high water at Umatilla where the flood waters would come down. So the floods would hit the gap by Umatilla Oregon and the waters would back up there as they went through the gap there. It created this temporary lake. There are a lot of silt layers and there is a lot of agriculture here. That was all happening 8,000 to 15,000 years ago. The ancestors were very deep because they were below some of the silt lines. They have been there for a very long time.*
>
> *There was a political thing about not building houses on top of burial site. The suggestion made was that kids were going to have the Down syndrome if houses were buried above burial sites. We followed our own process to asses this. We quartered off where the ancestors were buried and put a plaque there to recognize it as a burial site. It still continues to be discussed since we have some $2 million dollars' worth of infrastructure projects that have not been built because they are over the burial site.*
>
> *The reason this came up was when I was looking over some of the stored material and looking at some of the villages. There was burial site that had a building post right through it. They have these diagrams and it shows a body not in a fetal position. We have three types of*

> housing typologies and this was an older housing typology. It was at a newer housing type where the pole was put right through the body. At that time, they pushed the body out and put a post right through the middle of it. This was fascinating to me. Women were typically in charge of setting up the villages. There was a woman chief of the village. Historically their job was to move bodies when we had to build something new. This is challenging today because we don't do anything on that site and leave it just as it is. These are hypersensitive topics today.

Kristiina: Tribes are amazingly adaptive.

> **JD:** *I totally agree with that.*

Phil: Just going back to the hypersensitivity. Do you think the causation of that is what you talked about when you mentioned earlier about how some people are trying to tell Indians how to be Indians? Do you think it is the mixing of the cultures, languages or the media trying to tell Tribes what makes them Tribal?

> **JD:** *I think so. I can't think of any religion that is more ritualized than the Catholic Church. What is the purpose of the rituals? It is so you don't forget it. It is so ritualized, so it no longer has its original meaning, but you don't forget it. It has meaning but when you are doing 27 Hail Mary's what does that mean? What is the purpose of doing the Hail Mary's? It is to have forgiveness and pertinence. So people who do it are disconnected from the intent.*

Kristiina: It is controlling. I lived in Europe and saw this.

> **JD:** *But most of it is so people can remember it and don't forget it. So cultures end up doing that. This is especially when a culture is under attack. So we have to write it down to know how to do it. You have all these things to figure out that are now ritualized and they become hyper special. They were never intended to be this way.*

8.5 Stories in Navajo Lands*

Even though the traditional stories are ancient, they continue to have contemporary impact and application. Many current societal concerns can be positively

---

* By Dr. Nancy Maryboy and Dr. David Begay.

Chapter VIII Learning Indigenous People's Way to Tell Circular Stories

Dr. Nancy Maryboy (left) and Dr. David Begay (right).
Source: Nancy Maryboy.

addressed through sacred and secular recitation of oral narrative. Many stories deal with intrinsic relationships. They are not simplistic nor dualistic observations of positive and negative reactions. The flux of relationship is expressed by causality within causality, thus providing a continuum of complexity. A recognition and acknowledgement of the essential co-existence and inevitable interdependency of the positive and negative relationships expressed through SNBH is arguably the crux of the teaching (Box 17).

> **Box 17: Diné (Navajo) Paradigm of *Są'áh Naagháí Bik'eh Hózhóó* (SNBH)**
>
> Diné (Navajo) paradigm of *Są'áh Naagháí Bik'eh Hózhóó* (SNBH), a belief system that guides harmonious living, and demonstrate how the application of SNBH enhances understanding of Navajo principles for well-being.[1]

Stories and songs are a large part of the traditional way of life. The deep metaphysical and spiritual meaning and organization of knowledge reside within the stories and songs. This is somewhat difficult for the Western mind to accept since it is dissimilar to the Western scientific organization of knowledge. When traditionalists speak of story and song they are not referring to a story like Robin Hood nor a fable like Little Red Riding Hood. Navajo culture uses story and song to address the most serious societal problems and the most spiritual aspects of consciousness and humanity. Story and song are best used within the context of Navajo culture and society. When they are removed from the appropriate cultural context, they are trivialized, marginalized and relegated to the realm of myth.

---

1 Accessed June 15, 2018 at https://www.ncbi.nlm.nih.gov/pmc/articles/PMC5099227/

In the present American society, myth is seldom considered to be equal to science as an acceptable way of knowing. Thus, Western society, which largely lacks the cultural background to understand the significance of Indigenous story and song, continually associates Native myth with primitive paganistic fable.

A deeper analysis of many of the Navajo stories will show the biased narrow limits of popular western thinking. Western thinking disallow the deeper significance of story and song which, in fact, strongly correlate with the cutting edge of scientific thought, Quantum Physics, Chaos Theory and Systems Theory. Although Navajo and other Native people may discuss these concepts through cultural and metaphysical story and song, they nevertheless will recognize similar fundamental concepts in quantum physics and chaos theory; however they may understand them within a spiritual consciousness.

The organization of Navajo songs and prayers can be expressed visually through spherical diagrams. The example given in Fig. 8.1 is a simple illustration of organization in accordance with the interconnections of the mountains, placed and aligned through the four cardinal directions. As you can see by the diagram, the directional movement not only goes out, it also returns back to the center, creating in essence a recursive cyclical process. The two directions, according to the holistic thinking of traditional people, are one and the same. The diagramed process may seem abstract, but it has limitless applications.

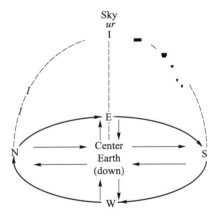

Fig. 8.1 A simple spherical diagram of organization.

To further illuminate the traditional Navajo thinking process, we have chosen to discuss a mountain song. This part of the song refers to Mount Taylor, *Tsoo Dzil*, the sacred mountain of the south. This is a free translation from the original song:

Tsoo dzil bi h6zh66go bee nashaadooleel

I will walk in harmony, in balance, with Mount Taylor

With the feet of Mount Taylor, in harmony I will walk
With the legs of Mount Taylor, in harmony I will walk
With the body of Mount Taylor, in harmony I will walk
With the breath, the voice of Mount Taylor, in harmony I will walk

With the thinking of Mount Taylor, in harmony I will walk

Dzil nanit'a, in harmonic movement I will walk

Aspects of psyche and body are manifested through the song. The song speaks about the mountain and the spiritual relationship with the mountain. Psyche, body, spirit, are all one and the same, expressed through the reality of the mountain. At the same time the mountain is part of a much larger and more comprehensive system, through intrinsic interconnections with a network of other mountains, united by a natural process of increasing complexity.

There is a fundamental difference in the use of the word mountain between English and Navajo. Webster's New Dictionary (1990) defines mountain as a land mass higher than a hill. A hill is defined as a "natural elevation" of the earth's surface, not as high as a mountain. Navajos use the word *dzil*. *Dzil* as understood by Navajo speakers goes far beyond the dictionary definitions of mountain, which remain exclusive to spatial attributes including elevation, soil, and stone mass. The root of the word *dzil* is *dziil* meaning approximately strength, firmness, energy, power and stability. The meaning of *dzil* in the Navajo language is relative. The use of *dzil* varies, depending on the circumstances. A small mound can be considered a hill or a mountain.

For the remainder of this discussion we will interchange the English term "mountain" with the Navajo term "*dzil*" but with the Navajo meanings attached, in order to avoid the limited western connotation of an elevated pile of soil. *Dzil* connotes an extension of Mother Earth as a living organism. *Dzil* is thus a living organism endowed with countless additional living organisms and natural phenomena. *Dzil* can be more fully understood in terms of a multiplicity of connections and interrelationships. The holistic quality of this relationship is usually expressed through ceremonial songs and prayers, where order and connections are meticulously defined.

A major characteristic of *dzil* is the organic process, an interconnective web encompassing the entire cosmic order. This process provides seasons, growth, reorganization and renewal, in relation to the growth process inherent in the complex relationships among plants and their environment. Ecological factors that determine these relationships can be expressed in scientific terms of climatic processes: Precipitation, evaporation, temperature, and photosynthesis. Additional factors could include edaphic relationships as well as biotic interactions with other organisms. The main point is that *dzil* is part of a systemic order, intrinsically connected to a vast universal order, best understood through the Navajo language and expressed through ceremonial organization.

The English language emphasizes a neutral gendered, but underlying male-oriented, dissociative process, whereby the speaker is disconnected from the action (verb) as well as from the object (direct object). Navajo language, on the other hand, is so intrinsically interconnected with subject, verb and object, that what would take an entire sentence, in English, can often be described in one or two words in the Navajo language.

Vine Deloria, Jr. argues that each cultural context has its own unity which should be acknowledged [212]:

> *"It is comforting to believe that we can transpose concepts and ideas from one cultural context to another and thereby gain new ways of understanding the world. But each cultural context has a unity of its own which gathers information into a particular set of interpretations. Removing an idea from one context to another is therefore a doubtful process because the methodologies by which each culture gathers information are different."*

Chapter IX
Medicine Wheel: Moving beyond Nature, People and Business Stereotypes

> "These Indians are fierce, they wear feathers and grunt. Most of us don't fit this idealized figure since we grunt only when overeating."
>
> "To be an Indian in modern American society is in a very real sense to be unreal and a historical."
>
> ~Vine Deloria, "Red Earth, White Lies: Native Americans and the Myth of Scientific Fact" (1997)~

Providing leadership to solve an environmental problem is very different from being a technician who solves a specific aspect of a problem. A technician is very good at keeping the status quo and stopping further degradation of a site. Both are needed even though you probably want fewer leaders and more technicians working on resolving an environmental problem. When the leaders make bad decisions, it impacts all the technicians who have to implement these decisions. Several historians have talked about how the foot soldiers win a war and the general loses it. Also, Napoleon Bonaparte was quoted as saying that *"Soldiers generally win battles; generals get credit for them."*[1] So, one cannot survive without the other. But the general is the leader that has to figure out the risks and strategies to win a war. An environmental leader is similar to a general but someone who knows when to step aside for someone else to lead when a different leadership is needed [3].

Environmental leadership also requires that the leader has not lost their connection to nature or the leader won't understand the repercussions of their actions. This resonates well with a quote attributed to Luther Standing Bear in 1933.[2]

---

1 Accessed March 16, 2018 at http://www.military-quotes.com/Napoleon.htm
2 Accessed on November 14, 2017 at http://www.indigenouspeople.net/standbea.htm

> "The old Lakota was wise. He knew that man's heart, away from nature, becomes hard; he knew that lack of respect for growing, living things soon led to lack of respect for humans too." He is also quoted as saying "The American Indian is the soil, whether it be the region of forest, plains, pueblos, or mesas. He fits into the landscape, for the hand that fashioned the continent also fashioned the man for his surroundings. He once grew as naturally as the wild sunflowers; he belongs just as the buffalo belonged."

Similarly, Quentin Regestein was quoted in a 2010 a Reuter's interview as saying,

> "...Patient should be made to understand that he or she must take charge of his own life, don't take your body to the doctor as if he were a repair shop."

Nature is not a repair shop that a leader (doctor) has all the tools to restore. So who is better to provide leadership than a person close to the land who does not see nature as being a washing machine where the parts are easy to replace?

Leadership is not learned when you are an adult. This has to start when you are young and you are already forming knowledge through interactions with elders. What is leadership? Can we decipher these by looking at the elements and differences or similarities of land management through different lens: Scientifically trained landscape ecologists, resource managers, and traditional practitioners? Leadership needs to be collaborative and not dictatorial. It isn't being a technician. It is not something that can play out at one institutional but includes multiple institutions, agencies and communities. Land-use and climate mitigation planning have to come from different levels and not just Congress. This requires that we should know how policy is formulated since they are the legal basis supporting why some decisions are made. In most cases policies are formed by the top administrative structures and do not consider Indigenous issues (*see* Section 6.2.2, 6.3.1) or we would not be experiencing the protests like what occurred at the Standing Rock Sioux tribe attempting to stop the Dakota Access pipeline potentially contaminating their water supplies from the Missouri River.[1]

We started this book talking about Leaders without Borders and how Dr. Mike Marchand exemplifies this (*see* Section 2.2). We will next include some thoughts written by Dr. Mike Marchand that reintroduces some of the ideas already mentioned but are qualities needed by a leader.

---

[1] Accessed on March 16, 2018 at https://www.theguardian.com/science/2017/may/15/those-are-our-eiffel-towers-our-pyramids-why-standing-rock-is-about-much-more-than-oil

9.1 When I Was a Young Boy*

Dr. Mike Marchand photo.
Source: Mike Marchand.

There are over 500 recognized Indian Tribes in the US, more in Canada and Mexico and South America, depending upon how you define the category. The various Nations are very diverse. Some are very urban. Some lived in small villages. Some had high levels of technical societies. Some preferred hunting and gathering and nomadic life styles. So, **there is no one monolithic block of people that you can define as Native American. When anyone speaks about Native American culture, keep these facts in mind.** But that is what makes the topic interesting. **The diversity is strength. Cultures adapted to their geographic regions. They lived in these regions for thousands of years. Usually but not always, they evolved into relatively stable societies and were synchronized with the ecosystems** which often were **sustained though centuries through culture and traditional practices, including proactive management of the environment. For anyone interested in the** interaction between humans and the environment, this diversity provides a rich treasure trove of multiple societies to observe and study and learn from.

My people live in the Pacific Northwest, anthropologists would categorize us into Plateau culture. We practiced hunting and gathering, but on a grand scale. The world perception is that hunters and gatherers lived on the precipice of survival, barely eking out a living on the edge of starvation. This is totally not true in the Pacific Northwest. Mother Earth provided and abundance of wild game and fish and plant foods. The world was like a super grocery store. Everything one needed merely had to be picked and gathered up or hunted or fished for. Over time the technologies and knowledge had become very efficient. So that time needed for "survival" was minimal. More time was spent on cultural activities. Less time was spent on getting rich. There was more emphasis on

---

* By Dr. Mike Marchand.

sharing and communally working together. Modern economies are often driven by scarcity and by supply and demand. But if nature is very abundant, the competition for scarce resources no longer is needed. So, for many Pacific Northwest Tribes, this was the situation for over 10,000 years. There were some down years with climate change and drought for example but in general life was very good.

Knowledge was passed down through non-written means. People learned through stories. People learned by doing. Ceremonies and traditions also transmitted knowledge. There were specialized societies and sub-groups. Some specialized in plants, often these were females. Some specialized in tools and weapons and war, often these were males. Some specialized in hunting and fishing. Some specialized in medicines. But these were not always segregated by sex, there were always exceptions.

Today, like all global societies, the Native Americans are being impacted by the digital Internet age. Whereas once there was a lot of elder-to-child contact, this is changing. Many children have access to smart phones and tablets and the Internet. So, the information technology is rapidly changing. Where this all goes is anyone's guess. Proponents of traditional Native culture are in the process of adapting to the new technologies as fast as possible.

Time in western society is generally perceived as linear in nature. Calendars start at year zero. Now it is 2018. Planning is usually done by calendar. Businesses commonly do 5-year plans or 20-year plans. There is often a perception that things improve over time. For example, one common iconic picture shows man's steady rise from apes and monkeys, to various types of cave men, all the way up to modern man. Civilizations are often similarly viewed. In western culture the Greeks are often seen as the founders of democracy.

Native Americans often have a different view of time. It is not so much linear. Often it is seen as cyclical. Each season and week have a relationship with Mother Earth. Certain plants are gathered at certain seasons. Big game animals migrated with different seasons. If your society depended upon these food, knowledge of these season was important. The concept of the past and present and future was also different non-linear. Western calendars divide things into days and weeks and months and years and centuries. The written calendar reflects this linearity. Native American calendar do have some variations. There are complex calendars in Central American societies which encompass thousands of years, but they tend to be more event oriented in my opinion. Like what are important galactic dates? When do important astronomical events happen, like eclipses, planetary and star alignment, etc.?

In Great Plains cultures, which were based on hunting buffalo, time and calendars tended to be different. Their calendars might focus on important Tribal events and individuals. This might include things such as battles, births of important leaders or Chiefs. On top of the Little Big Horn Mountains in Wyoming is a sacred site called a Medicine Wheel (*see* Section 1.2). It is essentially a site aligned with the sky and considered very sacred. I won't go into a lot of detail, but it is a large circular manmade formation of rocks. There are 28 spokes radiating from the center (28 is a common sacred number). There are 28 days in a lunar month, normally 28 days to a woman's menstrual cycle. So, there is a tie between the astronomy and human life. These are important connections to Native American people.

Modern science, particularly, Einstein, has changed the concepts of time. Einstein, in a nutshell, states the Western concepts of time are a human construct and has nothing to do with the reality of time. Past and present and future are really the same thing. This is more in line with Native American practice. **As a leader of 12 Tribes myself, when we make decisions we base those decision on three things. For any given issue, how would our ancient ancestors and culture view the issue? How do we view it today in the 2018 context? How does the issues affect countless future generations? Our existence is viewed as temporary in nature. But the tribe is more permanent. My life begins and ends. But if we do our jobs right, the tribe will continue on.** So, I personally put a lot of focus on my own grandchildren. What kind of world will I leave them? And, of course, they likely will have their own grandchildren someday long after I am gone. There is a photo of Einstein published wearing a headdress with a group of Native Americans. Some question the validity of the photo because he is visiting the Hopi pueblos but wearing a headdress and smoking a pipe from the Plains Indian culture. But Tribal people say that it is not uncommon to pick up headdresses and pipes from different tribes and then to pose with them at another tribe. It is somewhat humorous and contrived. For example, the eagle war bonnet is usually thought of as being from other Native American groups. But the juxtaposition of Einstein with Native people is worth pondering. It seems that the Native American concept of time is somewhat more in line with Einstein thinking.

When I was a young boy, my father and uncles would often take me deer hunting. On one such trip there was my father and his two brothers. They stopped at the base of a mountain and started walking. It was winter and as we walked higher the snow got deeper. One of my uncles put me on his shoulders and carried me. As the snow got deeper, they formed a single file. They were silent. No talking. I learned later that this was one of the family's old hunting trails, they knew it like a highway.

They would take turns breaking trail. About the second or third rotation, I asked *"When was it my turn to break trail?"* Uncle said *"You want to break trail?"* I said *"Yes, I think it's my turn."* They were big strong men, used to hard physical labor. I was just a spoiled toddler. But he took me down off his shoulders. I slung my little rifle on my shoulders like my uncles and led the way. It must have looked comical. My uncles and father were three big strong men, hardened by the labor of logging and ranching for years. Then me, a tiny tot, albeit very serious looking tot, was breaking trail for the three big men. Surely, I must have greatly slowed the hunt up. But they were patient. There was no talking. They would point to tell which way to go. Surely, I must not have gone too far, but I put my all into it and went as fast and as far as I could go. Finally my legs gave out and I was out of breath and sweating from the work. Then my uncle put me back on his shoulders and we continued.

They hunted as a team. They knew the trails, where the deer were likely to be. The eldest went ahead and stationed himself on a point. Another uncle went to another spot. My dad was the youngest and he carried me.

We got to the ridge top. By then the light was becoming visible, the sun was nearly rising. Golden light beams were piercing the cold. My father put me down as neared the top of the ridge. He peaked over a rock and spotted a herd of deer. He took a shot and killed a young buck. The rest took off running farther up the ridge. Soon there were more shots. My uncles were at their shooting stations and shot more deer as they tried to escape. That day we shot five deer. It was a successful hunt. They put a rope of one of the smaller deer and gave it to me. I was thinking I don't think I can pull a big deer, big relative to me anyway. But the snow was slick and the hill steep. The pull was easy and soon I was running to keep with the deer.

We loaded our deer into the truck and headed for home. The adults were happy and I could tell they were happy with my effort. This was a lesson. On how to hunt. On how to work together. On how to be patient with children. Now I am a grandfather myself and I try to treat my grandchildren with the same patience and put them into situations where they can succeed. They need to know the land. They need to understand how deer think.

Eventually we had a big family feast with the whole clan. A lot of home cooked food cooked on a woodstove. My grandmother would make a point of bragging on her hunters and would note that I was there too. Everyone was happy and well fed. Everyone was happy that the game was plentiful and that the hunt was successful. This was typical in my tribe and was replayed over and over again for thousands of years. My people have been hunters since the days of wooly mammoths. We worked together. We trusted each other. We respected life and Mother Earth.

The next summer I was on another hunting trip. It was late summer and much warmer. My dad was successful and shot a buck, a mule deer. While he took care of the deer he shot, I walked around the other side of the hill. I saw a little buck. I had to sneak closer because my rifle did not have a long range. I was able to get fairly close to him and shot. It was brain shot and he collapsed and died instantly. I ran back to get my dad and told him what happened. He smiled and was happy. We eventually got home, the deer were butchered and put into the freezer. My dad went behind the house where there was a big garden and some trees. I was spying on him. He did not know I was watching him. He took the

Chapter IX Medicine Wheel: Moving beyond Nature, People and...

deer heart and hung it in a tree to dry. A few weeks later he buried the heart in a corner of the yard, and he said some prayers. That was the custom for our people, the Sinixt nation. These customs and traditions were handed down for generation over the millennia. Our people say that ceremonies are important. That is we maintain ceremonies and do the proper prayers, that life will stay on course and life will be good. Game will be plentiful.

My wife has different stories. She is from the Okanogan Nation, they are neighbors to the Sinixt Nations, living in the next valley over in north central Washington State. Her mother and aunts would spend spring and summer gathering foods. These included various roots such as camas. To them the world was a huge garden. They knew where to go and gather foods and were constantly traveling. Berries were also very important. The elder woman had a vast knowledge of plants. They were taught that all things in life have a spirit. Animals have spirits. Plants have spirits. Even the earth and the heavens have spirits. One day they were looking for medicines, plants that had medicinal properties. The belief is that every ailment had a cure in nature and often these were plants. Her mother and aunt were busy hiking around. My wife was a toddler. She picked up what she thought was a pretty flower and proudly showed it to her mother. Her mother was horrified. She took the flower and slapped my wife. She was very angry and scared. My wife had accidentally picked a very powerful plant, a certain type of lily. It was deadly poisonous when properly prepared. Disturbing such plants was considered very bad luck and a danger to all. Her mother carefully took the plant back to where it came from. She prayed for forgiveness and sang certain songs and left a gift. All plants have a spirit. They must be respected.

My wife's family lived by the Okanogan River. Each morning her and her sister would be wakened very early and told to run up the hill adjacent to their house. It was a steep sandy hill and torture to run up. But they did this every morning. When they got done with the run, they were told to jump into the Okanogan River and go swimming. On sunny warm days this could be a nice experience. But they were expected to do this whether it was warm or cold. This was done to make children strong. These are warrior society customs developed over thousands of years. Being able to run and swim could be a matter of life and death if invaded by enemies. They also took sweat baths in their sweat lodge. The sweat lodge ceremonies are very important to Native Americans throughout North America.

The sweat lodge was constructed of willows and was a small dome type structure. Twenty-eight willows formed the ribs. Twenty-eight was a sacred number: The lunar month has 28 days; a woman's menstrual period is normally 28 days. A number that is tied to human existence everywhere. There is a large rock circle in the Little Bighorn Mountains called a Medicine Wheel. It also has 28 spokes. Numbers can be sacred. Circles are often sacred.

The first sweat lodge is called Coyote Sweat lodge. It is volcanic rock and is located in the middle of the Colville Tribes traditional territories in Washington State. Eons ago in the days of volcanoes, a bubble burped up out of the earth and formed a perfectly shaped sweat lodge. It is similar in shape to an igloo. Dome shaped. Hollow inside. Enough space for a man to fit inside. I mention this because it is one of many examples of sacred natural space. Native Americans do not generally build churches. The Earth itself and everything the Earth contains is considered sacred. For traditional Native Americans consider ancient sites as very sacred. Old can be good. Elders are to be respected and revered. In US society, new is often considered better. Elders are stored away in nursing home. Native Americans, if they are traditional, are guided by ancient traditions. This influences many things, but also time itself. Western thought sees time as linear. New is often seen as better. Often Native Americans see things the opposite. Ancient is better. So solid rock objects like the Coyote Sweat Lodge are very ancient. Rock has a long-life span. That rock was there long before humans. It is still there. Likely will be there for a long time to come.

9.2 Communicating Indigenous Knowledge to the Masses

So where do all our stories lead? It is the teaching of holistic ways of forming knowledge to non-Tribal people, so we can take better care of Mother Earth. We need to not leave this to politicians but the people who live or use resources from the land. That means all of us!

Today science needs to be understood not just by scientists but the general public who make decisions every day. With every storm or drought, it is apparent that climate change will affect your pocket book now! Many large businesses are planning for climate change, but how does a small business or individual cope? The need to plan for climate change is just as important for a small business or individual, maybe more important. Potential rising sea levels, drought, and abnormal storm activity can all raise the risk of your investments. Siting a business in the wrong location can spell disaster or at the very least make your hard-earned investment uninsurable. Current climate change planning tools are for trained scientists, do not include traditional knowledge/culture and are not scalable. Further science knowledge is fractured and science is used as a silo, and therefore unable to address climate change problems that include people and produce creative solutions to problems that persist for decades.

Scientists communicating with the masses is even more challenging today because technologies allow communication to occur in a manner of seconds. Today, social media translates or communicates science to decision-makers in many industrialized countries but rarely places the science in context of other drivers impacting the science being communicated. Today the line between communicating science to society is being distorted by the ease of communication and the availability of instant facts at everyone's fingertips, as long as you have a cell phone or iPad. Groshek and Bronda [213] recently wrote that the social media distorts and misinforms science facts when they communicate it to decision-makers — be it an individual or organization. They found that individuals, who are most active in contributing to social media, frequently propagate erroneous information because those facts are more news worthy. Groshek and Bronda wrote [213] that *"Once these items enter into the social media echo chamber, they're amplified. The facts become lost in the shuffle of competing information, limited attention or both."* Science misinformation is also propagated when legitimate information from some research is simplified by the media outlet to explain it to the public [213].

If one's culture is linked and formed through nature, it does not mean that one will make ethical decisions. The link to nature is important but more important is how it impacts your decision process and how you form knowledge. This means that a non-Tribal person does not begin to behave like a Tribal person by walking in woods or going "forest bathing". These are short excursions in the woods that do not result in making decisions that are socially and environmentally just. This gets back to the deep and surface culture we already wrote about (*see* Chapter III).

To accelerate our ability to make holistic decisions over shorter time scales will require getting tools need to get into the hands of the common people — the non-scientifically trained citizens — to actively participate in solving our environmental problems. It also needs knowledge of Native American practices that are holistic approaches to form knowledge related to nature. We need to produce the new generation of environmental decision-makers capable of comprehensively evaluating the impacts of resource consumption on the vulnerability of their lands to future land-uses and changes in the resource base as well as to identify any of its cultural impacts. Tribes are already dealing with climate change but are mostly excluded from the top-down and politically generated decisions that are made in the US today (*see* Chapter V).

Today, we need to build the tools and a platform to translate circular stories told by Indigenous people into a digital format or multi-media products. Circular stories do not have a fixed end point which means a digital story has to include knowledge of the interconnected nature held by Tribal members and ecosystem scientists in academic settings. This has to concurrently address the development of a roadmap or curriculum designed to teach youth how to write circular stories that include holistic knowledge and a morale. Today, youth can digitize or play linear game stories but these are mostly for entertainment. Circular stories based on holistic knowledge are needed to increase nature literacy. This form of storytelling has to become part of the popular media and culture if we are going to be able to increase nature literacy for the next generation of decision-makers, i.e., the youth of today. It will capture the attention of a younger audience, who are the future leaders of tomorrow, to build an easy-to-use mobile application where stories are integrated into a game.

9.3 Medicine Wheel and Not Case Studies

To achieve the concept described in the Medicine Wheel, we need to shift our focus from past "case study" approaches to define problems and their solutions for nature, people and the economy. Even business schools are debating whether the case study approach should be replaced by what they are calling the theoretical framework provided in class lectures. The latter approach is designed to give students the skills to become comfortable with plotting a roadmap to manage unpredictable and dynamics business environments.[1]

Harvard University pioneered the "case study" approach to teaching business practices where students analyze/debate real world management situations designed to provide students with reference models or existing frameworks based on their work experience to learn business strategies.[2] In a 2012 report by the Bloomberg BusinessWeek, 80% of Harvard's teaching uses the "case study" approach compared to about a third of the top 35 business schools.

Others suggest that the case study approach does not prepare students to react and adapt to the dynamic and unpredictable situations that a business will have to deal with and to innovate to beat the competition. What we have discussed in this book can be packaged in the dichotomy of teaching strategies that business schools are struggling with and have not developed consensus. The "case study" approach is similar to what is practiced by the Western science approach to identify factors needed to be addressed to respond to a problem. It is a linear approach to form knowledge. If the case study is an excellent reference model for the problem, this is an excellent approach to follow. Unfortunately, each problem we face is typically not a replicate of past problems. We also have only a few select models to draw on that were developed for specific site conditions with their site tailored criteria. The Indigenous knowledge framework is more similar to the theoretical approach that does not focus on "case studies". The theoretical approach provides students the theoretical knowledge which they adapt

---

[1] Accessed July 5, 2016 at http://www.topmba.com/why-mba/faculty-voices/debate-case-study-method-or-more-academic-approach

[2] Accessed on July 20, 2016 at http://www.topmba.com/why-mba/faculty-voices/debate-case-study-method-or-more-academic-approach

for future challenges introduced by dynamic social, economic and environmental conditions.

We allow stereotypes to continue to define our problems and solutions. We learn "facts" as we grow up that become firmly embedded in our psyche and become part of our core storehouse of knowledge. We used to believe that Native Americans were born as naturally good "trackers". When hunting was the main means of survival no doubt this was likely true for some Native Americans. This stereotype is reinforced in Western cowboy movies where a Native American character was always used to track down the bad guys. Since you saw this with your own eyes played out by the actors, it must be true. Even during the modern era, the US military used to purposefully recruit Native Americans because of their tracking skills. They were very disappointed to find out that they could not track any better than anyone else, and some could not track at all. Many Native Americans today live modern urban lives and are not working at refining their tracking skills.

We like to believe in some stereotypes because it makes life easier for us and they support our pre-conceived facts or values. A common stereotype believes science is too complex for any lay person to understand so they are not capable of making wise decisions, e.g., we need scientific experts to tell us the facts or the truth. These stereotypes persist even after they are no longer relevant. These stereotypes need to be made transparent since they influence our views of nature, our views of Tribes and what is their real knowledge, and the approaches we deem acceptable to develop our economies.

In contrast, when Tribes were forced to relocate into reservations, they had to work with what they were given even when the lands did not provide all of the cultural resources they used to collect. Even before the arrival of the European colonialists, Tribes already practiced a diversity of different approaches to making decisions related to land and water. They practiced a communal approach to land and resources. They did not rely only on "scientists" to develop or generate the knowledge needed by "decision-makers" to develop policy and make trade-off decisions. Tribes continued to work with what resources and lands European colonialists left for them and continued to make decisions following their holistic knowledge and practices. They attempted to work with the new government agencies to retain the lands they were relocated to but continued to lose parcels of land that were illegally appropriated by non-Tribal people [138]. Despite the loss of their lands, their access to cultural resources, and attempts to make American Indians to become American, they continued to retain a local knowledge of the land that was passed down inter-generationally. This knowledge is used to make decisions. Further challenging is that Tribes also have to live with the consequences of their decisions on the reduced land base they now own.

If there is going to be a climate change action plan, it will need to include the "common person" and not be relegated to flow through the hierarchical process for generating knowledge held by the Western European world. The only way the "common person" is going to be able to be part of the process is to link Native American knowledge-forming approach and how decisions are made for local land and water resources. This would provide an approach to shift from assuming that only scientists are capable of forming the knowledge needed to make sustainable decisions by decision-makers. For a climate change action plan to really work it needs individuals knowledgeable of the complex concepts and data relevant for each problem. It also needs to use a more communal approach to making decisions related to nature instead of leaving the decisions to a minority group of people with more disciplinary trained backgrounds. Forming knowledge holistically is not going to be easy — it will require a person to move from the surface culture to the deep culture as shown in the iceberg metaphor (*see* Fig. 3.1). We think global citizens have SISU and can move beyond the Western world traditions learn as they grow up. Mother Earth is depending on us!

9.4 "Fictional Tribe" as an Educational Tool to Teach How to Form Holistic Knowledge

Development of Tribal and non-Tribal governance capacity to regulate and to pursue resilient natural resource development in nature requires the creation of a new approach to education of youth in science, culture, engineering and business. This shift in educational approaches is not only a goal for non-Tribal youth. Even Tribal youth need to understood how self-governing Indian Nations can integrate into the American system of governance and participate in the American economy. All youth need to master technology for resource management, protection and development and use them to achieve holistic, environmental and socially just goals and objectives.

For Tribes, it is challenging to consider adopting environmental education programs at public schools. Historically education was used as a weapon used to break the ties Tribal youth had with their Tribal culture and identity [13]. Further, the public-school system is ill prepared to educate Indians youth about Tribal governance and holistic management. Tribal students learn the basics but as they progress in school, they find themselves caught between educational systems that are designed for non-Tribal Americans. Tribes' need people who are not only highly educated in management, science and engineering but they need to fit into the Tribal community and Tribe. The dissonance that Tribal students feel but do not understand, and is ignored by the public schools, is a major reason for the high failure rate public schools experience with their Native students. For Tribal youth, the education system should produce future decision-makers with the skills and vision of a modern self-governing Tribal economy founded on Tribal values. If these goals were met, it will preserve Tribal community cultural continuity while developing a prosperous, sustainable Tribal economy.

A novel approach to learning a Tribal model of environmental education is to have students function like a "Tribe". The heart of the academic and social program for the students should be that a class becomes the governing body of a "Fictional Tribe". Since they need to learn holistic knowledge, they need to be given the full political, social and economic indices along with the natural resource base of their "Fictional Tribe". This way they will factor in all of the complexities that exist for real Tribes in today's world. As the "Fictional Tribe", students could use methods of participatory facilitation where they develop a strategic vision for where the "Fictional Tribe" indicators would be in 20 years. They then would develop an action plan for achieving those standards, criteria and indicators of the desired development.

Students in the "Fictional Tribe" would form the committees they determined would be required within the Tribe's governance structure. Each committee would assist and motivate its members to develop, supervise and regulate the "Fictional Tribal" economy to achieve the economic goals set by the vision they originally developed. Each committee then is responsible for developing a plausible plan by researching how actual Tribes strive to achieve similar goals and how well their programs and different program approaches actually perform. They research natural resource protection, management, development and the extent to which differing approaches achieves differing results. The committees would coordinate with each other in order to create a consistent, coherent "Fictional Tribal" strategy for human, economic and natural resource development. This strategy should include the values and standards for protecting the fauna and flora that are important for sustaining cultural integrity for the people who are members of the "Fictional Tribe". The cultural integrity could consist of collecting traditional medicines, hunting or sustaining the Tribal ceremonial calendar. The "Fictional Tribe" would be required to develop a budget for allocating current general revenues existing in the "Fictional Tribe". They would also have to decide how they will coordinate to manage satisfying the regulations imposed by federal programs. As part of the "Fictional Tribe", they would also develop a coherent set of services that they will deliver to their community as they transition towards the Tribes desired future state.

A critical aspect of youth education programs should place priority on the study of science, engineering and business on a bedrock of holistic knowledge forming processes and nature cultural values. Such a program should be a tool that does not disaggregate the individual from their "Tribe" or non-Tribal cultural community. Education should be a tool to build the human resource capacity for youth to participate in our modern political and development economies but where their actions do not harm nature. If the program works it should produce the results that have already been discussed: High college retention and graduation rates for students in engineering, science and business; high degree of student confidence in their cultural values and identity as a strength in overcoming the challenges introduced by higher education for Native and non-Native students; and a means for students to have clearer understanding of their role as a professional in today's economy and environment.

References

[1] Corlett RT. The Anthropocene concept in ecology and conservation. TREE Jan 2015, 30(1): 36-41.

[2] Vogt KA, Patel-Weynand T, Shelton M, Vogt DJ, Gordon JC, Mukumoto C, Suntana AS, Roads PA. Sustainability Unpacked. Food, Energy and Water for Resilient Environments and Societies. Earthscan. UK. 2010.

[3] Gordon JC, Berry JK. Environmental Leadership equals Essential Leadership. Yale University Press. New Haven, CT., 192 pp., 2006.

[4] Stager C. The Silence of the Bugs. The New York Times, Sunday, May 27, 2018, p 7.

[5] Conniff R. Selling the Protected Area Myth. The New York Times, June 10, 2018.[1]

[6] Mis M. How to stop deforestation? Give indigenous people rights to land: U.N. expert. Planet Ark, July 20, 2016.[2]

[7] Couzin-Frankel J. When Mice Mislead. Science 2013, 342(6161): 922-925.

[8] Hook JB. The Alabama-Coushatta Indians. Texas A&M University Press. College Station, Texas. 152 pp., 1997.

[9] Deer L., Erdoes R. Lame Deer, Seeker of Visions. Simon & Schuster Paperbacks. New York. 1994.

[10] Rose C. Linear Versus Circular Thinking: Medicine Wheel Training for the Masses. Indian Country Today 2014.[3]

[11] Maryboy NC, D Begay, L Peticolas. Cosmic Serpent: Collaboration with Integrity. Indigenous Education Institute, Friday Harbor, WA. 2012.

[12] Santone S. Sustainability and Economics 101: A Primer for Elementary Educators (2011). Originally published in The Journal of Sustainability Education, 9 May 2010 under a Creative Commons Attribution License.[4]

[13] Marchand ME, Vogt KA, Suntana AS, Cawston R, Gordon JC, Miscawati M, Vogt DJ, Tovey JD, Sigurdardottir R, Roads PA. The River of Life: Sustainable Practices of Native Americans and Indigenous Peoples. With contributions by Wendell George, Cheryl Grunlose, John McCoy, Melody S Mobley, Jonathan Tallman, Ryan Rosendal. Higher Education Press (HEP) China and Michigan State University Press. 294 pp., 2016.

---

1 Accessed on June 15, 2018 at https://www.nytimes.com/2018/06/09/opinion/protected-area-myth.html
2 Accessed on Jule 15, 2017 at http://planetark.org/wen/74657#.V4_XpFRanaw.email
3 Accessed on January 4, 2018 at https://indiancountrymedianetwork.com/history/events/linear-versus-circular-thinking-medicine-wheel-training-for-the-masses/
4 Accessed on March 19, 2018 at http://www.pelicanweb.org/solisustv07n05page8.html

[14] Watt L. Tribes respond: Atlantic salmonid the Salish Sea. Ecotrust BLOG.[1]
[15] Lane P. Jr. Understanding and Leading Sustainable Development. Four Worlds International 2018.[2]
[16] Tzu Sun. The Art of War. Filiquarian. 68 pp., 2007.
[17] Jablon R. American Indian activist, poet John Trudell dies at 69. Orange County Register 2015.[3]
[18] Champagne D. In a Rational World, Can Tribal Knowledge Lead Us into the Future? Tribal knowledge could be the key to understanding earth and global climate change. Indian Country August 31, 2017.[4]
[19] Merriam-Webster Dictionary. 2016.[5]
[20] Drury F. Owning a library card, watching subtitled films and being skilled in the use of chopsticks are among the traits which make us cultured. Daily Mail 2015.[6]
[21] Dictionary. Culture. 2016.[7]
[22] Vogt KA, Gordon J, Wargo J, Vogt D, Asbjornsen H, Palmiotto PA, Clark H, O'Hara J, Keeton WS, Patel-Weynand T, Witten E. Ecosystems: Balancing Science with Management. Springer-Verlag. NY. 1997.
[23] INUIT (Inuit Circumpolar Council-Alaska) Alaskan Inuit Food Security Conceptual Framework: How to assess the Arctic from an Inuit perspective. Summary Report and Recommendations Report. Anchorage, AK. 2015.
[24] Edwards D. Artscience: Creativity in the Post-Google Generation. Harvard University Press. Cambridge. 2009.
[25] Ferguson E. Einstein, Sacred Science, and Quantum Leaps a Comparative Analysis of Western Science, Native Science and Quantum Physics Paradigm. Thesis. University of Lethbridge, Alberta, Canada. 135 pp., 2005.
[26] Barnhardt R. The Alaska Native Knowledge Network. Local Diversity: Place-Based Education in the Global Age, Greg Smith and David Gruenewald, eds. Hillsdale, NJ: Lawrence Erlbaum Associates. 2005.[8]
[27] National Park Service. National Park Service Policies, Chapter 5: Cultural Resources Management. 2001.[9]
[28] Turner NJ. The Earth's Blanket: Traditional Teaching for Sustainable Living. University of Washington Press, Seattle. 298 pp., 2005.

---

1 Accessed on February 15, 2018 at https://ecotrust.org/tribes-respond-atlantic-salmon-in-the-salish-sea/?utm_source=Sightline%20Institute&utm_medium=web-email&utm_campaign=Sightline%20News%20Selections

2 Accessed on March 19, 2018 at www.fwii.net, https://slideshare.net/4worlds/sustainable-development-4

3 Accessed on January 13, 2018 at https://www.ocregister.com/2015/12/08/american-indian-activist-poet-john-trudell-dies-at-69/

4 Accessed on February 12, 2017 at https://indiancountrymedianetwork.com/news/environment/rational-world-can-tribal-knowledge-lead-us-future/

5 Accessed on July 5, 2016 at http://www.merriam-webster.com/dictionary/culture

6 Accessed on July 5, 2016 at http://www.dailymail.co.uk/news/article-3240407/Owning-library-card-watching-subtitled-films-skilled-use-chopsticks-traits-make-cultured.html#ixzz59qDrzg84

7 Accessed on July 5, 2016 at http://www.dictionary.com/browse/culture

8 Accessed on February 27, 2018 at http://www.ankn.uaf.edu/curriculum/Articles/RayBarnhardt/PBE_ANKN_Chapter.html

9 Accessed on February 2009 at http://www.nps.gov/refdesk/mp/chapter5.htm

[29] Junger S. TRIBE. On Homecoming and Belonging. Hachette Book Group, New York. 168 pp. 2016.
[30] Gardner H. Leading and Learning Blog 2013.[1]
[31] WDFW. WA Department of Fish and Wildlife (2018) 2015—2021 Game Management Plan Development Process. 2018.[2]
[32] MacKendrick K. Climate Change Adaptation Planning for Cultural and Natural Resource Resilience: A look at planning for climate change in two native nations in the Pacific Northwest U.S. Unpublished Thesis. University of Oregon, 2009.
[33] Godoy M. How might Trump's food box plant affect health? Native Americans know all too well.[3]
[34] Allen N. Biggest American Indian tribe in the US introduces country's first junk food tax. The Telegraph 06 April 2015.
[35] Travis J. No Omnivore's Dilemma for Alaskan Hunter-Gatherers. Science Now 19 February 2012.[4]
[36] Keller SJ. How fish consumption determines water quality. April 1, 2013.[5]
[37] State of Washington Report. State of Washington. Department of Ecology. Fish Consumption Rates. Technical Support Document. A Review of Data and Information about Fish Consumption in Washington. January 2013. Publication No. 12-09-058.[6]
[38] Dunagan C. Skokomish Tribe sues state over hunting rights. Kitsap Sun February 20, 2013.
[39] Kawagley AO. A Yupiak Worldview. A Pathway to Ecology and Spirit. 2nd Ed. Waveland Press, Inc. 2006.
[40] Pew Research Center. Scientists and Belief 2009.[7]
[41] Einstein. Science, Philosophy and Religion, A Symposium, published by the Conference on Science, Philosophy and Religion in Their Relation to the Democratic Way of Life, Inc., New York, 1941.
[42] Taleb NN. Antifragile: Things that Gain from Disorder. Random House Trade Paperbacks, 544 pp., 2014.
[43] Taleb NN. Learning to Love Volatility. In a world that constantly throws big, unexpected events our way, we must learn to benefit from disorder. Wall Street Journal, November 16, 2012.[8]
[44] Macfarlane R. Landmarks. Penguin Books. 448 pp. 2016.
[45] Pinkham L. The seven levels of conscience in the River People: A native perspective of voice feeling, thought, land and lives. Dissertation Abstracts International, 59(11), 6119B. (UMI No. 9913239), 1998.

---

1 Accessed on January 13, 2016 at http://leading-learning.blogspot.com/2013/10/howard-gardner-on-creativity-are.html
2 Accessed on March 25, 2018 at https://wdfw.wa.gov/conservation/game/2015/
3 Accessed on June 8, 2018 at https://www.npr.org/
4 Accessed on January 1, 2013 at http://news.sciencemag.org/sciencenow/2012/02/no-omnivores-dilemma-for-alaskan.html
5 Accessed on November 12, 2017 at http://www.hcn.org/articles/how-fish-consumption-determines-water-quality?utm_source=wcn1&utm_medium=email
6 Accessed on April 7, 2018 at www.ecy.wa.gov/biblio/1209058.html
7 Accessed on March 3, 2018 at http://www.pewforum.org/2009/11/05/scientists-and-belief/
8 Accessed on February 5, 2017 at http://online.wsj.com/article/SB10001424127887324735104578120953311383448.html

[46] Carroll J B, Levinson SC, Lee P. Selected Writings of Benjamin Lee Wharf: Language. Thought and Reality. The MIT Press. 1964.
[47] Witherspoon G. Language and art in the Navajo universe. Ann Arbor: University of Michigan Press. 1977.
[48] Gaski H. Sami Culture in a New Era: The Norwegian Sami Experience. The University of Washington Press. 1998.
[49] Holm T. Strong Hearts Wounded Souls. Native American Veterans of the Vietnam War. University of Texas Press. 121 pp., 1996.
[50] Rosier PC. Native American Issues. Contemporary American Ethnic Issues. Greenwood Press. Westport, Connecticut, London. 2003.
[51] Newton DE. The Gale Encyclopedia of Fitness. Lacrosse. 2012.
[52] Vogt DJ, Vogt KA, Gmur SJ, Scullion JJ, Suntana AS, Daryanto S, Sigurdardottir R. Vulnerability of tropical forest ecosystems and forest dependent communities to drought. Environmental Research 2016, 144: 27-38.
[53] Russell S. Indigenous Ecology on the Half Shell: Native Methods Preserve Chesapeake Oysters. Indian Country Today Media Network.com July 11, 2016.[1]
[54] Alaska Rural Systemic Initiative. Sharing Our Pathways. A newsletter of the Alaska Rural Systemic Initiative. Jan/Feb 2005. Seeking Future Alaska Native PhDs![2]
[55] Gmur SJ, Vogt DJ, Vogt KA, Suntana AS. Effects of different sampling scales and selection criteria on modelling net primary productivity of Indonesian tropical forests. Environmental Conservation 2013, 41(2): 187-197.
[56] Suntana AS, Turnblom EC, Vogt KA. Addressing Unknown Variability in Seemingly Fixed National Forest. Estimates: Aboveground Forest Biomass for Renewable Energy. Energy Sources, Part A: Recovery, Utilization, and Environmental Effects (January 2012): 546-555.
[57] Diefenderfer HL, Thom RM, Johnson GE, Skalski JR, Vogt KA, Ebberts BD, Roegner GC, Dawley EM, Whiting AH. Assessing cumulative ecosystem response to coastal and riverine restoration programs. Ecological Restoration 2010, 29: 111-132.
[58] Vogt KA, Scullion JJ, Nackley LL, Shelton M. Conservation Efforts, Contemporary. 2012. In: Levin S, Ed. Encyclopedia of Biodiversity. 2nd Edition. Academic Press.
[59] Barone M. Why Political Polls Are So Often Wrong. Fewer landlines, fewer people willing to talk. And somehow conservatives tend to be undercounted. Wall Street Journal November 11, 2015.
[60] Miller RJ. Reservation Capitalism: Economic Development in Indian Country. Bison Books. 220 pp., 2012.
[61] Cordova VF. The Concept of Monism in Navajo Thought. University of New Mexico, unpublished dissertation., 1992.
[62] Ross R. Dancing With a Ghost: Exploring Indian Reality. Penguin, Canada. 2006.

---

1 Accessed on March 14, 2017 at http://indiancountrytodaymedianetwork.com/2016/07/11/indigenous-ecology-half-shell-native-methods-preserve-chesapeake-oysters-165094
2 Access on February 2, 2018 at http://ankn.uaf.edu/SOP/SOPv10i1.pdf

[63] Bonnicksen TM, Anderson MK, Lewis HT, Kay CE, Knudson R. Native American influences on the development of forest ecosystems. In: Szaro RC, Johnson NC, Sexton WT, Malk AJ, Eds. Ecological Stewardship: A Common Reference for Ecosystem Management. Vol. 2. Oxford, UK: Elsevier Science Ltd: 439-470. 1999.[1]

[64] WDFW. Final: Supplemental Environmental Impact Statement for the 2015—2021 Game Management Plan. 2014.[2]

[65] Cordero C. A Working and Evolving Definition of Culture. Canadian Journal of Native Education 1995, 21 supplement, 29: 28-41.

[66] Spretnak C. The Resurgence of the Real. Body, Nature, and Place in a Hypermodern World. Routledge. 292 pp., 1999.

[67] Bahti M, Joe EB. Navajo Sandpaintings. Rio Nuevo Publishers. 64 pp., 2009.

[68] Goosen IW. Dine Bizaad: Speak, Read, Write Navajo 1995. Salina Bookshelf, Inc, 81 pp.

[69] Prober SM, O'Connor MC, Walsh FJ. Australian Aboriginal peoples' seasonal knowledge: A potential basis for shared understanding in environmental management. Ecology and Society 2011, 6(2): 12.[3]

[70] Krosby M, Morgan H, Case M, Whitely BL. Stillaguamish Tribe Natural Resources Climate Change Vulnerability Assessment 2016. Climate Impacts Group, University of Washington.

[71] Ekins P, Folke C, De Groot R. Identifying critical natural capital. Ecological Economics 2003, 44(2): 159-163.

[72] EPA. Environmental Justice Timeline. 2014.[4]

[73] Bullard R. Dismantling environmental racism in the USA. Local Environment [serial online] 1999, 4(1): 5. Available from: Academic Search Complete, Ipswich, MA. Accessed April 22, 2017.

[74] Mellino S, Buonocore E, Ulgiati S. The worth of land use: A GIS-emergy evaluation of natural and human-made capital. Science of the Total Environment 2015, 506-507: 137-148.

[75] Coté C. Spirits of our whaling ancestors: Revitalizing Makah and Nuu-chah-nulth traditions. (1st ed., Capell family book). University of Washington Press, Seattle, and UBC Press Vancouver. 2010.

[76] Metzler M. Capital as Will and Imagination. Cornell University Press, 2013. ProQuest Ebook Central.[5]

[77] Tulalip Natural Resources Department. Climate Change Impacts on Tribal Resources. 2017.[6]

[78] Northwest Indian Fisheries Commission. Executive Summary CLIMATE CHANGE AND OUR NATUEAL RESOURCES. A Report from the Treaty

---

1 Accessed on March 3, 2014 at http://westinstenv.org/histwl/2008/01/04/native-american-influences-on-the-development-of-forest-ecosystems/

2 Accessed on March 25, 2018 at https://wdfw.wa.gov/publications/01657/

3 URL: http://www.ecologyandsociety.org/vol16/iss2/art12/

4 Accessed on April 3, 2017 at https://www.epa.gov/environmentaljustice/environmental-justice-timeline

5 Accessed on March 3, 2017 at http://ebookcentral.proquest.com.offcampus.lib.washington.edu/lib/washington/detail.action?docID=3138457

6 Accessed on September 12, 2017 at http://www.tulalip.nsn.us/

Tribes in Western Washington 2016.[1]
[79] O'Sullivan JL. The Great Nation of Futurity. The United States Democratic Review 1839, 6(23): 426-430.
[80] Mountjoy S. Manifest Destiny: Westward Expansion. Chelsea House Publishers, 2009.
[81] Lehrman Institute. The Founders and the Pursuit of Land. The Lehrman Institute 2017.[2]
[82] Sustainable Forestry Working Group. Menominee Tribal Enterprises: Sustainable Forestry to Improve Forest Health and Create Jobs. John D. and Catherine T. MacArthur Foundation. 1998.
[83] Onesmus DK. Sustainable Management of Forest by Menominee Tribe from Past to Present. University of Wisconsin Steven Point, 2017.[3]
[84] Nash R. Wilderness and the American Mind, 3rd. Yale University Press, 1982.
[85] Schama S. America's Verdant Cross. Wilson Quarterly Archives 1995, 19(2): 32-45.
[86] NPT PBS. (1998) American Buffalo: Spirit of a Nation. 1998.[4]
[87] Smits D. The Frontier Army and the Destruction of the Buffalo: 1865—1883. The Western Historical Quarterly Autumn 1994, 25(3): 312-338.[5]
[88] National Parks Service, U.S. Department of the Interior. Homesteading by the Numbers. 2017.[6]
[89] Ellis DM. The Forfeiture of Railroad Land Grants, 1867—1894. Mississippi Valley Historical Review June 1946, 33(1): 27-60.
[90] LaLande J. U.S. Bureau of Land Management. 1986.[7]
[91] Gordon J, Sessions J, Bailey J, Cleaves D, Corrao V, Leighton A, Mason L, Rasmussen M, Salwasser H, Sterner M. An assessment of Indian forests and forest management in the United States. Executive summary. 2013.[8]
[92] Morishima GS. Promises to keep: Paradigms and problems with coordinated resource management in Indian country. Evergreen 1998, 9: 22-25.
[93] Rigdon PH. Case 3.4. Indian Forest: Land in Trust. 105-109. In: Vogt KA et al. Forests and Society. Sustainability and Life Cycles of Forests in Human Landscapes. CABI International, United Kingdom, 2007.
[94] Moerman DE. Native American Medicinal Plants: An Ethnobotanical Dictionary. The medicinal uses of more than 3000 plants by 218 Native American tribes. Timber Press, Inc, Portland OR. 2009.

---

1 Accessed on October 12, 2017 at www.nwtreatytribes.org/climatechange
2 Accessed on November 15, 2017 at http://www.lehrmaninstitute.org/history/founders-land.html#ben
3 Accessed on November 1, 2017 at www.uwsp.edu/forestry/StuJournals/Documents/NA/donesmus.pdf
4 Accessed on November 2, 2017 at http://www.pbs.org/wnet/nature/american-buffalo-spirit-of-a-nation-introduction/2183/
5 Accessed on November 4, 2017 at doi:10.2307/971110
6 Accessed on November 3, 2017 at www.nps.gov/home/learn/historyculture/bynumbers.htm
7 Accessed on October 20, 2017 at https://oregonencyclopedia.org/articles/u_s_bureau_of_land_management/#.WqqqoWrwa70
8 Accessed on June 6, 2016 at http://www.itcnet.org/issues_projects/issues_2/forest_management/assessment.html

[95] Climate Adaptation Plan for the Territories of the Yakama Nation. 2016.[1]
[96] WA DNR Forest Health Program. 2017. Forest Health Highlights in Washington—2016. Argyropoulos M, Clark J, Dozic A, Fisacher M, Heath Z, Hersey C, Hof J, Kohler G, Omdal D, Ransey A, Ripley R, Smith B eds. Washington State Department of Natural Resources Forest Health Program. March 2017.[2]
[97] US Fish and Wildlife Service. Climate Change in the Pacific Northwest. 2017.[3]
[98] Franklin JF, Cromack K Jr, Denison W, McKee A, Maser C, Sedell JR, Swanson FJ, Juday G. Ecological characteristics of old-growth Douglas-fir forests. USDA Forest Service General Technical Report PNW-118. Pacific Northwest Forest and Range Experiment Station, Portland, OR. 48 pp., 1981.
[99] Sessions J, Gordon J, Rigdon P, Motanic D, Corrao V. 2017. Indian forests and forestry: Can they play a larger role in sustainable forest management? Journal of Forestry 2017, 115(5): 364-365.
[100] IFMAT I. First Indian Forest Management Assessment Team for the Intertribal Timber Council. Assessment of Indian Forests and Forest Management in the United States. Final Report. 1993.[4]
[101] IFMAT II. Second Indian Forest Management Assessment Team for the Intertribal Timber Council. Assessment of Indian Forest and Forest Management in the United States. Final Report. 2003.[5]
[102] IFMAT III. Third Indian Forest Management Assessment Team for the Intertribal Timber Council. Assessment of Indian forests and forest management in the United States Final report, Volume 1. 2013.[6]
[103] BIA. Report on the status of Indian forest lands Fiscal Year 2014. Submitted to Subcommittee on Indian, Insular and Alaska Native Affairs, US House of Representatives and Committee on Indian Affairs, US Senate, April 2015.
[104] Kenney B. Tribes as Managers of Federal Natural Resources. Natural Resources and Environment 2012, 27(1). American Bar Association, 2012.
[105] Ubelacker M, Wilson J. Time ball: A story of the Yakima people and the land: A cultural resource overview. 1984. Yakama Nation, Yakima, WA.[7]
[106] Fernandes R. Discovering our Story 2012.[8]
[107] Meisen P. Renewable Energy on Tribal Lands, National Congress of American Indians. Federally Recognized Tribes List. 2010.[9]

---

1 Accessed on June 8, 2017 at http://www.yakamanation-nsn.gov/final_yakama_nation_climate_adaptation_plan_version_1.pdf
2 Accessed on June 7, 2017 at http://file.dnr.wa.gov/publications/rp_fh_2016_forest_health_highlights.pdf
3 Accessed on June 9, 2017 at https://www.fws.gov/pacific/climatechange/changepnw.html
4 Accessed on May 5, 2018 at http://www.itcnet.org/issues_projects/issues_2/forest_management/assessment.html
5 Accessed on May 5, 2018 at http://www.itcnet.org/issues_projects/issues_2/forest_management/assessment.html
6 Accessed on May 5, 2018 at http://www.itcnet.org/issues_projects/issues_2/forest_management/assessment.html
7 Accessed on March 1, 2016 at http://www.worldcat.org/title/time-ball-a-story-of-the-yakima-people-and-the-land-a-cultural-resource-overview/oclc/10680372
8 Accessed on June 5, 2018 at http://www.wisdomoftheelders.org
9 Accessed on June 5, 2017 at http://www.ncai.org/index.php?ppgst_selectpro189=23&id=119#189

[108] Woodrow S. Growing Economies in Indian Country: Taking Stock of Progress and Partnerships, A Summary of Challenges, Recommendations, and Promising Efforts. 2012.[1]

[109] Prygoski PJ. From Marshall to Marshall: The Supreme Court's changing stance on tribal sovereignty. 2013.[2]

[110] Tungovia RF, Torbit S. Tribes can provide the key to clean energy. Arizona Republic: Viewpoints May 30, 2010.[3]

[111] Giago T. Native Sun: Indian gaming and tribal sovereignty. Native Sun November 19, 2009.[4]

[112] NCAI. Honoring the Promises: The Federal Trust Responsibility in the 21st century. Indian Country Budget Request FY 2012.[5]

[113] Atkinson KJ, Nilles KM. Tribal Business Structure Handbook. The Office of the Assistant Secretary — Indian Affairs U.S. Department of Interior. 2008.[6]

[114] Deloria V. Jr. Red Earth, White Lies: Native Americans and the Myth of Scientific Fact. Fulcrum Publishing. 288 pp., 1997.

[115] Regan S. 5 Ways the government keeps native Americans in poverty. Forbes Magazine March 13, 2014.[7]

[116] The Economist. Business in the Navajo Nation: Capitalism's last frontier. April 3 2008.[8]

[117] Gerdes J. Obama Administration Aims to Clear Roadblocks to Tribal Renewable Energy, Forbes Magazine Green Tech. January 31, 2012.[9]

[118] New York Times. Report: Agency hindering energy development on Tribal Lands. June 17, 2015.[10]

[119] Gates K&L LLP, Freedman EE, Freedman BJ, Moir AA. Department of Interior Streamlines Lease Approval Process for Developing Renewable Energy Projects on Tribal Lands. December 10, 2012.[11]

[120] DOE. Department of Energy, Office of Coal, Nuclear, Electric and Alternate Fuels. April 2011, Bureau of Indian Affairs, 25 CFR Part 162. [Docket ID: BIA-2011-0001], RIN 1076-AE73, Residential, Business, and Wind and Solar Resource

---

1 Accessed on March 3, 2016 at http://www.michiganbusiness.org/cm/files/tribal_business_development/growing-economies-in-indian-country-2012.pdf
2 Accessed on March 10, 2017 at http://www.americanbar.org/newsletter/publications/gp_solo_magazine_home/gp_solo_magazine_index/marshall.html
3 Accessed on May 30, 2015 at http://www.azcentral.com/arizonarepublic/viewpoints/articles/2010/05/30/20100530energy30.html#ixzz30DOUgHvQ
4 Accessed on April 10, 2016 at Tim Giago, http://www.indianz.com/News/2009/017461.asp
5 Access on December 1, 2016 at http://www.ncai.org/resources/ncai-publications/indian-country-budget-request/fy2012/FY2012_Budget_Energy.pdf
6 Accessed on February 8, 2017 at http://www.irs.gov/pub/irs-tege/tribal_business_structure_handbook.pdf
7 Accessed on May 15, 2016 at http://www.forbes.com/sites/realspin/2014/03/13/5-ways-the-government-keeps-native-americans-in-poverty/
8 Accessed on February 2, 2017 at https://www.economist.com/taxonomy/term/34/Financial?page=789
9 Accessed on March 30, 2017 at http://www.forbes.com/sites/justingerdes/2012/01/31/obama-administration-aims-to-clear-roadblocks-holding-up-tribal-renewables/
10 Accessed on February 5, 2017 at http://www.nytimes.com/aponline/2015/06/17/us/ap-us-tribes-energy-development.html?_r=0
11 Accessed on March 30, 2017 at http://www.jdsupra.com/legalnews/department-of-interior-streamlines-lease-33739/

Leases on Indian Land.[1]

[121] Kokodoko M. Making the New Markets Tax Credit work in Native communities. Community Dividend, October 2011 issue.[2]

[122] Woessner P. A super model: New secured transaction code offers legal uniformity, economic promise for Indian Country Community Dividend. March 1, 2006.[3]

[123] Stegman E. American Indian poverty and the potential of focusing on education. The Blog Article, September 5, 2012.[4]

[124] Bronin SC. The Promise and Perils of Renewable Energy on Tribal Lands. Ashgate Publishing Company, Burlington. 2012.

[125] Rave J. Renewable energy in Indian country, opportunities and challenges abound. Native Peoples Magazine 2010. 36-39.

[126] Shahinion M. The tax man cometh not: How the Non-Transferability of Tax Credits Harms Indian Tribes. The American Indian Law Review 2007, 32(1): 2007-2008.

[127] Anderson CA, Shibuya A, Ihori N, Swing EL, Bushman BJ, Sakamoto A et al. Violent video game effects on aggression, empathy, and prosocial behavior in eastern and western countries: A meta-analytic review. Psychological Bulletin 2010, 136: 151-173.

[128] Akin-Gump. IRS Ruling Creates Tax Opportunity for American Indian Tribes to Own Solar Projects. Global Project Finance Alert April 1, 2013.[5]

[129] Burton D, Nussbaum A, Gump A. Solar tax credit creates opportunity for Indian Tribes. Renewable Energyworld.com, May 24, 2013.[6]

[130] Depaul J. IRS to Reallocate $1.8 Billion of Tribal Economic Development Bonds. The Bond Buyer, July 17, 2012.[7]

[131] Bernstein D. Is the Tribal Economic Development Bond Program a failure? Policy Memos, February 13, 2013.[8]

[132] Lax A. A high-wire balancing act: Federal energy transmission corridors. Natural Resources and Environment 2008, 23(2).[9]

[133] Ailworth E. Fossil fuels remain at the forefront, energy expert says. The Boston

---

1 Accessed on May 2, 2017 at http://www.doi.gov/news/pressreleases/loader.cfm?csModule=security/getfile&pageid=331947

2 Accessed on June 6, 2017 at http://www.minneapolisfed.org/publications_papers/pub_display.cfm?id=4746&#f1

3 Accessed on January 5, 2008 at http://www.minneapolisfed.org/publications_papers/pub_display.cfm?id=2277

4 Accessed on February 6, 2015 at http://www.huffingtonpost.com/erik-stegman/american-indian-poverty-a_b_1855933.html

5 Accessed on July 8, 2017 at http://www.akingump.com/en/news-publications/irs-ruling-creates-tax-opportunity-for-american-indian-tribes-to.html

6 Accessed on March 13, 2017 at http://www.renewableenergyworld.com/rea/news/article/2013/05/solar-tax-credit-opportunity-for-indian-tribes

7 Accessed on July 7, 2017 at http://www.bondbuyer.com/issues/121_137/irs-to-reallocate-1-8-billion-of-tribal-economic-development-bonds-1041944-1.html?partner=sifma

8 Accessed on July 14, 2017 at http://policymemos.blogspot.com/2013/02/tribal-economic-development-bonds.html

9 Accessed on June 6, 2017 at http://www.jstor.org/stable/40924992?seq=1&Search=yes&searchText=increase&searchText=infrastucture&searchText=transmission&searchText=energy&list=hide&searchUri=%2Faction%2FdoBasicSearch%3FQuery%3Dincrease%2Benergy%2Btransmission%2Binfrastucture%2B%26amp%3Bacc%3Don%26amp%3Bwc%3Don%26amp%3Bfc%3Doff&prevSearch=&resultsServiceName=null

Globe, Business Section. Feburary 22, 2014.[1]
[134] Yirka B. REN21 renewable energy report shows healthy growth. June 28, 2011.[2]
[135] Wilkinson C. Messages from Frank's Landing: Story of Salmon, Treaties, and the Indian Way. University of Washington Press, Seattle. 2000.
[136] Royster JV, Blumm MC. Native American Natural Resources Law. Carolina Academic Press, North Carolina. 2008.
[137] Taylor JE, III. Making Salmon: An Environmental History of the Northwest Fisheries Crisis. University of Washington Press, Seattle. 1999.
[138] Cawston R, Frances C. Oral History with Frances Charles about Tse-whit-zen Village Site. 2008.
[139] JLARC. Review of port angeles graving dock project. Report 06-8. June 30, 2006.[3]
[140] Washington State. Washington State Senate Bill 6175. Establishing a government-to-government relationship between state government and federally recognized tribes. Washington State Legislature.[4]
[141] Cawston R. The Commissioner's Order for Tribal Relations. Washington State Department of Natural Resources. 2010.[5]
[142] Mukilteo Historical Society 2018.[6]
[143] Yuasa M. Tribal fisheries will see cutbacks on salmon seasons due to expected poor salmon forecasts. Seattle Times April 12, 2017.[7]
[144] Arctic Biodiversity Assessment (ABA). CAFF: Conservation of Arctic Flora and Fauna, Arctic Council. 2013.
[145] United Nations declaration on the Rights of Indigenous Peoples (UNDRIP). 2007.[8]
[146] Arctic Human Development Report (AHDR). Stefansson Arctic Institute. 2004.
[147] Shackeroff J, Campbell L. Traditional ecological knowledge in conservation research: Problems and rospects for their constructive engagement. Conservation and Society 2007, 5(3): 343-360.
[148] Alaska Federation of Natives. Subsistence Proclamation: To Achieve Subsistence Rights and Protection of Native Cultures. 2012.
[149] MEMA. Meaningful Engagement of Indigenous Peoples and Communities in Marine Activities. Workshop Report September 2016. Protection of the Arctic Marine Environment (PAME). Arctic Council.

---

1 Accessed on July 7, 2017 at http://www.bostonglobe.com/business/2014/02/22/fossil-fuels-will-continue-dominate-energy-mix-says-daniel-yergin/bXFiJqIyBocTTRiLsxsxcL/story.html
2 Accessed on March 6, 2017 at http://phys.org/news/2011-07-ren21-renewable-energy-healthy-growth.html#nRlv
3 Accessed on April 30, 2007 at http://leg.wa.gov/jlarc/AuditAndStudyReports/Documents/06-8_Digest.pdf
4 Accessed on January 21, 2014 at http://apps.leg.wa.gov/billinfo/summary.aspx?bill=6175&year=2011>2012
5 Accessed on December 12, 2011 at http://www.dnr.wa.gov/Publications/em_comm_tribalrelations_order_201029.pdf
6 Accessed on January 2, 2018 at http://mukilteohistorical.org/learn/history/city-of-mukilteo-noteworthy-dates-and-events/
7 Accessed on February 3, 2018 at https://www.seattletimes.com/sports/tribal-fisheries-will-see-cutbacks-on-salmon-seasons-ahead-of-expected-poor-salmon-forecasts/
8 Accessed September 30, 2017 at https://www.un.org/development/desa/indigenouspeoples/declaration-on-the-rights-of-indigenous-peoples.html

[150] Thornton T, Scheer A. Collaborative engagement of local and traditional knowledge and science in marine environments: A review. Ecology and Society 2012, 17(3): 8-12.
[151] State of the Arctic Marine Biodiversity Report (SAMBR). Conservation of Arctic Flora and Fauna (CAFF). Arctic Council. 2017.
[152] Paul-Wostl C, Tabara D, Bouwen R, Craps M, Dewulf A, Mostert E, Ridder D, Taillieu T. The importance of social learning and culture for sustainable water management. Ecological Economics 2008, 64(3): 484-495.
[153] Inuit Circumpolar Council – Alaska. Expected 2018. Conservation through Use.
[154] Alaska Federation of Natives. 2016 Priorities. 2016.
[155] Andrews R. Progressivism: Conservation in the public interest. In: Managing the Environment, Managing Ourselves: A History of U.S. Environmental Policy. Second Edition. Yale University Press. 2006.
[156] Northwest Arctic Borough. Inuunialiqput Ililugu Nunannuanun: Documenting Our Way of Life through Maps. Northwest Arctic Borough Subsistence Mapping Project. Volume 1. Northwest Arctic Borough. 2016.
[157] Jensen A. Nuvuk, Point Barrow, Alaska: The Thule Cemetary and Ipiutak Occupation. Dissertation. Bryn Mawr College. 2009.
[158] Langlois K. Alaska wilds lose out. An industry-friendly White House helps Sen. Lisa Murkowski score long-sought goals. High Country News 2018, 50(5).
[159] Morris BJ, Croker S, Zimmerman C, Gill D, Romig C. Gaming science: The "Gamification" of scientific thinking. Front. Psychol. 09 September 2013.[1]
[160] Plutzer E, McCaffrey M, Hannah A Lee, Resenau J, Berbeco M, Reid AH. Climate confusion among U.S. teachers. Science 2016, 351: 664-665.
[161] Johnson CV. New ways scientists can help put science back into popular culture. The Conversation. January 18, 2018.[2]
[162] Alberts B. Prioritizing science education. Science 2013, 340: 249.
[163] STEM. STEM-overview. 2017.[3]
[164] Ferrini-Mundy J. Driven by diversity. Science 2013, 340: 278.
[165] Biswas AK, Kirchherr J. Prof, no one is reading you. The Straits Times. April 11, 2015.[4]
[166] Barclay DA. The challenge facing libraries in an era of fake news. The Conversation January 4, 2017.[5]
[167] Field CB, Mortsch LD, Brklacich M, Forbe DL, Kovacs P, Patz JA, Running SW, Scott MJ. North America. Climate Change 2007: Impacts, Adaptation and Vulnerability. 617-652 pp. Contribution of Working Group II to the Fourth

---

[1] Accessed on January 14, 2018 at http://journal.frontiersin.org/article/10.3389/fpsyg.2013.00607/full. Accessed 1/5/2017
[2] Accessed on January 28, 2018 at https://theconversation.com/new-ways-scientists-can-help-put-science-back-into-popular-culture-84955
[3] Accessed on January 30, 2017 at https://www.ed.gov/sites/default/files/stem-overview.pdf
[4] Accessed on January 16, 2017 at http://www.straitstimes.com/opinion/prof-no-one-is-reading-you
[5] Accessed on January 16, 2017 at https://theconversation.com/the-challenge-facing-libraries-in-an-era-of-fake-news-70828?utm_medium=email&utm_campaign=The%20Weekend%20Conversation%20-%206436&utm_content=The%20Weekend%20Conversation%20-%206436+CID_2fe394bcfd38e9893f46dad924e37755&utm_source=campaign_monitor_us&utm_term=The%20challenge%20facing%20libraries%20in%20an%20era%20of%20fake%20news

[168] Tulalip Tribes. Climate change impacts on tribal resources. Tulalip Tribes. 2016.[1]
[169] Shapiro J. Duolingo for schools is free, and it may change the edtech market. Forbes Janurary 8, 2015.[2]
[170] Werbach K. Gamification harnesses the power of games to motivate. The Conversation March 19, 2015.[3]
[171] Gonzalez HB, Kuenzi JF. Science, Technology, Engineering, and Mathematics (STEM) Education: A Primer 2012. Congressional Research Service, 7-5700, www.crs.gov, R42642.[4]
[172] ACT. The Condition of STEM 2016. MS501. 2016.[5]
[173] Root-Bernstein M, Root-Bernstein R. From STEM to STEAM to STREAM: Writing as an Essential Component of Science Education. Writing mastery is essential to STEM learning and innovating. Psychology Today, Posted March 16, 2011.[6]
[174] Mistry J, Berardi A. Bridging Indigenous and scientific knowledge. Local ecological knowledge must be placed at the center of environmental governance. Science 2016, 352(6291): 1274-1275.
[175] Conniff R. People or Parks: The Human Factor in Protecting Wildlife. 2013 Yale Environmental Forum Report 2013.[7]
[176] Leonard S, Parsons M, Olawsky K, Kofod F. The role of culture and traditional knowledge in climate change adaptation: Insights from East Kimberley, Australia. Global Environ Change 2013, 23: 623-632.
[177] Russell-Smith J, Cook GD, Cooke PM, Edwards AC, Lendrum M, Meyer CP, Whitehead PJ. Managing fire regimes in north Australian savannas: Applying Aboriginal approaches to contemporary global problems. Front Ecol. Environ. 2013, 11 (Online Issue 1): e55-e63.
[178] Rick TC, Reeder-Myers LA, Hofman CA, Breitburg D, Lockwood R, Henkes G, Kellogg L, Lowery D, Luckenbach MW, Mann R, Ogburn MB, Southworth M, Wah J, Wesson J, Hines AH. Millennial-scale sustainability of the Chesapeake Bay Native American oyster fishery. National Academy of Sciences 2016, 113(23): 6568-6573.
[179] von Winterfeldt D. Bridging the gap between science and decision making. National Academy of Science 2013, 110 Supplement 3: 14055-14061.

---

1 Accessed on April 12, 2016 at www.tulalip.nsn.us/pdf.docs/FINAL%20CC%20FLYER.pdf
2 Accessed on June 5, 2017 at https://www.forbes.com/sites/jordanshapiro/2015/01/08/duolingo-for-schools-may-change-the-edtech-market/#6ff96c9740d2
3 Accessed on May 12, 2017 at https://theconversation.com/gamification-harnesses-the-power-of-games-to-motivate-37320
4 Accessed on January 16, 2017 at https://fas.org/sgp/crs/misc/R42642.pdf
5 Accessed on January 30, 2017 at http://www.act.org/content/dam/act/unsecured/documents/STEM2016_52_National.pdf
6 Accessed on January 16, 2017 at https://www.psychologytoday.com/blog/imagine/201103/stem-steam-stream-writing-essential-component-science-education
7 Accessed on April 4, 2014 at http://e360.yale.edu/feature/people_or_parks_the_human_factor_in_protecting_wildlife/2707/

[180] Bang M, Medin D. Cultural processes in science education: Supporting the navigation of multiple epistemologies. Science Education 2010, 94: 1008-1026.

[181] NSF. Women, Minorities, and Persons with Disabilities in Science and Engineering 2017.[1]

[182] Houde N. The six faces of traditional ecological knowledge: Challenges and opportunities for Canadian co-management arrangements. Ecology and Society 2007, 12(2): 34.

[183] Harvard GSE. 2016. Harvard Graduate School of Education study led by Dr. Mehta entitled In Search of Deeper Learning.

[184] NASA/Design Squad. The Design Process in Action 2017.[2]

[185] Friedman TL. Need a Job? Invent It. New York Times March 30, 2013.[3]

[186] National Research Council. Exploring the Intersection of Science Education and 21st Century Skills. A Workshop Summary 2010. Washington DC, National Academics Press.

[187] Reyes-Garcıa V, Aceituno-Mata L, Calvet-Mir L, Garnatje T, Gómez-Baggethun E, Lastra JJ, Ontillera R, Parada M, Rigat M, Vallès J, Vila S, Pardo-de-Santayana M. 2013. Resilience of traditional knowledge systems: The case of agricultural knowledge in home gardens of the Iberian Peninsula. Global Environ. Change 2014, 24: 223-231.

[188] Smith D, Schlaepfer P, Major K, Dyble M, Page AE, Thompson J, Chaudhary N, Salali GD, Mace R, Astete L, Ngales M, Vinicius L, Migliano AB. Cooperation and the evolution of hunter-gatherer storytelling. Nature Communication 2017, 8(1853): 1-9.

[189] De Jong T, MC Linn, ZC Zacharia. Physical and virtual laboratories in science and engineering education. Science 2013, 340: 305-308.

[190] Iosup A, Epema DHJ. An experience report on using gamification in technical higher education. 2014.[4]

[191] Fleischmann K, Ariel E. Gamification in science education: Gamifying learning of microscopic processes in the laboratory. Contemporary Educational Technology 2016, 7(2): 138-159.

[192] Matthews R. How green gamification can change the world. The Green Market Oracle. August 29, 2016.[5]

[193] Susskind L, Kim E. Playing 'serious games', adults learn to solve thorny real-world problems. The Conversation January 7, 2016.[6]

---

1 Accessed on June 4, 2017 at https://www.nsf.gov/statistics/2017/nsf17310/static/downloads/nsf17310-digest.pdf

2 Accessed on January 15, 2017 at http://pbskids.org/designsquad/parentseducators/workshop/welcome.html

3 Accessed on July 12, 2016 at http://www.nytimes.com/2013/03/31/opinion/sunday/friedman-need-a-job-invent-it.html?_r=0

4 Accessed on January 28, 2017 at https://www.researchgate.net/punlivsyion/262395542

5 Accessed on August 4, 2017 at http://globalwarmingisreal.com/2016/08/29/how-green-gamification-can-change-the-world/

6 Accessed on January 15, 2016 at https://theconversation.com/playing-serious-games-adults-learn-to-solve-thorny-real-world-problems

[194] Green Games. 2017.[1]
[195] Wagner K. How many people are actually playing Pokémon Go? Here's our best guess so far. Just a little back-of-the-napkin math. JUL 13, 2016.[2]
[196] The Associated Press and NORC. 2015. Public opinion and the environment: The nine types of Americans. December 2015.[3]
[197] Hegg J. Darwin's Demons: Better video games through natural selection. Posted September 30, 2016. PLOS Ecology Community 2016.[4]
[198] IPCC. Managing the risks of extreme events and disasters to advance climate change adaptation. In: Field CB, Barros V, Stocker TF, Qin D, Dokken D J, Ebi KL et al. A Special Report of Working Groups I and II of the Inter- governmental Panel on Climate Change. Cambridge University Press, Cambridge, UK; New York, USA. 2012.
[199] Agrawal A. Studying the commons, governing common-pool resource outcomes: Some concluding thoughts. Environ. Sci. Policy 2014, 36: 86-91.
[200] Chuvieco E, Martínez S, Román MV, Hantson S, Pettinari ML. Integration of ecological and socio-economic factors to assess global vulnerability to wildfire. Glob. Ecol. Biogeogr. 2014, 23: 245-258.
[201] Downing AS, van Nes EH, Balirwa JS, Beuving J, Bwathondi POJ, Chapman LJ et al. Coupled human and natural system dynamics as key to the sustainability of Lake Victoria's ecosystem services. Ecol. Soc. 2014, 19: Article 31.
[202] Mumby PJ, Chollett I, Bozec YM, Wolff NH. Ecological resilience robustness and vulnerability: How do these concepts benefit ecosystem management? Curr. Opin. Environ. Sustain. 2014, 7: 22-27.
[203] Starkey D. More than shamans and savages: American Indians and game development. 2014.[5]
[204] United States. Ethnologue Languages of the World 2017.[6]
[205] Jackson T. TMI launches Cherokee language video game. 2013.[7]
[206] Tulalip Tribes. Tulalip Lushootseed. 2017.[8]
[207] LaPensée E. Techno-Medicine Wheel: An Indigenous Alternate Reality Game. Posted on December 9, 2009.[9]
[208] Scimeca D. Why Never Alone is so much more than a video game. The State of Gaming 2015. March 1, 2015.[10]

---

1 Accessed on February 3, 2017 at http://meetthegreens.pbskids.org/games/
2 Accessed on February 3, 2017 at http://www.recode.net/2016/7/13/12181614/pokemon-go-number-active-users
3 Accessed on February 3, 2017 at http://www.apnorc.org/PDFs/Global%20Issues/12-2015%20Segmentation%20Report_D10_DTP%20Formatted_v2b-1a.pdf
4 Accessed on January 31, 2017 at http://blogs.plos.org/ecology/2016/09/30/darwins-demons-better-video-games-through-natural-selection/
5 Accessed on February 3, 2017, Usgamer.net. http://www.usgamer.net/articles/more-than-shamans-and-savages-american-indians-and-game-development
6 Accessed on February 3, 2017 at https://www.ethnologue.com/country/US
7 Accessed on February 3, 2017 at http://www.cherokeephoenix.org/Article/Index/7130
8 Accessed on February 3, 2017 at http://www.tulaliplushootseed.com/stories.htm
9 Accessed on January 15, 2017 at http://www.abtec.org/blog/?p=436
10 Accessed on December 8, 2016 at http://kernelmag.dailydot.com/issue-sections/features-issue-sections/11965/never-alone-alaska-native-video-game/#sthash.hQyJs7b5.dpuf

[209] Ferro S. New Brazilian Video Game Teaches Players About Indigenous Culture 2016.[1]

[210] Gottschall J. The Story Telling Animal. How Stories Make Us Human. Houghton Mifflin Harcourt. Boston. New York, 2012.

[211] Dove M (Hu-mis'-hu-ma). Coyote stories. The CAXTON Printers, Ltd., Caldwell, Idaho. 1934.

[212] Deloria V. Custer Died for Your Sins. University of Oklahoma Press. 1988.[2]

[213] Groshek J, Bronda S. The Conversation. How social media can distort and misinform when communicating science. June 30, 2016.[3]

---

[1] Accessed on February 4, 2017 at http://mentalfloss.com/article/76686/new-Brazilian-video-game-teaches-players-about-indigenous-culture (UTUBE link: https://www.youtube.com/watch?v=f5m88A4oRHo)

[2] http://www.feminish.com/wp-content/uploads/2012/08/Custer_Died_for_Your_Sins.pdf

[3] Accessed on March 3, 2017 at https://theconversation.com/how-social-media-can-distort-and-misinform-when-communicating-science-59044

Authors

Dr. Michael [Mike] E. Marchand is former Chair and Council Member of the Confederated Tribes of the Colville, President of the Affiliated Tribes of Northwest Indians Economic Development Corporation, PhD University of Washington Seattle.

Dr. Kristiina A. Vogt is professor of ecosystem management and holistic assessment of ecosystems at the University of Washington Seattle.

Rodney Cawston is an enrolled member of the Colville Confederated Tribes, he is the Chair of the Confederated Tribes of the Colville, previously he was the Tribal Relations Manager for WA DNR, PhC University of Washington Seattle in Holistic Management of Nature.

John J. Tovey is Director of Planning for the Confederated Tribes of the Umatilla Indian Reservation in Oregon, PhC in Urban Design and Planning at the University of Washington Seattle, he was an NSF IGERT Fellow.

John McCoy is a Tulalip Tribal member, Senator in Washington's 38th Legislative District, former General Manager of Quil Ceda village and White House Computer Technician.

Dr. Nancy Maryboy is a Cherokee/Navajo, President and Founder of the Indigenous Education Institute at Friday Harbor, Affiliate Professor at the University of Washington Seattle, PhD California Institute of Integral Studies in Indigenous Science (Astronomy) in San Francisco.

Calvin [Cal] T. Mukumoto has a Forest Management and a U. W. Master of Business Administration from the Executive MBA Program. He has his own company "Mukumoto Associates" and provides forestry/business services in Indian Country and to minority small business owners.

Daniel [Dan] J. Vogt is a Soils and Ecosystem Ecologist at the University of Washington Seattle, formerly a Soils and Ecosystem Ecologist and Director of the Greeley Analytical Laboratory at FES at Yale, formerly he was a co-founder and coordinator of the Forest Systems and Bioenergy program in SEFS at U. W.

Melody Starya Mobley is a Cherokee, first Black woman hired as a Wildlife biologist by the U.S. Forest Service, Independent Consultant and author.

Contributing Authors

Dr. David Begay is an enrolled member of the Navajo Nation (*Navajo*: Naabeehó Bináhásdzo), Navajo Council Member, has a PhD from the California Institute of Integral Studies in Indigenous Science (Astronomy) in San Francisco.

Victoria Buschman is an Alaska Native of the Iñupiat People, she has a MS in Wildlife Sciences from the University of Washington Seattle, and she is a PhD Student at U. W.

Vincent Corrao is the President, Northwest Management Inc in Idaho.

Dr. Phil Fawcett is the U. W. iSchool Capstone Program Coordinator, former Microsoft Technology Research Developer, PhD in Information Wayfarers in the iSchool at the University of Washington Seattle.

Dr. John Gordon is emeritus Dean at FES at Yale, he Chaired or co-Chaired IFMAT I, II, III.

Jessica Hernandez is an Indigenous Scientist, Zapotec & Ch'orti, Climate/Environmental Justice Scholar & Scientist, she is working on a PhD degree at U. W.

Donald Motanic is an enrolled member of the Confederated Tribes of the Umatilla Indian Reservation and a Technical Specialist (Forester) in the Intertribal Timber Council.

Mark Petruncio is a Yakama Nation forester, Affiliate Faculty at SEFS at the University of Washington Seattle.

Philip [Phil] Rigdon is an enrolled member of the Yakama Nation, he is the President of the Intertribal Timber Council Board of Directors, has a MS degree in forestry from Yale.

Jonathan Tallman is a Yakama Nation member, he is the Yakama Nation Fuels Manager, MS from the University of Washington in forestry and climate change.

Dr. Michael [Mike] Tulee is a Yakama Nation member, Executive Director United Indians of all Tribes Foundation in Seattle; he is a former NSF IGERT Fellow, has a PhD from the University of Washington Seattle in Energy.

Alexa Schreier is a Masters Student of Western Land-Use History and Policy at University of Washington Seattle.

John Sessions is a University Distinguished Professor and Strachan Chair of Forest Operations Management at Oregon State University.

Jaskaran [Jesse] Singh, Student of History at the University of Washington Seattle.

John D. Wros is employed by the Alaska Conservation Trust, is working on his PhD degree at the University of Washington Seattle in Conservation Policy and Private/Public Partnerships.

Index

9/11, 94, 95

A

a historical, 389
AAAS scientists, 117
Aandeg, 58
Aboriginal
    approaches, 362
    Media Lab (AML), 370
    Peoples, 160
abstract thinking, 356
academic(s)
    knowledge, 361
Adams, 186
adapt, 54, 59, 87, 151, 183, 293, 359, 361, 362, 398
adaptability, 79, 355
adaptive ecosystem management, 301, 302
Affiliated Tribes of the Northwest Indians (ATNI), 276
Afghanistan, 167
African American, 93
agrarian
    society, 187
Agricultural
    Act of 1949, 109
    and Consumer Protection Act of 1973, 109
Ahwahneechee, 188
air, 21, 30, 63, 111, 168, 182, 193, 197, 249, 296, 379
Air Force, 89, 345, 347
airplane science, 135
Alaska
    Native Corporations (ANCs), 325
    Native Institute of Higher Learning, 157
    Native Practice, 300, 302
    Rural Systemic Initiative, 144
Alaska Inuit Food Security, 137

Alaska National Interest Lands Conservation Act (ANILCA), 309
Alaska Native Claims Settlement Act (ANCSA), 309
Alaskan Inuit, 104
Alaskan Native Corporation, 215
Alcatraz, 39, 40
alcoholism, 77, 81, 84, 94
Alexander the Great, 34
Alford, Dan Moonhawk, 126
Algonquians, 187
altered
    lands, 359
    waterways, 359
Alternative Conservation Philosophies, 300
American
    Bar Association, 214
    democracy, 88
    Indian movement, 39
    Recovery and Reinvestment Act, 246
Ancestor, 6, 21, 383
Anchor Forests Initiative, 222
ancient heritage, 179
Angwusngyam, 58
animal rights, 174
Annual Allowable Cuts (AACs), 217
anthropocentric, 158
anthropogenic disasters, 295
Anti-fragile, 212
antibacterial, 7
Antioch, 76, 82
apps, 368, 369
aquatic lands, 3, 269, 270, 282, 283, 290
Arapaho chiefs, 189
Archaeology and Cultural Resources, 194
architectural design, 287
Arctic
    ecosystem, 104, 301
    National Wildlife Refuge, 310, 321

Peoples, 294
Arctic Coast, 310
argument, 92, 224
Arian Nations, 74
Army Corps of Engineers, 175, 350
Arrow Lakes people, 45
Arsenic, 112
art, 12, 22, 117, 118, 128, 147, 158, 209, 210, 356, 357, 359, 363, 370, 381
Artifacts, 288, 289
Artist, 118, 371
Artscience, 118
asbestos, 320
Asian
    Gypsy Moth, 205
    philosophy, 99
assessment panels, 204
assimilation, 61, 73, 87, 151, 352
astronomical, 4, 8, 392
astronomy, 393
Atlantic salmon, 14
ATNI, 68, 227, 289
auction, 289
Australia, 168, 362
Austria, 28
autochthonous, 149
awareness, 149–151, 153, 154, 159, 288

B

bacon, 109
Bad River Chippewa, 261
bag limits, 106, 107
Bakken oil fields, 341
balance
    in life, 80, 87
ball-striking, 129
Bandwidth, 376
Banks, Dennis, 39, 42
barnacles, 108, 339
Barracks, 168
baseline data, 299
basket(s)
    weaver, 170
bear grass, 273
Bear Paw, 90
bee yajoltiih, 154
beef, 109
Beers, W. George, 131
Beetle, 236

beeweed greens, 110
Begay, David, 11
behavioural finance, 14
Bellon, Maia, 15
Bend River Mall, 239
berries, 3, 24, 28, 56, 62, 65, 71, 108, 115, 116, 194, 275–277, 339, 340
Berry, Joyce, 200
Bible, 29
Big hART, 370
Bighorn
    Mountains, 8
billboard, 102
bio-pirating, 55
biodiverse ecosystem dynamics, 303
biodiversity, 148, 175, 211, 295, 317
biological diversity, 183
biomass plant, 234
biophilia, 161
birds, 127, 179, 182, 249, 321, 338
Bismarck, 343
bison, 107, 188–191
black
    sheep, 93
    widows, 15
Blackfoot, 90
blankets, 187
BLM, 184, 200, 213, 214, 378
Bloomberg Businessweek, 398
Blue Mountains, 168
BMW, 236
Board of Natural Resources, 203, 254, 283
boarding schools, 83, 109
Boeing, 347
Boldt
    , Judge, 281
    decision, 263, 281, 283
bond issuance, 246
Bonneville Power, 47, 340
boom-and-bust cycles, 6, 110, 359, 373
bottles of oil and juice, 109
bows, 187
Boylston Street, 39
Brazil, 55, 371
brick of commodity cheese, 109
bridging the gap, 377
Bristol Bay, 291, 321, 322
British

Empire, 30
    rule, 30
brittle, 16
brown bear, 370
Brundtland Report, 219
buck, 394
Buddha, 118
Buffalo
    Bill, 102
    hunters, 31
building communities, 370
bullying, 369
Bureau of Indian Affairs (BIA), 192, 244
bureaucracy, 244, 302, 336
burials, 252, 383
Burke Museum, 21, 252
Burroughs, 352
Buschman, Victoria, 293, 294
Bush, George W, 243
business community, 347
butter clams, 108

C

Cabazon Band of Mission Indians, 132
Cabela's, 354
Caddo, 58
California
    Carbon Offset Program, 315
Camas, 105
Cambridge Hilton, 39
Camry, 236
Canada, 4, 31, 48, 49, 63, 125, 131, 153, 210, 272, 370, 391
cancer-causing toxins, 112
canned meats, 109
Canoe
    Journey, 3, 288
    tree, 273
capitalistic
    ideologies, 172, 176
    society, 170
carbon
    dating, 8
    footprint, 52, 367
    sequestration, 217
caretakers, 379
caribou, 303, 310, 321
Cartesian-Newtonian, 158
Cascades, 276

casino, 94, 97, 99, 132, 235, 237, 240, 343, 349, 354
CAT drivers, 238
Catholic
    Church, 384
    education, 28
Catholicism, 69
catlinite, 187
Caucasian, 69, 85
Cawston, Rodney, 1, 2, 61, 64, 115, 250, 253, 257, 267
Cayuse, 90, 339, 378
cedar
    bark, 4, 182, 274
    trees, 271, 274
    veneer, 233
ceded lands, 115, 192
celebrities, 88
Celilo Village, 289
cell phones, 369
cemetery sites, 310
Census data, 89
Centennial Accord, 252, 255, 290, 291
Central America, 74
ceremonial, 2, 3, 52, 56, 61, 65, 106, 119, 121, 126, 129, 291, 387, 388, 401
ceremonies, 3, 8, 24, 55, 59, 62, 70, 73, 104–106, 112, 128, 273, 275, 292, 340, 395
Champagne, 52
chaos, 30, 343, 386
Chaos Theory, 125, 386
Charles, Francis, 250, 287
Che Guevara, 38
chemical contamination, 111, 291, 320
Cherokee, 154, 155, 157, 370
Chesapeake, 136, 362
Chief Seattle, 44
Chiefs, 44, 392
children, 8, 24, 25, 27, 70, 74–76, 83, 96, 109, 127, 151, 251, 281, 353, 367, 392, 394, 395
China Town, 95
Chinese, 33, 34, 79, 215, 251
Chinook salmon, 175
Chippewa, 58
Christian majority, 88
Christmas, 93
cigarettes, 34

cigars, 34
circles, 9, 10
circular
    stories, 365, 371, 397
    thinking, 10, 12
citizen scientists, 141
Civil Rights Act, 75
civil unrest, 38
Civil War, 189
clamming, 108
clams, 108
clan animals, 58
Climate
    Adaptation Plan, 193
    change, 194
        impacts, 59, 110, 138, 180, 183, 184, 193, 198, 292, 294, 323
        projections, 307
    mitigation plans, 179
    patterns, 183
    refuges, 183
cloudberries, 56
coal mine, 90, 180
coal-fired power plants, 111
Coast Salish Tribes, 171, 173
cockles, 108
Cody, Wyoming, 8
Coeur d'Alene, 267
coffee
    stand, 239
cognitive scaffolds, 367
cohogs clams, 108
cold remedies, 193
Cold War, 38
Coldwater Creek, 239
collaboration, 297, 304, 306, 316, 326, 329, 332, 334, 335, 340, 361, 371
collaborative inertia, 316
College of
    Engineering, U.W., 22
    Forest Resources, U.W., 22, 33, 283
Collier, John, 244
Collins, JW, 174
Columbia
    River, 26, 31, 46–48, 112, 119, 169, 250, 289, 340
    University, 187
Colville
    Council Room, 44
    Indian Reservation, 9, 20, 28, 126
    reservation, 65, 67, 267, 281
commercial
    catch, 283
    fisheries, 172
Commissioner's Order on Tribal Relations, 268, 269, 285
commodity program, 109
communication, 151, 268, 284, 318, 321, 358, 361, 396, 397
Communism, 38
communities, 2, 4, 7, 25, 42, 44, 54, 55, 69, 108, 109, 128, 132, 148, 149, 153, 161, 163, 170, 171, 180, 181, 183, 193, 205, 212, 230, 240, 242, 244, 291, 292, 294–297, 299–304, 306, 307, 322, 323, 335, 336, 360, 370, 390
Community Development Financial Institution (CDFI), 246
completer and non-completer, 207
complex
    communication/social skills, 355
    real-world problems, 356
conceptual thinking, 356
Confederated Tribes of the Colville, 1, 31, 63, 156
Conflict, 314
conflicts of interest, 204, 293, 297
Confucius, 118
Congress, 21, 29, 30, 62, 132, 175, 183, 186, 199, 225, 245, 246, 263, 282, 319–321, 334, 346, 390
Congress of American Indians (NCAI), 380
conquest, 370
consilience, 161
Constitution, 88, 263
constrained learning, 367
constraints, 147, 211, 223, 300, 380
Cook Inlet Tribal Council, 371
Cooke Aquaculture, 15
cookies, 110
cooks, 187
cooperating agency status, 329
Cooperative Endangered Species, 316
Coos Bay Wagon Road, 214
Coquille, 102, 213, 214, 217, 218

Cordero, Carlos, 157
Cordova, Viola, 152
cornerstones, 88
cornmeal, 109
cornucopia, 102
Corrao, Vincent, 215
corvette, 237
cosmic
    order, 122, 387
    relationships, 119
cosmovision, 123
costumes, 56
Council on Environmental Quality, 330
Cow Creek Band of Umpqua Indians, 218
cowboys, 8, 102, 103
coyote — a trickster, 374
Coyote Sweat lodge, 396
crab, 108, 241
create knowledge, 360, 366
Creative Commons Attribution, 60
creator
    for Native Americans, 105
crisps, 110
Critical
    Analysis, 359
    thinking, 359–362, 366, 368
crocus, 87
crow, 58, 156, 377
Crow Clans, 58
Cuba, 38
cucumber, 381
cultural
    interpretation, 53
    norms, 70, 162, 300
    resilience, 183
    Resources, 61, 191, 253, 326, 351
    and Natural Resources, 344
    sites, 182, 289
    theme, 52
cultural foods, 59, 104, 105, 107, 109–113, 155, 186, 197
culturally
    -based decisions, 359
    insensitive statements, 337
    relevant environment, 363
culture, 2, 14, 16, 17, 24, 32, 35, 44, 51–60, 194, 230, 400
culture-based planning, 179
culture/art, 59, 160
cultured, 51, 53–55, 58

Cuna Tribe, 15

D

D.C., 29, 30
Daily Mail, 53, 54, 58
Dakota
    Access Pipeline (DAPL), 341
    Tribe, 187
Dale American Horse, 342
dam removal, 171, 175, 340
dancing, 70, 104
Dartmouth, 42
Darwin's Demons, 367
data collection, 299, 304, 306
Daybreak Star, 227
de Brebeuf, Jean, 131
de-contextualized, 363
dead zones in the ocean, 46
decision-making, 14, 53, 59, 129, 245, 272, 295, 323, 362
Declaration of Independence, 88
Deep Culture, 61, 66, 117, 118
deer, 3, 8, 24, 65, 66, 70, 107, 110, 127, 155, 156, 339, 378, 393–395
Deer, John Lame, 10
deforestation, 136, 187
degraded
    habitats, 180
    soils, 368
degree, 22, 42, 68, 77, 79, 123, 125, 152, 158, 225, 227, 257, 285, 356, 357, 402
DeLoria, Vine, 135, 179, 355, 388, 389
demographic groups, 356, 357, 363, 364
Dena'ina people, 310
Denver, 348
Department of
    Archaeology and Historic Preservation, 286, 288
    Ecology, 272, 350
    Fish and Wildlife, 107, 254, 255, 290
    Forestry in Oregon, 203
    Natural Resources, 249, 254, 257, 267–269, 313, 315
    Transportation, 251, 276, 286, 287
depression era, 96
Design Process, 360
Designated Hunter, 107

destruction, 109, 249
developmental damage, 111
diabetes, 110, 339
diesel mechanic, 235
digital
    preservation, 377, 380
    technology, 368
Diné
    College, 122
    cosmology and spirituality, 121
dioxin/furans, 112
disagreements, 87
discharge papers, 90
disciplinary
    -based knowledge, 358
    focused, 139, 357
discriminatory profiling, 88
disease, 5–7, 51, 182, 193, 194, 207, 361
dissertation, 152, 382
dissociative genesis, 158
dissolved oxygen, 181
distortion, 377
distributive justice, 171
District
    of Columbia, 185
    Ranger, 326
distrust of government, 316
ditch diggers, 187
DNA, 21
DNR Lands Commissioner, 203
Dodge, Richard Irving, 189
dog kennels, 343
donkeys, 93
double decker, 349
Douglas-fir (*Pseudotsuga menziesii*), 193
dried beans, 109
drinking water, 197, 264, 320
drones, 343
dropouts, 237, 238
drought
    severity, 196
drumming, 70
Duolingo, 364
dzil
    nanit'a, 387

E

E-Line Media, 371
eagle, 379, 393
Earth, 394
earth, 8, 10, 11, 51, 53, 63, 102, 149, 150, 154, 243, 372, 395, 396
earth's resources, 53
easements, 309, 313, 314
Eastern
    Seaboard, 186
    Washington, 156, 271, 280, 281, 290
    University, 42, 43
Eastern border, 348
eBay, 288
ecological
    calendars, 4, 155
    function, 183
Economic
    and Community Development, 97
    barriers, 246
    development activities, 226
    self-sufficiency, 212, 223
ecosystem
    health, 6
    Management, 366
    services, 147, 172, 176, 180, 191, 216, 217, 307, 316, 317, 330
ecotones, 180
Ecotrust BLOG, 15
Eddie Bauer, 234
Edenic, 188
Edison, Thomas, 118
education, 28, 29, 35, 46, 67–69, 79, 82, 83, 86, 125, 127, 141, 149, 152, 158, 183, 225, 238, 239, 262, 263, 266, 272, 273, 278, 344, 346, 352, 355–357, 361, 363, 400–402
educators, 36, 162, 339, 352, 356, 358, 360
eels, 108
Einstein
    , Albert, 355
Eisenhower, 94
Eklutna, 312
elastic, 15, 16
elections, 88, 189
elitist process, 204
Elk, 3, 65, 70, 106, 107, 127, 156, 232, 292
Elliott State Forest, 219
Elwha
    Dam, 175

River Ecosystem and Fisheries Restoration Act, 175
embedded reasoning skills, 367
Emerson, Elaine, 373
empires clams, 108
enchantment interpretation, 53
energy production, 22, 243
engineering
　department, 349
　model, 43
England, 29, 37, 123, 162, 349, 353, 354
English, 11, 27–29, 31, 58, 75, 79, 98, 120–123, 126, 150, 151, 159, 160, 187, 237, 353, 371, 379, 382, 383, 387, 388
Entiat, 26
environmental
　assessments, 319
　conditions, 183, 295, 399
　economic model, 12, 13, 15
　economics, 16, 170, 171, 174–176
　education, 400, 401
　issues, 87
　justice, 148, 170, 171, 174, 176, 319, 320
　laws, 87, 316, 320
　managers, 160, 359
　policies, 170
　pollution, 170
　problem, 6, 7, 136, 145, 146, 356, 389
　Protection Agency (EPA), 112, 170
　sciences, 11, 372
　trade-offs, 358
　trends, 183, 307
equal employment opportunity (EEO), 337
equitable distribution, 170
equity investor, 246
ESP, 33
Essential Leader, 136
ethical behavior, 4, 6, 373
ethnic, 88, 369
ethnocentric foundations, 119
Euro Canadian culture, 153
Europe, 30, 38, 57, 102, 166, 168, 187, 189, 376, 384
European
　colonialists, 57, 58, 128, 129, 319, 399

settlers, 131, 174, 337, 380
U.S. planning model, 43
Everett
　Community College (EVCC), 347
Evergreen, 266, 344
Executive Director United Indians of all Tribes Foundation, 66
Executive Order, 31, 325, 328, 330, 331
expansionism, 188
extended family relationships, 107
exurban, 100
Exxon Valdez Oil Spill, 313

F

fake news, 358
false manner, 370
famine, 300
far reaches, 120
farm lands, 35, 99
Father Sky, 11
Fawcett, Phil, 90, 113, 165, 375
feast, 24, 62, 105, 276, 300, 394
feathers, 227, 389
federal
　authority's jurisdiction, 247
　forests, 215, 218, 222, 334
　income tax, 246
　Land Policy and Management Act, 326
federal tax credit, 246
Federal Universal Conservation Easement Act, 317
Federally
　mandated assessments, 198
　Recognized Tribes and Indigenous Peoples, 320
fedora, 25
fee lands, 217
Ferguson, Elizabeth, 59
fiduciary responsibilities, 322, 323
fight, 33, 40, 74, 88, 92, 100, 103, 128, 275, 294, 341
film, 356
financial
　analysts, 246
　incentives, 245
　markets, 14
Finland, 31, 56, 57, 78, 84, 86, 87
Finnish values, 57
Fire, 191, 221

first foods, 217, 339
Fish
    and Wildlife Commission, 107, 254
    hatchery, 46
    ladders, 47
fishing, 24, 61, 108
fishing season, 106, 283, 291
fixed land base, 46, 180
Flathead, 237
Fleming, Alexander, 7
floral industry, 273–275
Florida, 98, 100, 132
flounder, 108
flour, 109
flowers, 87, 141
Food
    , Conservation, and Energy Act, 330, 331
    Distribution Program on Indian Reservations (FDPIR), 109
    security, 105
    Stamp Act, 109
    web, 110
forces of nature, 9
Forest
    bathing, 397
    Ecology Interest Group, 309
    engineers, 209
    fires, 163, 164
    fuels, 184
    genetics, 210
    Legacy Program, 316
    loss, 185
    Management Plan (FMP), 194
    productivity, 223
    products, 226, 332
    Service Directives System, 332
    Supervisor, 326
forgotten peoples, 179
Fort McDowell, 42
FORTRAN, 209
Four Sacred Mountains, 154
fox, 156, 370
fracking, 247, 342
fragmented data, 358
framework, 12, 45, 55, 117, 118, 136–139, 142, 146, 216, 245, 296, 299, 323, 346, 359, 360, 365, 373, 398
Franklin, Benjamin, 186
free translation, 386

Freedom
    of Information Act, 330, 331
French Academy of Languages, 382
fried potatoes, 110
Friedman, Thomas, 361
frost kills, 210
fruit, 109, 275
fruiting of berries, 182
Fukushima, 46
Funerals, 65, 106, 292

G

gambling, 97, 128, 129, 132
Game
    Management Plan (GMP), 155
    protagonists, 367
gamified, 380
gaming, 132, 216, 237, 245, 368, 376, 377, 381
Gandhi, 118
garbage, 46, 270, 367
Garden of Eden, 45, 188
gas station, 238
Gaski, Harold, 123
gay rights, 95
GED, 260
gender roles, 167
generalized models, 303
generational inheritance, 242
geneticists, 210, 211
geoducks, 270
geographic information systems, 307
geography, 11, 68, 126
geology, 12, 357
geotagged digital images, 369
Germans, 74, 187
gerrymandering, 88
GHGs, 362
Gifford Pinchot National Forest, 277
Glass Mountains, 168
goals, 43, 52, 53, 200, 212, 213, 217, 219, 222, 247, 273, 292, 297, 318, 335, 341, 400, 401
Gobin, Hank, 351
Goldmark, Peter, 269
Goodman, Dr. Robert, 172
Google Earth, 369
Gordon, John, 136, 143, 198, 215, 359
Government

Accountability Office (GAO), 244
politics, 240
Governor, 252, 287, 290
Graduate Record Examination (GRE), 22
Grand Coulee Dam, 46, 47
grandmother, 10, 24, 45, 89, 95, 96, 98, 103, 278, 378, 379, 394
grandparents, 23, 27, 28, 95, 107, 151, 361
graves, 278, 286, 289
gravesite, 286
Great
    Dying, 378
    Lakes, 186
    Promise, 378, 379
    Recession, 217
Greek, 28, 29, 126, 151
Green Games, 367
greenhouse
    gas emissions, 180, 362
Greenpeace, 299
grizzly bears, 33
groundbreaking, 289
group socialization, 120
grunt, 389
Gulf of Alaska, 310

H

Habitat
    enhancement, 107
    fragmentation, 183
    loss, 172, 173
    Service, 172
Hail Mary's, 384
half breed, 94
Hancock, General Winfield Scott, 189
hardware, 347, 350
harmonic cosmic laws, 122
harmonious, 15, 385
Harvard
    Graduate School of Education, 360
harvesting oysters, 362
Haskell University, 344
Head Start to kindergarten, 382
health, 4, 104, 107, 110–112, 170, 192–194, 219, 221, 225, 294, 301, 303, 320–322, 338, 339, 362, 366, 371
healthy diets, 110

heat stress, 182
Heathrow Airport, 349, 350
heritage, 88, 89, 150, 158, 194, 330, 373
Hernandez, Jessica, 170
hero, 369
Hewlett Packard, 347
hidden half, 355
hiding, 129
High Country News, 111, 321
high-tech, 260
Hispanics, 89, 356, 357
historical recordings, 370
Hitler, Adolf, 74
hitting the bones, 129
hogan, 11
Hoh
    tribe, 226
holidays, 24, 39, 56
holistic
    approach, 4, 5, 136, 152, 155, 161, 176, 186, 194, 363, 368, 371
    management, 187, 190, 256, 400
    management practices, 155
Hollywood, 234
Home Depot, 354
Homelands, 3, 57, 63, 108, 181, 216, 222, 223
homeownerships, 280
Homestead Act of 1862, 186, 188
homework, 92, 382
hooghan, 11
Hook, Jonathan, 9
horses, 8
house
    pits, 310
housewives, 240
Hubbard Brook Experimental Forest, 162
huckleberries, 24, 113, 114, 167, 276, 277
Human
    activity, 13, 375
    economy, 172
    Resources and Planning Departments, 346
    systems, 13
Humboldt State, 224
Hummer, 236
Huni
    Jui, 371

Kuin, 371
hunter-gathers, 110
hunting
    and gathering, 4, 115, 116, 391
    season
Hupa tribal members, 224
husband, 86, 98
hydraulic system, 301

I

Iñupiat whaling communities, 322
IBM, 347
iceberg
    metaphor, 16, 59, 400
Idaho, 31, 63, 89, 90, 97, 168, 267, 376
identity, 70, 88, 89, 154, 187, 295, 337, 400, 402
    adoption, 367
idiots, 100
IFMAT, 192, 198–202, 204–207, 211, 218, 221, 222, 323
Imagination, 147, 355
immersion classes, 382
immigrants, 87, 88, 186
inalienable human rights, 88
inappropriate joking, 337
Indian
    Awakening, 94, 97
    clothes, 32
    Country, 171, 193, 224, 242
        Media Network, 370
        Today, 10, 175
    food distribution programs, 108
    Forest Management Assessment Team (IFMAT), 192
    Forestry, 204, 211
    Land Tenure Foundation, 97
    politics, 29
    Removal Act, 108
    Tribal, 109, 245
        Energy Development and Self-Determination Act (ITEDSA), 245
        Organizations (ITOs), 109
        Trust Asset Reform Act of 2016 (ITARA), 221
    Village, 103
Indigenous
    business model, 12, 13, 15, 16
    Knowledge, 1, 104, 137, 304, 396
    peoples, 185, 294
Indonesia, 30
information
    -based economies, 356
    scientists, 381
infrastructure, 151, 217, 245, 265, 322, 332, 354, 383
innovative solutions, 183
insights, 137, 147, 257, 366, 368
institutional agendas, 299
Integrated Conservation and Development projects (ICDPs), 145
integrator houses, 347
intellectual property, 55
interdisciplinary team, 219
intergenerational
    exchange, 304
    knowledge, 20, 161
    transfer, 141
intergovernmental, 306
International
    agreements, 299
    conservationists, 145
    Covenant on Civil and Political Rights, 296
    Game Developer's Association, 369
    Renewable Energy Agency (IRENA), 247
Intertribal Timber Council (ITC), 199, 200
intra-governmental, 345
Inuit Circumpolar Council, 299, 300, 306
Inupiaq Eskimos, 310
invasive
    species, 182
        encroachment, 182
investing partners, 245
IRA, 235
irrigation systems, 47
IRS (Internal Revenue Service) policy, 246
island biogeography, 160
Italians, 187
Itchoak, Karlin, 310
Izembek National Wildlife Refuge, 321

J

Jackson, Martha, 122

Jamestown
    Klallam, 226
    S'Klallam, 253, 340
Jeannie, 353
Jefferson, 115, 186
Jesus, 118
Jicarilla Apache, 152
job
    "find" a, 361
    "invent" a, 361
Johnson, Norm, 200
Joint Legislative Audit and Review Committee (JLARC) Study, 286
Jones, Stanley G., Sr., 343
Joseph Band Nez Perce, 90

K

K-12 Education, 352
Kaleb, 212
Kawagley, Oscar, 151, 157
Kaxinawá, 371
Kennewick Man, 20, 21
Kettle Falls, 101
Kindergarten, 36
King George, 30
knowledge
    framework, 4–7, 117, 118, 136, 137, 142, 145, 373, 374, 398
Ku Klux Klan, 88
Kuskokwim, 301

L

La Push, 226
Lacrosse, 131
Lake
    Chelan, 26
    Roosevelt, 112, 289
    Superior, 187
Lakota, 10, 390
Land
    and Water Conservation Fund, 316
    commissioner, 269, 276, 285
    scarcity of, 137, 142
    speculators, 232
land-use activity, 289
landscape
    -level management, 223
    -scale ecosystem analysis, 305
    scales, 4, 6

Landscape Architecture, 99, 100
landslide, 280
language, 2, 29, 32, 69, 75, 76, 79, 85, 88, 92, 98, 104, 109, 118–123, 150, 256, 268, 294, 295, 327–329, 352, 357, 364, 377, 382, 387
lawyers, 29, 47, 94, 251
lead, 3, 6, 53, 111, 112, 126, 207, 219, 222, 305, 320, 330, 389, 396
Leaders without Borders, 390
leadership models, 136
legal counsel, 91
legislation, 35, 63, 91, 216, 243, 244, 254, 261, 264, 266, 319, 330
Lewis and Clark, 378
life, 10, 11, 15, 24, 27, 36, 42, 52, 67, 87, 96, 100, 104, 118, 122, 135, 149, 150, 154, 159, 182, 187, 194, 207, 249, 258, 276, 294, 317–319, 344, 346, 361, 372, 385
lifegiving force, 122
lifestyles, 322
Lightning, 8
Linear thinking, 10, 11
Lip service, 336
liquor, 188
Little
    Bighorn Mountains, 395
    Red Riding Hood, 385
LLC, 235
ln'chi wana (the Great River, the Columbia River), 119
lobbyist, 91, 92
log brokerage company, 226
Logan, Utah, 96
logging contractors, 238
long houses, 69, 71, 166
Los Angeles Times, 174
loss of
    biodiversity, 3, 293
    habitat, 107
Lower
    Elwha Klallam, 171, 175, 340
        Tribe, 175, 250, 253
    Kuskokwim School District, 59
lumber, 190, 273, 280
Lummi, 340
Lushootseed language, 370

Luther Standing Bear, 389

M

Madagascar, 16
mainstream society, 69, 74, 77
Makah
    Indian tribe, 224, 225
    Tribe, 65, 173, 174, 225
Malad, 90
Manifest Destiny, 58, 185, 186, 188
mankind, 179
Marchand, Edwin, 342
Marketplace, 361
marriages, 87
Marsh Hall (at FES, Yale University), 209
Maryboy. Nancy, 11, 15, 80, 110, 119, 120, 149, 151, 153, 157, 384, 385
Marysville, 258, 348
mass extinctions, 172
mathematical equations, 58
Mayflower, 30, 41, 103
Mazda, 237
McCain, Senator John, 201
McCoy, John, 86, 257, 258, 341, 343
Means, Russell, 39, 42
meatballs, 56
medicinal plants, 55, 192, 194, 197, 201
Medicine
    man, 10, 70
    Wheel, 1, 4, 9–11, 118, 119, 142, 389, 392, 395, 398
melting pot
    theory, 87, 88
Memorandum of Understanding, 270
Menominee, 58, 186, 187
mental, 4, 5, 14, 36, 57, 77, 150, 158, 372
mental bifurcation, 158
mercury poisoning, 111
meta-files, 366
metacognition, 367
metacognitive scaffolding, 367
methylmercury, 111
Metlakatla, 313
Metzler, Dr. Mark, 176
Mexican-American War, 189
microbially contaminated, 111
Microsoft, 90, 375, 376

military
    battalion, 188
Miller, Jay, 374
mills, 238, 251
Mills, Peter, 348
Mineral Block, 277
mining camp, 187, 188
Miriwoong people in Australia, 362
Mississippi River, 108, 186
MIT, 36
mitigation, 2, 6, 184, 297, 301, 302, 311, 390
Mobile
    Learning Incubator, 369
Mobley, Melody, 88, 324, 333, 334
Mohave Desert, 98
mole, 370
molybdenum, 231
Monroe, 186
moose, 127, 156, 370
Morishima, Gary, 200, 202
Motanic, Donald, 215
Mother Earth, 9, 11, 170, 172, 176, 243, 387, 391, 392, 396, 400
motivational scaffolds, 367
motorcycles, 236
Mount
    St. Helens National Volcanic Monument, 277
    Taylor, 386, 387
mountain song, 386
Muckleshoot, 129, 340
    Tribe, 217
Muir, John, 188
Mukilteo, 288
mule deer, 394
Murphy's Law, 351
music, 25, 356
Muslim
    -majority, 88
mussels, 108, 339

N

naming ceremonies, 65, 106
Napoleon Bonaparte, 389
nation of immigrants, 87, 89
National
    Academy of Science, 205
    energy crisis, 243

Environmental Policy Act, 330
Forest System, 331, 333, 334
Forests, 106, 215, 220
Geographic, 369
grasslands, 334
Indian Forest, 199
Indian Forest Resource Management Act (NIFRMA) of 1990 (NIFRMA), 215
Park, 254–256
Parks Services, 175
policy, 35
politics, 87
Research Council, 197, 355
Native
  allotments, 310, 312
  American Graves Protection and Repatriation Act (NAGPRA), 20
  American Studies program, 224
  Corporation, 312, 314
    lands, 310, 313, 314
  eyes, 157
  language, 119, 150
  people, 274
  tongue, 119
Native Alaska Corporations, 291
natural
  capital, 147, 170, 175, 176
  elevation, 387
  gas, 247, 312
  physical processes, 183
  resource, 64, 113, 172, 201, 211, 215, 241, 250, 267, 272, 285, 290, 303, 304, 334, 336, 357, 358, 362, 363, 379, 400, 401
  Resources and Environment Law Review, 214
  resources management, 66, 183, 340
  systems, 13, 314
  world, 4, 5, 66, 82, 374
NatureServe's Climate Change Vulnerability Index (CCVI), 164
Navajo
  Area Indian Health Service, 110
  Code Talkers, 125
  Nation, 110, 244
Navy, 72
Neah Bay
  , WA, 173
  Field Station, 226

Neo-Nazis, 88
Nespelem, 72
net-pen failure, 15
networking, 345, 346, 348, 350
neurological problems, 111
neutral assessment, 206
Never Alone, 368, 371, 376, 377, 380
New
  Hampshire, 37, 162
  Markets Tax Credits (NMTC), 245
  Mexico, 58, 110, 152, 240
Nisqually, 338, 340
Nobel Prize in Economics, 14
non-native fisher people, 322
non-verbal communication, 166
Nooksack, 340
norm, 78, 91, 117, 161, 187
Norman Conquest, 123
North
  American Indians, 4
  American Wetlands Conservation Grants, 316
  Central Forest Experimental Station, 210
  Dakota, 31, 341, 343
  Island County, 259
  Star, 11
Northern Plains, 8
Northwest
  Arctic Borough, 305, 307
  Forest plan, 214
  Indian Fisheries Commission, 106, 111, 172, 292, 339, 340
  U.S., 198, 338
NSF IGERT, 21, 113
nurse, 89, 104
nutrient cycles, 6
nutrition, 106, 144, 182

O

Oak Harbor, 258
obnoxious, 370
ocean chemistry, 182
odor, 353
Office of
  Indian Affairs, 108, 109, 290
  the Special Trustee, 221
offshore drilling, 322
Ogden Utah, 96

Okanogan
    Nation, 395
    River, 48, 49, 395
old-growth forests, 181
Olympic National Park, 175
Omak, Washington, 32
Omnibus bills, 35
one-stop-shop, 282
Ontario, Oregon, 168
open
    and unclaimed, 106, 115, 116
Operations Manager, 226, 227
opportunities, 15, 155, 182, 193, 216, 223, 245, 317, 318, 322, 335, 356–358, 375
opportunity, 89, 218, 242, 251, 294, 296, 316, 317, 328, 329, 333
oppression, 95, 172
oral tradition, 304
orchards, 47
Order on Tribal Relations, 269
Oregon
    Coast, 107, 108, 236, 240
    State Land Board, 219
Oregon Board of Forestry, 204
Our Common Future, 219
outlet mall, 240
overeating, 389
overharvesting, 107, 172, 173, 276, 279
ownership, 14, 57, 114, 115, 202, 244, 250, 280, 311, 312, 316, 370
Oxford University, 29

P

Pacific Islanders, 112
pack of wolves, 33
Paiute, 101
paleontologist, 21
palm, 115
Papua New Guinea, 16
parcel, 280, 281, 311, 313
Pasco, Washington, 260
pass through, 191, 246
pasta, 109
PCB, 112, 320
pea soup, 56
Peace Corps, 230
Pendleton
    Round-up, 103
Penicillium, 7

Penstone, James, 60
per-capita
    allocation, 246
    income, 246
pesticide, 282
pests, 182, 194, 195
Petruncio, Mark, 191, 230
Pettit
    Engineering, 349
Pew Forum, 117, 146
pharmaceutical companies, 55
PhD, 21–23, 33, 61, 68, 78, 81, 85, 95, 127, 257, 376
Philippines, 278
Philips Exeter Academy, 39
philosophy, 22, 57, 73, 99, 100, 120, 152, 187, 203, 317, 379
physical world, 7
physics, 12, 98, 99, 162, 357, 363, 386
pickles, 381
Pike Place Market, 30
pilgrim, 24, 30, 41
Pinkham, Lloyd, 119
Pinnguaq, 370
pinyon nuts, 110
pipeline, 266, 341–343, 390
pipes, 34, 393
place-based
    knowledge, 11, 164, 256, 358, 364, 368
    reality, 367
Plains, 131, 188, 189, 392, 393
planetary history, 135
plant
    gathering areas, 182
    productivity, 182
plaque, 383
Plateau, 65, 69, 129, 276, 391
Plymouth Rock, 40, 41
Pokémon Go, 367
polar bears, 33, 321
policy recommendations, 299
political suicide, 317
Polybrominated diphenyl ethers (PBDEs), 112
Polycyclic aromatic hydrocarbons (PAHs), 112
Ponderosa pine, 195, 213
Pontoons, 286
porcupine, 127, 370

Port
    Angeles, 251–253, 286, 287
    Gamble S'Klallam, 253, 340
    Townsend, 233
Portuguese, 371
potatoes, 56, 339
Potlatches, 108
poverty, 94, 101, 153
POW WOWS, 70
powdered milk and eggs, 109
Power of the World, 10
pre-1492, 166
pre-statehood actions, 309
Premium Outlet Mall, 354
Prep Schools, 36
Presbyterian Church, 96
preservation of language, 377
President
    Obama, 343
    Trump, 343, 346
President Andrew Jackson, 108
prey availability, 182
prickly pears, 110
Prince William Sound Eskimos, 310
Prisoner, 10, 143
pro-social skills, 369
proactive, 48, 285, 287, 391
problem-solving skills, 7, 355
procedural justice, 171
production tax credits, 245
professions, 68, 71
professor, 42, 68, 91, 153, 169, 187, 224
programmatic environmental impact statement (PEIS), 325
progression, 370
Project Noah, 369
protein, 115
Protestant, 69
psyche, 119, 387, 399
public
    -private partnerships, 309
    protests, 88
Pueblo, 58
Puerto Rico, 16, 55, 278, 334
Puget Sound, 3, 15, 111, 112, 172–174, 271, 290, 338, 352
pulp
    and paper industry, 210
Purépecha, 369
Puyallup, 319, 340

puzzle
    -platformer game, 368

Q

Quantum Physics, 125, 151, 386
Quil Ceda Village, 239, 258, 261, 265, 346, 351, 354
Quileute, 340
Quillenchooten, 9
Quinault, 200, 340

R

raccoon, 156, 370
racial, 53, 88, 337
Rafeedie decision, 281
ranchers, 240
range shifts, 180, 359
razor clams, 108
reactive, 48, 256, 285, 304
real estate, 97, 309, 310, 314, 316
reciprocal relationship, 301
recognition justice, 171
recreation, 129, 217, 316
Red
    alder (Alnus rubra), 193
    Earth, White Lies, 135, 179, 355, 389
    sports car, 237
reductionistic method of thinking, 11
Regestein, Quentin, 390
regulating activities, 87
REI, 233, 234
reindeer, 123, 295
religion, 53, 85, 88, 89, 117, 118, 129, 131, 146, 158, 356, 384
religious
    freedom, 88
    monoculture, 96
renewable energy, 242–247
repayment period, 246
Request for Proposal, 347
resilience, 148, 170, 183, 366, 372
resource
    and environmental problems, 136
    exploitation, 293, 295
    extraction, 279, 303
Retaliation, 337
Reuters, 96
reverence, 75, 243

revolutionary literature, 38
rice, 109
Riegle Community Development and Regulatory Improvement Act, 246
Rigdon, Philip, 215
ritualized, 384
River
    People, 119
roadmap, 374, 397, 398
Robin Hood, 385
Rock Chairman Archambault, 342
Rocky Reach Dam, 26
Roosevelt Administration, 244
Ross, Rupert, 153
royalty, 24
rules, 17, 52, 54, 87, 100, 105, 113, 131, 167, 202, 205, 217, 220, 245, 283, 303
rural
    character, 99
    spaces, 99, 100

S

sacred
    mountain, 386
    site, 8, 26, 175, 392
    symbol, 8
SAI Cuentos Mágicos de los Ancestros, 369
salad bowl theory, 88
salal (*Gaultheria shallon*), 274
salicylic acid, 193
Salish
    Networks, 346, 348
Salmon
    Chief, 6, 113
    fishermen, 31
    habitat restoration, 292
    Recovery Office, 111
salt
    of the earth, 92
saltwater intrusion, 182
Sami people, 123
San Juan Islands, 14
San Simeon Bay, 174
sandpainting, ikaah, 159
Santa Fe, 288
Sauk-Suiattle, 291, 340
Savings and Loan (S&L), 97
scarcity of
    knowledge, 6, 118, 137, 143–145, 161, 323, 357
    land, 137, 143
    resources, 147, 167, 300
Schama, Simon, 187, 188
schismogenesis, 158
Schreier, Alexa, 21, 22, 66, 89, 90, 113, 165, 224, 225, 257, 267, 375
Science
    based assessments, 200
    literacy, 355, 356
scientific
    jargon, 86
    method, 11, 12, 20, 117, 125, 139, 306, 360
    practices, 11
scientists, 7, 11, 33, 43, 80, 117, 118, 136, 160–162, 205, 304–307, 317, 396, 400
Scottish ancestors, 91
Sea
    level rise, 2
    Shepherd, 299
seal, 306
Seattle
    Times, 173, 287
secondary disturbance agents, 195
Secretary of the Interior, 199, 221
sediment, 175, 182
self-
    determination, 218, 295, 345, 346, 354
    Act, 212
    governance, 221, 244, 345, 346, 354
    management/self-development, 355
semi-nomadic, 128, 166, 168
sequential thinkers, 209
sergeant, 67, 89
service-oriented staffs, 337
Sessions, John, 202, 209, 215
seven generations, 154, 213
Sewage Plant, 346, 348
sewer system, 348, 350
shellfish beds, 181
Sherman, Pastel, 29
shikeyah, 154
shingles, 273
Shinny, 131
Shoshone
    -Bannock Tribes, 99

Index —— 439

Language Program - University of
    Utah's, 370
Sicangu Lakota, 342
Siccama, Tom, 162, 163
Silcosasket, 27
silt layers, 383
silver bullets, 227
silvicultural
    practices, 193, 197
silviculture, 218, 220, 335
simulations, 367
singing, 70, 129
Sinixt nation, 395
SISU, 57, 78, 84, 400
Skagit, 264
Skokomish, 115, 116, 288, 340
Smithsonian Museum, 21
Smokey the Bear, 163, 164
Snohomish County, 350, 354
snowpack, 193, 197
social
    change, 10
    ecologist, 284
    mechanism, 95, 303
    movements, 95
    pressures, 95
    sciences, 117, 138–140, 210, 366
    scientists, 136, 139, 140, 145
    triggers, 145
social bonds, 108
socio-ecological system, 364, 368
Socioeconomic, 170
sockeye, 250, 322
software, 347, 350
solar development, 246
soldier, 72, 168
Soul of applied science, 211
South
    Africa, 167
    America, 158, 369, 391
    Dakota, 8, 31
Southeast Alaska, 322
sovereign nations, 106, 132, 133, 218,
    262, 263
sovereignty, 4, 42, 147, 172, 180, 223,
    224, 242, 243, 264, 270, 294,
    295, 302, 319, 323, 338, 341
space, 6, 10, 11, 95, 96, 99, 188, 253,
    347, 369, 372, 378, 396
Spam, 110
spatial and temporal patterns, 307

spears, 187
Special Forest Products Policy, 332
species, 4, 15, 16, 106, 110, 111, 156,
    219, 282, 293–295, 304–307,
    333, 338, 358, 359, 369
Sperry-UNIVAC, 352
spider web, 12, 16, 138
Spinoza, Bernard, 152
spiritual, 4, 21, 62, 64, 65, 82, 117, 118,
    121, 122, 125, 137, 143, 146,
    192–194, 218, 306, 329, 330,
    333, 342, 357, 359, 366, 368,
    385, 387
spiritual consciousness, 120, 154, 159,
    386
spirituality, 12, 125, 158, 294
split estate rights, 309
Spokane Indian reservation, 67
sport fishermen, 173
sports activity, 25
sportsmen, 283
squares, 9
Squaxin Island, 340
St. Mary's Mission, 28
stakeholders, 6, 204, 256, 318, 338, 363
Standing
    Cloud, 27
    Rock, 24, 341–343, 390
        Indian Reservation, 341, 342
STEAM, 357, 368
STEM
    -related fields, 363
    education, 356
stereotype, 102, 167, 399
Stick Game, 129
Stillaguamish Tribe, 164
stone circle, 4
strategic planning, 246
Strawberry Island, 169
STREAM, 357, 368
stream temperatures, 197
strengths
    , weaknesses, opportunities and
        threats (SWOT), 216
submission, 109
Subsistence
    fishing, 295
    gathering, 111
    herding, 295
    hunting, 295

Mapping Project, 305, 307
   practices, 295, 307
   quotas, 299
substance abuse, 94
sugar, 109
sugary drinks, 110
suicidal, 167
sumac berries, 110
supermarket, 211
Supervisory Control and Data Acquisition system (SCADA), 350
Supreme
   Court, 79, 106, 115, 116, 132, 133, 318, 321, 338
   law of the land, 106, 319, 321
Suquamish, 113, 340
surplus Food, 108, 109
Sustainability U, 369
Sustainable
   Development, 15
   forest management, 185, 216, 218
swearing, 90
sweat lodge, 150, 395, 396
Swinomish Tribe, 112, 172
symptom, 5–7, 361
Systems
   Theory, 125, 386
   thinking, 355

T

Tallman, Jonathan, 191
tax
   -free status, 245
   codes, 245
   credit, 246
   exempt, 246
Taylor, Mike, 345
teacher, 35, 80, 81, 91, 98
teacher's certification program, 68
team sports, 25
technical universities, 367
technology, 46, 47, 137, 144, 162, 172, 188, 347, 349, 360, 362, 392, 400
Teck Cominco lead-zinc smelter, 112
tempo and rhythms of life, 17
tepee, 9, 26
termination period, 94
Thaler, Richard, 14
Thanksgiving
   dinner, 103
The
   Alabama-Coushatta Indians, 9
   Art of War, 33, 34, 38, 215
   Cosmic Serpent, 11
   Great Land, 309
   Spirit of Spring, 369
   Tribes, 180, 211, 272, 273, 275, 276, 279, 286, 338
theater, 356
Theory of Relativity, 11
thick skin, 337
Thornton Media, Inc. (TMI), 370
threat, 107, 173, 217
timber tribes, 201
Tlingit, 58
toilet paper, 359
Tongass National Forest, 321
toothpicks, 381, 382
top predator, 110
tortillas, 110
tourism, 297, 322, 330
towers, 232, 348
Traditional
   Conservation Philosophies, 300
   foods, 3, 4, 24, 62, 65, 105, 182, 268, 273, 280, 339
   homelands, 63, 272
   hunting grounds, 182
Trail of Tears, 154, 155
Transcontinental Railroad, 186, 191
transformation, 53, 121, 219, 221
transit, 91
transmission
   development, 247
   infrastructure, 247
transportation, 91, 193, 332, 369
treasure trove, 391
treaties, 4, 30, 31, 63, 105, 106, 109, 181–183, 245, 249, 256, 292, 318–322, 329, 331
Treaty
   Hunting Rights, 106
   of Neah Bay in 1855, 173
   rights lands, 220
trees, 2, 7, 47, 71, 72, 100, 102, 141, 162, 179, 194, 195, 197, 203, 205, 212, 214, 220, 236, 243, 250, 268, 270, 271, 274, 321, 346, 361, 368, 377, 394

Index — 441

Tri-Cities, 168, 169
Tribal
    business enterprise, 244
    citizen engagement, 183
    Council, 22, 103, 226, 257, 288
    economic development (TED) bonds
    elders, 2, 62, 65, 339, 370, 373
    forest management, 185, 193
    Forest Protection Act (TFPA) 2004, 218
    history, 100, 261, 262, 290
    liaison, 249, 255, 267, 271, 289, 290
    logger, 235
    resource, 218, 222
        access, 280
    sovereignty, 132, 171, 241, 242, 264, 289, 320, 321
    timber management, 211
    visions, 192
tribally managed forests
tribe, 14, 22, 31, 45, 48, 70, 97, 116, 127–129, 138, 143, 166–168, 188, 204, 205, 230–235, 285, 319, 373, 374, 383, 390, 393, 394
triple bottom line approach, 220
true fir species, 198
trust
    lands, 193, 199, 201, 217, 222
Tsoo Dzil, 386
Tulalip
    reservation, 67
tunnel vision, 59, 209
turkey, 24, 156
Turku, Finland, 239
Two Bulls, 10
Tyonek, 313
Tzu, Sun, 34

U

U.S.
    Coastal Program, 316
    Constitution, 106, 318, 321
    Department of Agriculture, 110, 190, 333
    Drought Monitor, 195
    Marines, 124
    National Forest Planning, 323
    Westward Expansion, 185

U.W. (UW)
    Climate Impacts Group, 164, 180, 196, 197
    Digital Collections, 174
Ukpeagvik Iñupiat Corporation, 310
ultimate adapters, 59
unclaimed lands, 106, 115
Under Secretary for Indian affairs, 221
undercutting, 247
undergraduate capstone, 360
Ungwish-wungwa, 58
uniqueness, 88
UNISYS, 347
United
    Declaration on the Rights of Indigenous peoples, 296
    Nations, 295, 296, 306
        Declaration on Indigenous Peoples, 295
Universal Declaration of Human Rights, 296
University of
    Idaho, 98, 276
    São Paulo, 371
    Tromso, Norway, 123
    Washington, 9, 22, 23, 68, 100, 127, 175, 197, 291, 309, 373
    Bothell, 347
    Wisconsin, 369
University of British Columbia, 276
untrodden spaces, 185
Upper Skagit, 291, 340
urban landscapes, 6
urbanization, 173

V

VA (Veteran Affairs), 168
Vamos a aprender Purépecha, 369
Vancouver, British Columbia, 370
vegetables, 109
vendors, 226, 227
venture capital sources, 219
vests, 239
video games, 367, 370
Vietnam War, 38, 77, 79
Village, 250–252, 312, 314, 315
Villains, 58
VIP, 27
Virgin Islands, 334

Virtanen, 31
vision, 8, 32, 43, 150, 205, 211, 216, 336, 344–346, 351, 354, 401
vision loss, 111
volcanic rock, 396
voluntary goals, 179
voting rights, 88

W

WA DNR Forest Health Program 2017, 195
Wales, 90
Wall Street Journal, 261
Walla Walla, 339, 382
Wallowa Band Nez Perce, 90
Wallowa Valley, 90
Walmart, 229, 354
war dancing, 70
Warm Springs, 201, 212, 267, 340
warrior, 27, 70, 73, 74, 395
Washburn General Store, 229
Washington
    D.C., 29, 30, 258, 282, 291
    Department of Ecology, 15
    State legislators, 15
    Territory, 106
waste
    incinerators, 111
Water
    cannons, 49, 343
    Protectors, 341, 342
    wheeling, 348
watershed protection, 217
weaknesses, 137, 218, 222, 290
wellbeing, 294
Wellpinit, 72
Welsh, 90
Wenatchee, 26
Western, 376
western
    clothes, 32
    Practice, 300, 302
    science, 5–7, 9, 11, 45, 48, 55, 59, 77, 117, 118, 135–139, 160, 162, 297, 299, 302, 357, 368, 398
    Territories, 185
    trained scientist, 137, 143

Western Australia, 370
western civilization, 158
wetland mitigation, 315
Weyerhaeuser, 201, 206, 213
Whale Hunt, 174
whaling, 173–175, 225
Wheeler-Howard Act, 244
White
    , Justice Byron, 132
    House, 321
    man, 66
    papers, 200
    people, 10, 95, 96, 378
    spruce, 210
whores, 187
wild potato, 110
Wilderness Society, 206
Wildernesses, Research Natural Areas, 277
wildland fire management, 280
wildlife
    biologists, 208
    management, 107, 155, 217, 295, 311
    monitoring, 299
    resources, 107
    science, 296, 299, 307
Williams,
    Daryl, 349, 351
    Herman, Jr., 349, 353
    Wayne, 343–345
Wilson, EO, 160
Winchell, 43
Wind, 8
    power, 232
windmills, 232
Wisconsin, 186, 240
Wishusen, Fran, 172
Witherspoon, Gary, 120
wolf packs, 156, 157
wolverines, 33, 128
Womer, Willie, 342
wood processing
    infrastructure, 223
woodstove, 34, 394
wooly mammoths, 394
World Wildlife Fund, 299
world's problems, 87
Wyoming, 7, 8, 31, 392

X

Xenon, 348

Y

Yakama
    Nation, 66, 87, 191–194
    reservation, 67, 76
Yale's School of Forestry and Environ-
    mental Studies, 198
Yellowstone, 90, 107
Yochemate, 188
Yosemite National Park, 187, 188

YouTube, 33
yucca fruit, 110
Yukon Inter-Tribal Fish Commission,
    301

Z

Zinke, Ryan
    Interior Secretary, 321

#

21st Century trading post, 354